Crop Production, Diversity and Improvement

Crop Production, Diversity and Improvement

Editor: Lancaster Mason

www.callistoreference.com

Callisto Reference,
118-35 Queens Blvd., Suite 400,
Forest Hills, NY 11375, USA

Visit us on the World Wide Web at:
www.callistoreference.com

ISBN: 978-1-64116-256-2 (Hardback)

Trademark Notice: Registered trademark of products or corporate names are used only for explanation and identification without intent to infringe.

Cataloging-in-Publication Data

Crop production, diversity and improvement / edited by Lancaster Mason.
    p. cm.
Includes bibliographical references and index.
ISBN 978-1-64116-256-2
1. Crop science. 2. Crops. 3. Field crops--Varieties. 4. Crop yields. 5. Agrobiodiversity.
6. Crop improvement. I. Mason, Lancaster.
SB91 .C76 2020
631--dc23

Table of Contents

Preface

This book aims to highlight the current researches and provides a platform to further the scope of innovations in this area. This book is a product of the combined efforts of many researchers and scientists, after going through thorough studies and analysis from different parts of the world. The objective of this book is to provide the readers with the latest information of the field.

A crop is a plant or animal product which is harvested for profit. It is usually expanded to include alga or macroscopic fungus. Crop may refer either to the harvested parts or to the harvest in a more refined state. Most of them are harvested as fodder for livestock or food for humans. Industrial crops, horticulture and floriculture are the important non-food crops. Industrial crops include the crops produced for biofuels, clothing or medicine. Horticulture refers to the production of the plants used for other crops such as fruit trees. Floriculture is associated with the production of houseplants, cut flowers and cultivated greens, flowering garden and pot plants, and bedding plants. Potato, wheat, maize, sugarcane and rice are the important food crops in terms of produced weight. This book strives to provide a fair idea about crop production and to help develop a better understanding of the latest advances within this field. It will also provide interesting topics for research, which interested readers can take up. Scientists and students actively engaged in this area of study will find this book full of crucial and unexplored concepts.

I would like to express my sincere thanks to the authors for their dedicated efforts in the completion of this book. I acknowledge the efforts of the publisher for providing constant support. Lastly, I would like to thank my family for their support in all academic endeavors.

Editor

Effect of nickel toxicity on growth, photosynthetic pigments and dry matter yield of *Cicer arietinum* L. varieties

Saima Batool*

Department of Botany, Govt. Degree College for Women, Samanabad, Faisalabad, Pakistan

*Corresponding author email: saimabatoolbhatti@gmail.com

Abstract

Effects of nickel toxicity on photosynthetic pigments and dry matter yield of *Cicer arietinum* L. (Chickpea) varieties were observed. Nickel as $NiCl_2$ was applied to the soil in solution form @ 0mgL^{-1}(Control), 25mg L^{-1}, 50mg L^{-1}, 100mg L^{-1} and 150mg L^{-1}. All the treatments were replicated six times. Experiment was laid down in Completely Randomized Design (CRD) with two factor-factorial arrangement. A significant decrease in growth, chlorophyll a, b, total chlorophyll and carotenoid contents was observed with increasing concentrations of nickel application. A similar decreasing trend was also noted for dry matter yield of the plants which may be attributed to decreased pigment contents and decreased photosynthetic activity. Chickpea was found to be an indicator of nickel toxicity showing its inhibitory effects on various growth and biochemical parameters.

Keywords: Nickel, Chlorophylls, Carotenoids, Dry Matter, Chickpea

Introduction

Bioaccumulation of heavy metals in the environment has become a danger for plants and other living organisms (Emamverdian et al., 2015). Nickel, a heavy metal is an essential plant nutrient (Harasim and Filipek 2015; Chen et al., 2009) and is needed in very low quantities for normal growth of plants (Chen et al., 2009). In legume plants, it is required by hydrogenase enzyme during nitrogen fixation (Gerendás et al., 1999; Seregin and Kozhevnikova, 2006), and is the component of urease enzyme which brings about hydrolysis of urea (Seregin and Kozhevnikova, 2006). Deficiency of nickel results in inhibited activity of urease which leads to accumulation of toxic levels of urea in shoots (Yusuf et al., 2011) causing necrosis of leaf tips or chlorosis of older leaves (Seregin and Kozhevnikova, 2006; Yusuf et al., 2011).

At higher levels, nickel is highly toxic and causes alterations in antioxidant enzymes (Dubey and Panday 2011; Bhalerao et al., 2015) and may form reactive oxygen species which can cause oxidative stress (Bhalerao et al., 2015).

Its toxicity adversely affects the growth of plants causing inhibition of shoot growth (Gajewska et al., 2006), and reduction of fresh biomass of root and shoot (Lu et al., 2010). Dry matter yield is also reduced due to the accumulation of higher concentrations of the nickel (Rathor et al., 2014).

Nickel at elevated levels causes reduction in photosynthetic activity (Hussain et al., 2013). It causes destruction of photosynthetic apparatus as it damages the chloroplast grana structure and reduces the size of grana (Chen et al., 2009). Chlorophyll a, chlorophyll b, and total chlorophyll contents decrease due to nickel toxicity (Singh, 2011; Lu et al., 2010) which results in chlorosis of leaves (Singh, 2011; Lu et al., 2014). Carotenoid contents also decrease due to nickel toxicity (Hassanpour and Rezayatmand, 2015; Lu et al., 2010; Singh et al., 2012).

Cicer arietinum L. commonly called chickpea is a legume crop of family Fabaceae. It is mostly used as seed food (Ibricki et al., 2003), as a rich source of proteins and other nutrients (Ibricki et al., 2003; Al-Snafi, 2016). Dried seeds contain 20% proteins, 61% carbohydrates and seed coat serves as the source of crude fiber. Seeds also contain certain minerals such as Ca, P and K alongwith vitamins A, B and C (Al-Snafi, 2016).

It has been reported that 10ppm concentration of nickel is damage threshold for chickpea and elevated levels resulted in reduced seed germination, growth, yield and pigment contents (Khan and Khan, 2010).

The experiment aims at providing some basic mechanisms regarding toxic effects of varying concentrations of nickel on plants.

Material and Methods

The present study was conducted in Botanical Garden of University of Agriculture, Faisalabad, Pakistan. Sixty earthen pots (30x30cm) were used allocating six pots for each treatment. 8kg homogeneously mixed and sun dried soil was used for each pot. Eight seeds of each variety of Cicer arietinum L. (Punjab 2000 and Bittal 98) were sown in each pot separately. After germination, the plants were thinned to maintain five seedlings in each pot. The plants were irrigated with tap water at alternate days.

Nickel as $NiCl_2$ was applied to soil in solution form 30 days after the germination of plants. There were five treatments viz., T_0, T_1, T_2, T_3 and T_4 @ 0 mg L^{-1}(Control), 25 mg L^{-1}, 50 mg L^{-1}, 100 mg L^{-1} and 150 mg L^{-1}, respectively. The experiment was laid down in Completely Randomized Design (CRD) with two factor-factorial arrangement.

Growth parameters were studied at three successive harvests at an interval of two weeks each starting from treatment application. At the time of sampling, three plants from each treatment were taken to determine mean values.

Growth analysis

Values of growth parameters were subjected to growth analysis using Radford's formulae (1967).

1. Relative growth rate (RGR) (g/g/day)

RGR= (log W_2-log W_1) / t

Where

W_1 = Dry weight of preceding harvest

W_2 = Dry weight of following harvest

t = Time interval between two harvests

Using above formula, following were also calculated:

2. Relative increase in plant height
3. Relative increase in number of branches
4. Relative increase in number of leaves

These parameters were not subjected to statistical analysis.

The chlorophyll (chl. a, chl. b and total) and carotenoid contents were determined when plants were 87 days old. 0.5 g fresh leaves were chopped in small pieces and extracted with 10 ml of 80% acetone. The chlorophyll contents were determined by the method of Arnon, (1949) and carotenoid contents following Davies, (1979) by a spectrophotometer.

At maturity of the crop, dry biomass of the plants was determined. For this purpose, the sample plants were kept in an oven at 70°C for 4 days and then mean values were calculated.

The data collected were analyzed statistically by applying ANOVA. Duncan's Multiple Range Test (Steel and Torrie, 1986) was also applied to find out significant differences among treatment means.

Results

Data for various growth parameters are presented in Fig.1, Fig.2, Fig.3 and Fig.4 which showed a gradual decrease with increasing levels of nickel application.

ANOVA for different parameters are presented in Table.1 while Table.2 and Table.3 showed comparison of means for variety Punjab 2000 (V_1) and variety Bittal 98 (V_2) respectively. For chlorophyll 'a', T_1 showed maximum contents of chl.'a' (0.615) and (0.628) for (V_1) and (V_2) respectively whereas T_4 showed the minimum (0.570) for (V_1) and (0.591) for (V_2). Chlorophyll 'b' contents revealed the gradual decrease of 2.49%, 4.53%, 6.90% and 9.17% in T_1, T_2, T_3 and T_4 respectively for variety Punjab 2000. Similar trend was noted for variety Bittal 98 showing a decrease of 1.84%, 3.68%, 6.09% and 7.24% for T_1, T_2, T_3 and T_4 respectively.

For total chlorophyll, there was a little decrease at T_1 and T_2 but then more pronounced at T_3 and T_4. The maximum decrease of 17.79% and 7.29% was observed for (V_1) and (V_2) respectively at 150 mg L^{-1} of nickel treatment. For carotenoids, (V_1) showed maximum decrease of 16.83% at T_4 while minimum of 6.57% was noted for T_1. T_2 and T_3 showed intermediate values. (V_2) also had maximum decrease of 12.80% at T_4 and minimum decrease of 3.20% at T_1. Maximum dry matter yield was recorded for control which gradually decreased in all other treatments of nickel for both varieties.

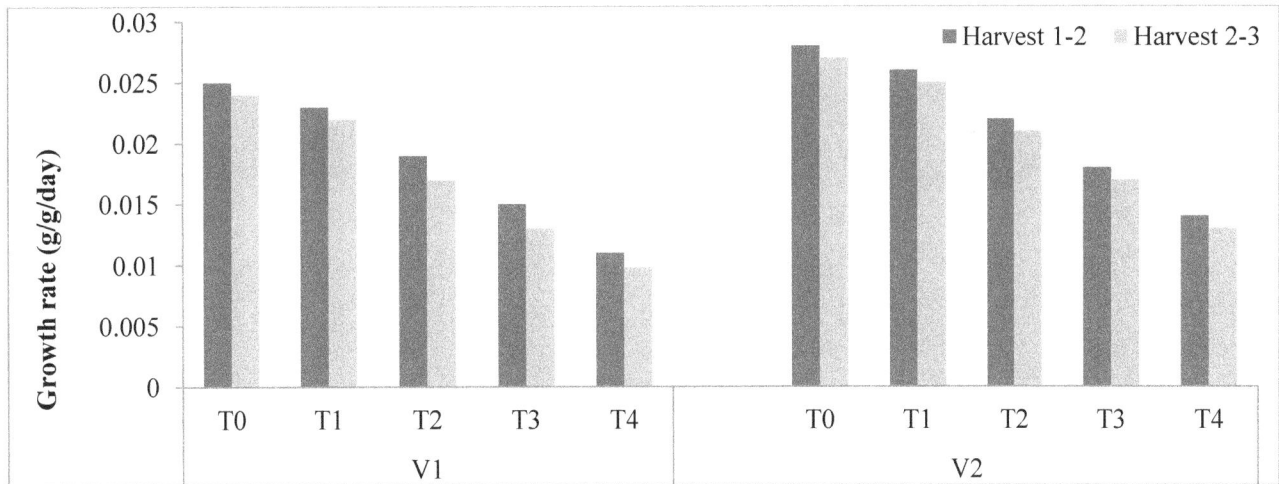

Fig.1. Relative growth rate as influenced by nickel of two chickpea varieties

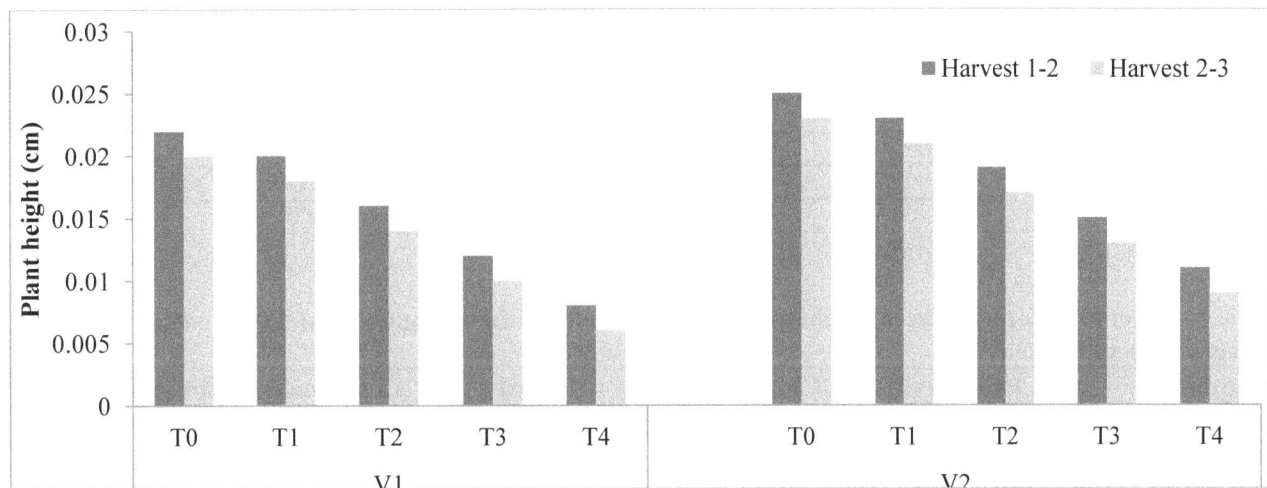

Fig.2. Relative increase in plant height (cm) as influenced by nickel of two chickpea varieties

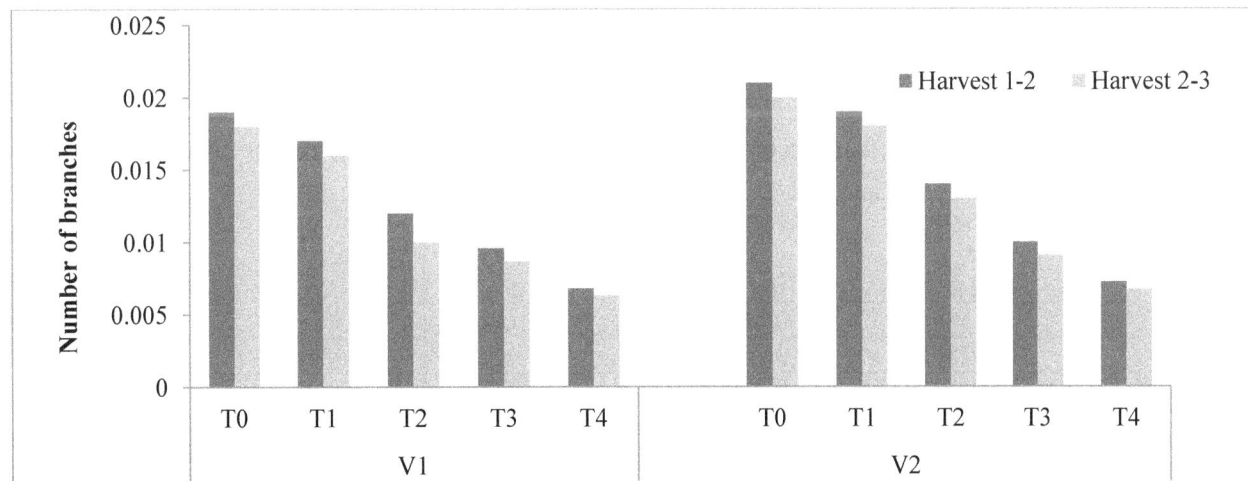

Fig.3. Relative increase in number of branches as influenced by nickel of two chickpea varieties

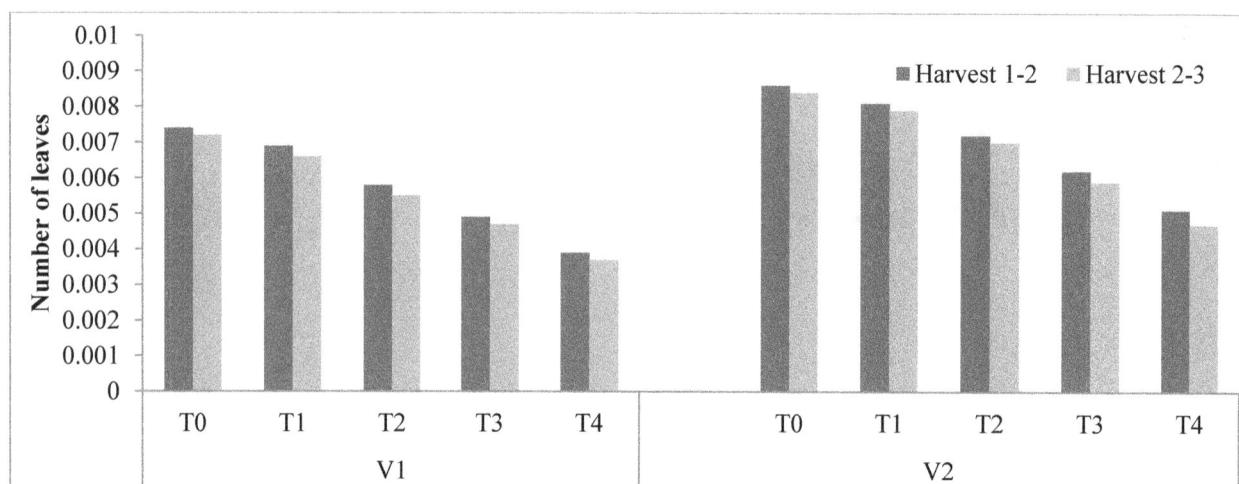

Fig.4. Relative increase in number of leaves as influenced by nickel of two chickpea varieties

Table. 1. ANOVA for effect of nickel on various measured parameters of two chickpea varieties.

Source of variation	df	Chlorophyll 'a' (mg g^{-1})	Chlorophyll 'b' (mg g^{-1})	Total chlorophyll (mg g^{-1})	Carotenoids (mg g^{-1})	Dry matter yield (g)
Varieties (V)	1	0.002**	0.000**	0.004ns	0.007**	0.172**
Treatments (T)	4	0.002**	0.005**	0.036**	0.005**	0.847**
Interaction (V x T)	4	0.000**	0.000**	0.009ns	0.000*	0.004*
Error	20	0.000	0.000	0.008	0.000	0.001

**= Highly Significant   *= Significant   ns= Non Significant

Table. 2. Effect of different concentrations of nickel on photosynthetic pigments and dry matter yield of chickpea variety Punjab 2000.

Treatments	Chlorophyll 'a' (mg g^{-1})	Chlorophyll 'b' (mg g^{-1})	Total chlorophyll (mg g^{-1})	Carotenoids (mg g^{-1})	Dry matter yield (g)
T_0	0.620a	0.883a	1.669a	0.487a	3.31a
T_1	0.615b	0.861b	1.480ab	0.455b	3.07b
T_2	0.608c	0.843c	1.452b	0.435c	2.87c
T_3	0.598d	0.822d	1.411b	0.421d	2.64d
T_4	0.570e	0.802e	1.372b	0.405e	2.41e

Values in columns followed by same letters indicate non-significant difference according to DMR Test.

Table. 3. Effect of different concentrations of nickel on photosynthetic pigments and dry matter yield of chickpea variety Bittal 98.

Treatments	Chlorophyll 'a' (mg g^{-1})	Chlorophyll 'b' (mg g^{-1})	Total chlorophyll (mg g^{-1})	Carotenoids (mg g^{-1})	Dry matter yield (g)
T_0	0.638a	0.869a	1.507a	0.500a	3.51a
T_1	0.628b	0.853b	1.481ab	0.484b	3.24b
T_2	0.616c	0.837c	1.453b	0.472c	3.07c
T_3	0.608d	0.816d	1.424b	0.459d	2.74d
T_4	0.591e	0.816d	1.397b	0.436e	2.51e

Values in columns followed by same letters indicate non-significant difference according to DMR Test.

Discussion

The present study was an attempt to evaluate the toxic effects of nickel on photosynthetic pigments and dry matter yield of Cicer arietinum L. Experiment clearly showed that principal photosynthetic pigments, chlorophyll 'a' and chlorophyll 'b' as well as total chlorophyll contents gradually decreased with increasing levels of nickel toxicity (Rastgoo et al., 2014; Shafeeq et al., 2012). However, chlorophyll 'b' was found to be more sensitive to nickel than chlorophyll 'a' (Aldoobie and Baltgai, 2013), showing more decrease over control than that of chlorophyll 'a'. This reduction in chlorophyll contents might be due to inhibition of enzymes for chlorophyll biosynthesis (Kaveriammal and Subramani, 2013), such as protochlorophyll-dereductase and δ-aminolevulinic acid dehydratase (Younis et al., 2015), and also due to decrease in Mg uptake (Chen et al., 2009). It has been reported that excess nickel resulted in decreased nitrogen contents in chickpea plants (Athar and Ahmad, 2002) which might be a contributing factor towards reduced cholorophyll contents. Carotenoids, which are the accessory photosynthetic pigments also reduced due to higher concentrations of nickel (Sai Kachout et al., 2015; Flemotomou et al., 2011).). Photosynthetic activity has been reported to be inhibited due to decrease in stomatal conductance and destruction of mesophyll tissue (Hussain et al., 2013), decrease in cholorophyll contents (Chen et al., 2009) and reduction in number of leaves (Hussain et al., 2013; Chen et al., 2009). This reduction in photosynthesis adversely affected the growth and fresh biomass (Younis et al., 2015; Ishtiaq and Mahmood, 2011) as well as dry matter yield of plants (El-Enany et al., 2000).

References

Aldoobie NF and Beltagi MS, 2013. Physiological, biochemical and molecular responses of common bean (*Phaseolus vulgaris* L.) plants to heavy metal stress. Afr. J. Biotechnol. 12: 4614-4622.

Al-Snafi E, 2016. The medical importance of Cicer arietinum- A Review. IOSR J. Pharmacy. 6: 29-40.

Arnon DI, 1949. Copper enzymes in isolated chloroplasts: polyphenol oxidase in Beta vulgaris. Plant Physiol. 24: 1-15.

Athar R and Ahmad M, 2002. Heavy metal toxicity in legume - microsymbiont system. J. Plant Nutr. 25: 369-386

Bhalerao SA, Sharma AS and Poojari AC, 2015. Toxicity of Nickel in Plants. Inter. J. Pure App. Biosci. 3: 345-355.

Chen C, Haung D and Liu J, 2009. Functions and Toxicity of Nickel in Plants: Recent Advances and Future Prospects. Clean. 37: 304-313.

Davies BH, 1976. Carotenoids. In chemistry and biochemistry of plant pigments. (Goodwin T.W. Ed.) Academic Press, London. 2: 38-165.

Dubey D and Panday A, 2011. Effect of Nickel (Ni) on chlorophyll, Lipid peroxidation and Antioxidant enzyme activities in black gram (*Vigna mungo*) leaves. Int. J. Sci. Nature. 2: 395-401.

El-Enany AE, Atia MA, Abd-Alla MH and Rmadan T, 2000. Response of bean seedlings to nickel toxicity: role of calcium. Pak. J. Bio. Sci. 3: 1447-1452

Emamverdian A, Ding Y, Mokhberdoran F and Xie Y, 2015. Heavy metal stress and some mechanisms of Plant Defense Response. The Sci. World J. 2015: 18pp.

Esfahani H, and Rezayatmand Z, 2015. Evaluation of some physiological and biochemical parameters of variety of sunflower sanbaro (Helianthus annuus L.) under nickel toxicity. Ind. J. Fund. Appl. Life Sci. 5: 88-99

Flemotomou E., Molyviatis T and Zabitakis I, 2011. The Effect of Trace Elements Accumulation on the Levels of Secondary Metabolites and Antioxidant Activity in Carrots, Onions and Potatoes. Food Nut. Sci. 2: 1071-1076

Gajewska E, Sklodowska M, Slaba M and Mazur J, 2006. Effect of nickel on antioxidative enzyme activities, proline and chlorophyll contents in wheat shoots. Bio. Plant. 50: 653-659

Harasim P and Filipek T, 2015. Nickel in the Environment. J. Elementol. 20: 525-534

Hussain MB, Ali S, Azam A, Hina S, Farooq MA, Ali B, Bharwana SA and Gill MB, 2013. Morphological, physiological and biochemical responses of plants to nickel stress: A review. Afr. J. Agri. Res. 8: 1596-1602

Ibricki H, Knewtson SJB and Grusak MA, 2003. Chickpea leaves as a vegetable green for humans: evaluation of mineral composition. J. Sci. Food Agri. 83: 945-950

Ishtiaq S and Mahmood S, 2011. Phytotoxicity of nickel and its accumulation in tissues of three Vigna species at their early growth stages. J. Appl. Bot. Food Qual. 84: 223-228

Kaveriammal S and Subramami A, 2013. Toxic effect of nickel chloride (NiCl$_2$) on the growth behavoiur and biochemical constituent of groundnut seedlings (*Arachis hypogeae* L.). Int. J. Res. Bot. 3: 48-52

Khan MR and Khan MM, 2010. Effect of varying concentrations of nickel and cobalt on plant growth and yield of chickpea. Aust. J. Basic Appl. Sci. 4: 1036-1046

Lu Y, Li X, He M, Wang Z and Tan H, 2010. Nickel effects on growth and antioxidative enzymes activities in desert plant *Zygophyllum xanthoxylon* (Bunge) Maxim. Sciences in Cold and Arid Regions. 2: 436-444

Radford PJ. 1967. Growth analysis formulae their use and abuse. Crop Sci. 7: 171-175

Rastgoo L, Alemzadeh A, Tale AM, Tazangi SE and Eslamzadeh T, 2014. Effects of copper, nickel and zinc on biochemical parameters and metal accumulation in gouan, *Aeluropus littoralis*. Plant Knowledge J. 3: 31-38

Rathor G, Chopra N and Adhikhari T, 2014. Effect of variation in nickel concentration on growth of maiz plants: A comparative overview for pot and Hoagland culture. Res. J. Chem. Sci. 4: 30-32

Sai Kachout S, Ben Mansoura A, Ennajah A, Leclerc JC, Ouerghi Z and Karray Bouraoui N, 2015. Effects of Metal Toxicity on Growth and Pigment Contents of Annual Halophyte (A. hortensis and A. rosea). Int. J. Environ. Res. 9: 613-620

Shafeeq A, Butt ZA and Muhammad S, 2012. Response of nickel pollution on physiological and biochemical attributes of wheat (*Triticum aestivum* L.) Var.Bhakkar 02. Pak. J. Bot. 44: 111-116

Singh G, Agnihotri RK, Reshma RS and Ahmad M, 2012. Effect of lead and nickel toxicity on chlorophyll and proline content of Urd (*Vigna mungo* L.) seedlings. Inter. J. Plant Physiol. Biochem. 4: 136-141

Singh K, 2011. Effect of nickel-stresses on uptake, pigments and antioxidative responses of water lettuce, Pistia stratiotes L. J. Environ. Bio. 32: 391-394

Steel RGD and Torrie JH, 1986. Principles and Procedures of Statistics.2nd Ed. (McGraw Hill Book Inc. New York) pp. 336-354

Younis U, Athar M, Malik SA, Raza Shah MH and Mahmood S, 2015. Biochar impact on physiological and biochemical attributes of spinach *Spinacea oleracea* (L.) in nickel contaminated soil. Global J. Environ. Sci. Manage. 1: 245-254.

Combined application of sorghum and mulberry water extracts is effective and economical way for weed management in wheat

Shahbaz Khan[1], Sohail Irshad[2]*, Faisal Mehmood[1], Muhammad Nawaz[3]

[1]Department of Agronomy, University of Agriculture, Faisalabad, Pakistan.
[2]In-Service Agricultural Training Institute, Rahim Yar Khan, Pakistan.
[3]Department of Agronomy, College of Agriculture, BZU, Bahadur Campus Layyah, Pakistan.

Abstract

Wheat (*Triticum aestivum* L.) is of prime importance being staple food of the masses of Pakistan. Weed infestation in wheat not only reduce the yield but also affects the efficiency of other production factors. Allelopathy is sustainable and ecofriendly method for the management of weeds and diseases. A field trail was carried out to assess the allelopathic potential of mulberry and sorghum water extracts against invasive winter weeds like *Phalaris minor*. Retz, *Chenopodium album* L., *Avena fatua* L. and *Convolvulus arvensis* L. in wheat at Agronomic Research Farm, University of Agriculture, Faisalabad. Experiment was laid out in randomized complete block design (RCBD) with four replications. All the treatment application showed better weed management and enhanced final yield of the crop. Application of sorghum water extract (SWE) at the rate of 18 L ha^{-1} and mulberry water extract (MWE) at the rate of 18 L ha^{-1} indicated better weed management (51-55%) and improved grain yield (28%) as compared to control. SWE at the rate of 27 L ha^{-1} and MWE at the rate of 9 L ha^{-1} in combined application also showed better weed management and improved the grain yield while Atlatis (Standard herbicide) had 66-68% weed control and 32% increase in final yield but it was not found economical. Combined application of SWE at the rate of 18 L ha^{-1} and MWE at the rate of 18 L ha^{-1} was found best combination with maximum net return while hand weeding and herbicide application were extravagant because of high cost and low net return.

Keywords: Allelopathy, Wheat, Water Extracts, Weeds

*Corresponding author email:
sohail_99uaf@yahoo.com

Introduction

Wheat is of prime importance being the main source of food due to its nutritional value since long. Most of the nutritional requirements of human body are met through this crop. It has main role in national economy. The final yield of the crop is not enough to meet dietary requirements of ever increasing population. Multiple factors including weed infestation, late sowing, improper nutrition and water use efficiency are responsible for limiting the yield (Jabran et al., 2011). A number of constraints hinder wheat production like poor quality seed, poor storage conditions, high prices of fertilizers and inadequate measures of plant protection principally weed infestation. Noxious weeds of wheat like *Phalaris minor* and *Avena fatua* reduce the yield up to 30%. Reduction in the wheat yield up to 20-30% was observed in case of high infestation of field with *Avena fatua* L. (Hassan and Khan, 2007).

Most of the weeds have established resistance against herbicides (Heap, 2008) and use of these agrochemicals have many health concerns (Kudsk and Streibig, 2003) with the problems like their residual effects and degradation in soil. Application of Mesotrione (Selective herbicide) to soil affects the population of microbes (Crouzet et al., 2010). Allelopathy is a novel method for the management of weeds and diseases. It refers to positive or negative impact of one plant to other including both weed and crop species by the release of some allelochemicals from plant parts through leaching, residue decomposition, root exudates and other ways in agro ecosystem (Ferguson and Rathinasabapathi, 2003). It has stimulatory along with inhibitory effect on plants as well as on microorganism. Number of studies depicted the effectiveness of allelochemicals against crop pests while working on the effect of sorghum, sunflower and brassica water extracts along with reduced dose of glyphosate for the management of purple nutsedge in cotton, Iqbal et al. (2009) investigated a reduced density and biomass of purple nutsedge.

Allelochemicals can interrupt the normal functioning of plant and can affect the physiological functions like respiration, photosynthesis and cell elongation (Field et al., 2006). There is mounting problem of pesticide and herbicide resistance in modern agriculture. Owing to pesticide resistance number of health related problems are increasing, unselective use distressed the ecosystem with soil and environment concerns, in this scenario allelopathy can play its role to cope these problems (Zhu and Li, 2002; Roseveld and Bretveld, 2008; Dayan et al., 2009). It is common observation of reduced vegetation under mulberry tree and its leaf defoliation in autumn suppress the growth of neighboring weeds. Use of mulberry water extracts reduced germination, root and shoot length with reduced dry weight of raddish (Hong et al., 2003). Wheat yield is affected by different broad and narrow leaf weeds including broad leaf dock, sweet clover, swine cress, field bind weed, fumitory, wild medic, little seed canary grass and wild oat (Shamsi and Ahmed, 1984).

Hong et al. (2004) reported decrease in weed density (65%) and dry weights (70%) when mulberry leaves were incorporated in soil at the rate of 2 t ha^{-1} in rice crop which ultimately increased the yield of the paddy by 23%. Some of the studies report the stimulatory effect of mulberry e.g. mulberry water extract of 50% concentration promoted the growth and development in broad bean and pea while in case of lentil, germination and growth showed positive impact by the single application of 25% concentration of mulberry water extract (Mughal, 2000).

Present study was conducted to assess the combined and alone application of mulberry and sorghum water extracts for the management of weeds in wheat and in particular to endorse an ecofriendly approach to reduce environmental and human health hazards by indiscriminate use of agrochemicals.

Material and Methods

A field trail was conceded to assess the allelopathic impact of sorghum and mulberry water extracts on weeds of wheat during the season 2014-2015 at Agronomic Research Area, University of Agriculture, Faisalabad, using randomized complete block design (RCBD) with four replications. Wheat cultivar AARI-2011 was sown by hand drill with R*R of 22 cm with gross plot size of 7 m × 2.2 m and 10 rows in each plot. Sowing was done in well pulverized soil (soil type: Lyallpur soil series (Aridsol-fine silty, mined, hyperthermic, ustalfic, Haplagrid) in USDA classification). Fertilizers were applied at the rate of 100-90-75 kg ha^{-1}. Irrigations were given according to the crop requirement. Treatments combination plan has been shown in table 1.

Preparation of extracts and herbicidal solution

Water extracts of sorghum and mulberry were prepared according to Cheema and Khaliq (2000a). Briefly, mulberry leaves were collected from the Agronomic Research Area, University of Agriculture, Faisalabad, Pakistan. The leaves were washed three times under running tap water and were dried in an oven at 70°C for 48 h till a constant weight was attained. The dried leaves were soaked in tap water at 10% (w/v) at room temperature (25 _ 3°C) for 24 h. After that, the material was filtered to obtain a 10% extract. Then, the extract was boiled to concentrate it by 95%. This concentrated extract was named as 100% (stock solution). Mature sorghum crop was used for all these treatment preparations. It was placed in the shade for couple of days to avoid leaching by rain. After drying in the shade , it was chopped by ordinary fodder cutter ,thereafter, then this material was soaked in water in 1:10 ratio for 24 hours and concentrated to reduce its volume by 95% via boiling at100∘C (Cheema and khaliq, 2000a). The volume of the spray was 330 L ha $^{-1}$. Recommended doses of iodosulforun + mesosulforun at the rate of 14.4 g ha^{-1} a.i. (Atlantis 3.6 WG); product of Bayer crop sciences, was used.

Application of treatments

There was no application of treatment in weedy check plot. It was left untreated throughout the crop growth period. Hand weeding was done twice at 30 and 45 days after sowing the crop. Mixture of Iodosulfuran and Mesosulfuron (Atlantis 3.6WG) at the rate of 14.4 g a.i. ha^{-1} was sprayed after 30 days of sowing of crop. Similarly all other combinations of mulberry and sorghum water extracts were applied once in crop growth period after 30 days of sowing the wheat crop.

Weed Parameters

Weed density was calculated by counting all weeds present in 1 m^2. Fresh weight of all weeds was recorded by pulling out all of weeds plants and weighing them by ordinary scientific balance. After recording the fresh weight, weed samples were first sun dried, thereafter placed in an oven at 70°C for 72 hours till then all of the moisture was removed. These samples were weighed by weighing balance to get dry weight.

Yield and yield components

Total number of tillers was calculated by counting the number of tiller from an area of 1 m^2. Regarding the data of number of spikelets/spike, five spikes were selected at random from each plot and number of spikelets was calculated and means of the calculated values was determined. For the estimation of final grain yield, an area of 1 m^2 of fully mature crop was harvested from each plot. After threshing, the weight of grains was taken.

Economic analysis

Economic analysis was carried to determine which treatment was much cost effective. Economic analysis was done through method prescribed by Byerlee et al. (1986).

Statistical analysis

Data were analyzed statistically by using Statistix 8.1 version (a computer software package for statistical analysis) and the difference between treatments' means was compared by employing LSD test at 5% level of probability (Steel et al., 1997).

Results

Weed Parameters
Total weed density/m^{-2}

Significant impact was observed on growth parameters of weed when applied with allelopathic water extracts. A number of weed growth parameters like fresh and dry biomass of weeds and weed density provides the information of weed infestation in the field. In the present study, weed density was calculated after 45 and 60 after sowing of the crop. Different broad and narrow leaf weeds like *Phalaris minor, Avena fatua, chenopodium album* and *convolvulus arvensis* infestation in the field was observed. Most of treatments applied showed reduced weed population (Table 3) but combined application of SWE at the rate of 18 L ha^{-1} and MWE at the rate of 18 L ha^{-1} reduced extreme weed density by 55% at 45 days after sowing (DAS) and 60% at 60 DAS while combined application of SWE at the rate of 27 L ha^{-1} and MWE at the rate of 09 L ha^{-1} depressed weed density by 41% at 45 DAS and 54% at 60 DAS. Total fresh and dry weed biomass was also affect significantly by the phytotoxic effect of allelochemicals of extracts when compared to control. Fresh biomass of weeds was also reduced by the application of mixture of Iodosulfuran + Mesosulfuron (Atlantis 3.6 WG) at the rate of 14.4 g a.i. ha^{-1} up to 65-75% in comparison to other treatments this trend was followed by combined application of water extracts of both sorghum and mulberry at the rate of 18 L ha^{-1} each by 51-56% while alone application of both the extracts found less effective in reducing the fresh biomass of weeds. In case of dry weights of the weeds similar observations were recorded and it was reduced up to 66-68% in case of application of (Atlantis 3.6WG) at the rate of 14.4 g a.i. ha^{-1}) this trend was followed by combined application of both the water extracts at the rate of 18 L ha^{-1} of each with the reduction in dry weight up to 51-55%. It was observed that the combine application of both water extracts reduced the dry weight of weeds as compared to sole application but alone application of MWE was found less effective as compared to SWE in reducing dry weight of weeds.

Yield parameters

Total tillers, number of spikelets/spike and grain yield were expressively affected by the application of definite treatments in comparison with control. In case of hand weeding, number of tillers and spikelets/spike were more and this trend was at par where herbicide application was done. In case of crop water extracts, combined application of both the extracts at the rate of 18 L ha^{-1} of both produced maximum number of tillers which were statistically similar to application of herbicide (Table 2). Least number of tillers and spikelets/spike were observed in case of weedy check which was found statistically similar to alone application of MWE at the rate of 18 L ha^{-1}. Same was the case with the grain yield; maximum grain yield was recorded in hand weeding which was similar to herbicide application while minimum grain yield was observed in case of weedy check (Table 2).

Economic analysis

Effectiveness of the production scheme can be assessed for its economic fitness for the farmers.

Present study was evaluated by conducting economic analysis along with marginal rate of return to determine best treatments applied. Results of the analysis depicted increase in income of Pakistani rupees 2570 to 25803 ha^{-1} in different combination of treatments as compared to control treatment. Maximum net benefits (Pak Rs. 25803) (236.72$) were obtained with combined application of both crop water extracts at the rate of 18 L ha^{-1} each while it was followed by application of herbicide (Pak Rs. 25463) (233.60$) and hand weeding with Pakistani rupees of 24850.5. Minimum net return (Pak Rs. 2570) (23.57$) was obtained in case of sole application of MWE at the rate of 18 L ha^{-1}.

Table 1: Treatment Combinations

	Description
T$_1$	Weedy check (Weed was not removed throughout growth period)
T$_2$	Hand weeding (30 and 45 DAS; days after sowing)
T$_3$	Iodosulfuran + Mesosulfuron (Atlantis 3.6WG) at the rate of 14.4 g a.i. ha^{-1} (30 DAS)
T$_4$	Sorghum water extract18 L ha^{-1} (30 DAS)
T$_5$	Sorghum water extract 27 L ha^{-1} (30 DAS)
T$_6$	Sorghum water extract 36 L ha^{-1} (30 DAS)
T$_7$	Mulberry water extract 18 L ha^{-1} (30 DAS)
T$_8$	Mulberry water extract 27 L ha^{-1} (30 DAS)
T$_9$	Mulberry water extract 36 L ha^{-1} (30 DAS)
T$_{10}$	Sorghum water extract 27 L ha^{-1} + Mulberry water extract 09 L ha^{-1} (30 DAS)
T$_{11}$	Sorghum water extract 18 L ha^{-1} + Mulberry water extract 18 L ha^{-1} (30 DAS)
T$_{12}$	Sorghum water extract 09 L ha^{-1} + Mulberry water extract 27 L ha^{-1} (30 DAS)

Table 2: Effect of sorghum and mulberry water extracts on yield parameters of wheat

Treatments	Number of tillers m^{-2}	Number of spikelets per spike	Grain yield (t ha^{-1})
Weedy check	331 g	11 e	3.88 f
Hand weeding	382 a	18 a	5.31 a
Iodosulfuron + Mesosulfuron	374 ab	17 ab	5.17 ab
SWE at the rate of 18 L ha^{-1}	347 ef	13 de	4.08 de
SWE at the rate of 27 L ha^{-1}	350 de	13 de	4.10 de
SWE at the rate of 36 L ha^{-1}	357 cd	14 cd	4.44 d
MWE at the rate of 18 L ha^{-1}	337 fg	13 de	4.02 e
MWE at the rate of 27 L ha^{-1}	343 ef	13 de	4.08 de
MWE at the rate of 36 L ha^{-1}	352 de	14 cd	4.27 d
SWE at the rate of 27 L ha^{-1} + MWE at the rate of 09 L ha^{-1}	358 cd	16 bc	4.70 c
SWE at the rate of 18 L ha^{-1} + MWE at the rate of 18 L ha^{-1}	368 bc	16 abc	5.00 b
SWE at the rate of 09 L ha^{-1} + MWE at the rate of 27 L ha^{-1}	361 cd	15 bc	4.50 c
LSD at 5% probability level	11.695	2.2781	0.2166

Different lettering shows the statistically significant difference among the performance of treatments ($P <$ 0.05) DAS = Days after sowing; SWE = Sorghum water extract; MWE = Mulberry water extract; LSD = Least significant difference.

Table 3: Effect of sorghum and mulberry water extracts on total weed density (m^{-2})

Treatments	Weed density at 45 DAS (m^{-2})	Weed density at 60 DAS (m^{-2})	Fresh weight at 45 DAS (g m^{-2})	Fresh weight at 60 DAS (g m^{-2})	Dry weight at 45 DAS (g m^{-2})	Dry weight at 60 DAS (g m^{-2})
Weedy check	35.25 a	38.25 a	47.25 a	56.07 a	5.97 a	7.09 a
Hand weeding	0.00 i	0.00 i	0.00 h	0.00 h	0.00 i	0.00 h
Iodosulfuran + Mesosulfuron	8.25 h	7.250 h	15.00 g	14.00 g	1.89 h	2.37 g
SWE at the rate of 18 L ha⁻¹	27.00 bc	23.75 bc	40.25 bc	48.18 bc	5.09 bc	6.09 bc
SWE at the rate of 27 L ha⁻¹	25.25 d	24.25 bc	39.00 bc	45.01 bcd	4.93 cd	5.94 bc
SWE at the rate of 36 L ha⁻¹	23.75 e	21.75 de	33.50 d	38.23 d	4.23 e	5.09 d
MWE at the rate of 18 L ha⁻¹	28.25 b	25.50 b	44.00 ab	50.37 ab	5.48 b	6.62 ab
MWE at the rate of 27 L ha⁻¹	26.75 c	23.50 cd	42.00 b	49.65 ab	5.34 b	6.53 ab
MWE at the rate of 36 L ha⁻¹	26.50 cd	23.00 cd	35.75 cd	40.87 cd	4.52 de	5.42 cd
SWE at the rate of 27 L ha⁻¹ + MWE at the rate of 09 L ha⁻¹	20.75 f	17.50 f	24.50 ef	31.42 ef	3.09 f	3.92 ef
SWE at the rate of 18 L ha⁻¹ + MWE at the rate of 18 L ha⁻¹	15.75 g	15.00 g	20.75 f	26.90 f	2.63 g	3.44 f
SWE at the rate of 09 L ha⁻¹ + MWE at the rate of 27 L ha⁻¹	22.00 f	20.75 e	27.50 e	32.97 e	3.42 f	4.17 e
LSD at 5% probability level	1.3849	1.8073	5.0765	6.9199	0.4145	0.7176

Different lettering shows the statistically significant difference among the performance of treatments ($P < 0.05$) DAS = Days after sowing; SWE = Sorghum water extract; MWE = Mulberry water extract; LSD = Least significant difference

Discussion

Weed density is the key parameter that gives index of weed infestation in the field. In present study, weed density was recorded after 45 DAS and 60 DAS. Infestation of different braod and narrow leaf weeds like *convolvulus arvensis*, *Avena fatua*, *chenopodium album* and *Phalaris minor* was observed during the study in the field. Combined application of SWE and MWE at the rate of 18 L ha⁻¹ of both showed maximum weed control as compared to other treatments as depicted in Table 3. There was substantial variation in weed control between 45 and 60 DAS; this might be due to natural variation of weed seed bank in soil. Findings of our study depicted that sole or combine application of both crop water extracts had more or less effect on weed density in field. These results are in line with the findings of Cheema and Khaliq (2000) who reported less population of weeds when applied with sorghum water extract. Our findings are quite similar to the findings of Cheema et al. 2000 who reported weed control up to 16-55% by spraying sorghum water extract. Hong et al. (2004) reported 65% reduction in weed density in paddy field when applied with mulberry water extract. Parameters like fresh and dry weight are also of prime importance as it regulates the uptake of different nutrients from soil profile by altering it into micro and macro molecules. Allelopathic extracts when applied to the crop, affected the fresh as well as dry weight (Table 3).

Alone application of MWE was less effective in suppressing weed fresh weight as compared to SWE. It might be possible due to the presence of allelochemicals present in water extracts that possessed inhibitory effect rather than killing weeds. Our results are in line with the findings of Putnam and Defrank (1979) who reported that crop plants release some of the chemicals that have inhibitory effect which could be used for the management of weeds. Our findings also confirm the study of Purvis et al. (1985) who stated inhibitory effect of sorghum that suppressed the dry biomass of weeds due to these allelochemicals.

Improvement in the yield parameters of wheat was observed due to less wheat density by the application of water extracts. Present study revealed that presence of allelochemicals in water extracts have promoting effect on wheat while negative impact on weed growth and development. Application of water extracts significantly affected yield parameters like number of spikelets/spike, total tillers and grain yield. Maximum number of tillers, spikelets/spike, total tillers and grain yield were observed in case of hand weeding which was statistically at par with herbicide application. In case of water extracts, combined application of water extracts of SWE and MWE at the rate of 18 L ha^{-1} produced maximum number of tillers and other yield parameters (Table 2). Improvement in spikelets/spike may be due to better weed management. Our results are supported the findings of Einhelling and Rasmussen (1989) who reported inhibitory effect of sorghum on weeds resulted in improvement in the yield of wheat. Likewise, findings of Hejl and Koster (2004) and Jabran et al. (2008) also supports our findings who reported inhibitory effects of sorghum and mulberry on weeds and their growth. Increase in grain yield might be due to suppressing ability of water extracts to weeds which ultimately caused in less competition of nutrients between crop and weeds resulted in more number of tillers, more spikelets/spike, increase in spike length, more number of grains/spike and grain weight. These results are in accordance with Haq et al. (2010) who reported better weed management in wheat by the application of MWE resulted in increase in grain yield.

Conclusion

It is concluded from present study that combined application of SWE at the rate of 18 L/ ha and MWE at the rate of 18 L/ha at 45 and 60 days after sowing of wheat helped to achieve maximum weed control. It also produced highest grain yield by improvement in yield contributing parameters with maximum net benefit returns. Sorghum and mulberry water extracts can be used effectively to control weed in wheat crop. It is recommended to use both these water extracts as tool for better weed management approach in wheat as these are safe and ecofriendly.

References

Cheema ZA and Khaliq A, 2000. Use of sorghum allelopathic properties to control weeds in irrigated wheat in a semi-arid region of Punjab. Agric. Eco. Envt. 79: 105-112.

Cheema ZA, Sadiq HMI and Khaliq A, 2000. Efficacy of Sorgaab (Sorghum water extract) as a natural weed inhibitor in wheat. Int. J. Agri. Biol. 2: 144-146.

Crouzet O, Batisson I, Hoggan PB, Bonnemoy FEE, Bardot C, Poly F, Bohatier J and Mallet C, 2010. Response of soil microbial communities to the herbicide mesotrione: A dose-effect microcosm approach. Soil Biol. Biochem. 42: 193-202.

Dayan FE, Cantrell CL and Duke SO, 2009. Natural products in crop protection. Bioorg. Med. Chem. 17: 4022-4034.

Einhelling FA and Rasmussen JA, 1989. Prior cropping with grain sorghum inhibits weeds. J. Chem. Ecol. 15: 951-960.

Ferguson JJ and Rathinasabapathi B, 2003. Allelopathy: How plants suppress other plants. online. Internet. Available from URL: http://edis.ifas.ufl.edu./ Accessed October 2011.

Field B, Jordan F and Osbourn A, 2006. First encounters – deployment of defence-related natural products by plants. New Phytol. 172: 193-207.

Haq RA, Hussain M, Cheema ZA, Mushtaq MN and Farooq M, 2010. Mulberry leaf water extract inhibits bermuda grass and promotes wheat growth. Weed Biol. Manage. 10: 234-240.

Hassan G and Khan AI, 2007. Yield and yield components of wheat affected by wild oat *Avena fatua* L densities under irrigated conditions in: Afri. Crop Sci. Proceed. 8: 33-36.

Heap I, 2008. The international survey of herbicide resistant weeds. Online. Internet. Available from URL: http://www. Weedscience.com/. Accessed August 2011.

Hejl AM and Koster KL, 2004. The allelochemical sorgoleone inhibits root H+-ATPase and water uptake. J. Chem. Ecol. 3: 2181-2191.

Hong NH, Xuan TD, Tsuzuki E and Khanh TD, 2004. Paddy weed control by higher plants from Southeast Asia. Crop Prot. 23: 255-261.

Hong NH, Xuan TD, Tsuzuki E, Hiroyuki T, Mitsuhiro M and Khanh TD, 2003. Screening for allelopathic potential of higher plants from Southeast Asia. Crop Prot. 22: 829-836.

Iqbal J, Cheema ZA and Mushtaq MN, 2009. Allelopathic crop water extracts reduce the herbicide dose for weed control in cotton (*Gossypium hirsutum*). Int. J. of Agri. Biol. 11: 360-366.

Jabran K, Cheema ZA, Farooq M and Khan MB, 2011. Fertigation and foliar application of fertilizers alone and in combination with canola extracts enhances yield in wheat crop. Crop Environ. 2: 42-45.

Jabran K, Cheema ZA, Farooq M, Basra SMA, Hussain M and Rehman H, 2008. Tank mixing of allelopathic crop water extracts with pendimethalin helps in the management of weeds in canola (*Brassica napus*) field. Int. J. Agric. Bio. 10: 293-296.

Kudsk P and Streibig JC, 2003. Herbicides – a two-edged sword. Weed Res. 43: 90-102.

Mughal AH, 2000. Allelopathic effect of leaf extract of *Morus alba* L. on germination and seedling growth of some pulses. Range Manage. Agroforest. 21: 164-169.

Purvis CE, Jessop RS and Lovett JV, 1985. Selective regulation of germination and growth of annual weeds by crop residues. Weed Res. 25: 415-421.

Putnam AR and Duke WO, 1979. Allelopathy in agro ecosystem Ann. Rev. Phytopathol. 16, 431-451 in Rice, E.L. Allelopathy 2nd ed. pp. 72.

Roseveld N and Bretveld R, 2008. The impact of pesticides on male fertility. Curr. Opin. Obstet. Gyn. 20: 229-233.

Shamsi SRA and Ahmed B, 1984. Ecophysiological studies on some important weeds of wheat. Final technical report, Pakistan Science foundation Research. P-PU/agriculture (64). Dept. Bot. Univ. Punjab, Lahore, Pakistan.

Steel RGD, Torrie JH and Dickey D, 1997. Principles and Procedures of Statistics: A Biometrical Approach. 3rd Ed. McGraw Hill Book Co. Inc. New York, USA. pp: 172-177.

Zhu Y and Li QX, 2002. Movement of bromacil and hexazinone in soils of Hawaiian pineapple fields. Chemosphere. 49: 669-674.

3

Uptake, Translocation of Pb and Chlorophyll Contents of *Oryza Sativa* as Influenced by Soil-Applied Amendments under Normal and Salt-Affected Pb-Spiked Soil Conditions

Muhammad Mazhar Iqbal[1,2,3,4,5]*, Ghulam Murtaza[2], Tayyaba Naz[2], Wasim Javed[2], Sabir Hussain[6], Muhammad Ilyas[3,4], Muhammad Ashfaq Anjum[3,4], Sher Muhammad Shahzad[5], Muhammad Ashraf[5] and Zafar Iqbal[7]

*[1]Soil and Water Testing Laboratory for Research, Chiniot, Department of Agriculture, Government of Punjab.
[2]Institute of Soil and Environmental Sciences, University of Agriculture, Faisalabad.
[3]Provincial Pesticide Reference Laboratory, Kala Shah Kaku, Sheikhupura.
[4]Institute of Soil Chemistry and Environmental Sciences, Ayub Agriculture Research Institute, Faisalabad.
[5]Department of Soil and Environmental Sciences, University of Sargodha, Sargodha.
[6]Department of Environmental Sciences and Engineering, Government College University Faisalabad, Faisalabad.
[7]Department of Plant Pathology, University of Sargodha, Sargodha.

Corresponding author email:
mazhar1621@gmail.com

Abstract

Heavy metal contamination of the soil environment has become a major source of concern and has posed serious human health related problems in many developing countries particularly Pakistan. Chemical immobilization of heavy metals can be accomplished by the addition of amendments to reduce contaminant solubility and ultimately uptake by the plants. However, a very scarce information is available on the immobilization of Pb with the application of different Ca, S and P sources (gypsum i.e., gyp, rock phosphate i.e., RP and Di-ammonium phosphate i.e., DAP) on rice grown normal and salt-affected Pb-spiked soils. Therefore, a pot trial was conducted to investigate the uptake, translocation of Pb and chlorophyll contents of rice as influenced by soil applied amendments (gyp, RP and DAP) and their variable amounts in normal and salt-affected Pb-spiked soils. The results showed that the Pb and salinity stress induced decrease in chlorophyll contents of rice were significantly ($p \leq 0.05$) counteracted by the applied gyp, RP and DAP. Application of 7.5 g gyp kg^{-1} soil was found the most effective in improving chlorophyll contents, and reducing Pb uptake and translocation both in normal and salt-affected Pb-spiked soils.

Keywords: Rice, lead accumulation, transport, photosynthetic pigments, amendments, saline Pb stressed soil.

Introduction

During the past few years, the rapid industrialization and urbanization have resulted in serious complications of environmental pollution. The natural ecological system has been severely affected due to rise in heavy metals status in soils (Meenakshi et al., 2006). The toxic metals marked a substantial impact on the environmental pollution due to increased human activities including mining and smelting operations, electroplating and energy processes, fuel production from fossils, power transmission setups, intensification of agriculture, sewage sludge dumping operations and military actions (Britto et al., 2011,

Rizwan et al., 2016). The presence of heavy metals is a menace for flora and fauna and ultimately to humans. Among heavy metals, lead (Pb) is one of the most abundant toxic elements that cause serious concerns to human health. It affects growth and metabolism of plants, also having visible symptoms such as stunted growth and lesser leaves, as well as membrane disorder and decrease in photosynthetic rate (Sharma and Dubey, 2005). Moreover, inhibition of chlorophyll synthesis is one of the most Pb-sensitive plant physiological feature (Li et al., 2012).

Rice (*Oryza sativa* L.) is the most dominant staple food in the world. According to an estimate, rice denotes 30% of the global cereal food production and will be needed for 4.6 billion people for their daily nutrition in 2025 (Gnanamanickam, 2009). In Pakistan, rice is grown on an area of 2.89 mha and its total production is 6.79 mt by average yield of 2423 kg ha^{-1} (GoP, 2015). Despite higher yield potentials, the average yield of rice in Pakistan is lower than that of other rice-growing countries of the world. Rice varieties of Pakistan are found susceptible to different environmental stresses including soil salinity and toxicity of heavy metals that are responsible for decreasing the rice yield and quality in Pakistan (Ahmad, 2007). A soil containing an excess amount of soluble salts and/or exchangeable Na$^+$ affect most of the crops adversely (Ghafoor et al., 2004). Rice was marked sensitive to salinity and its sensitivity to salinity differs with stages of growth and development (Maas and Grattan, 1999).

The continuous use of city and industrial effluents is creating soil salinization, sodication and builds up of toxic metals in the surface soils (Murtaza et al., 2008; Abd-Elrahman et al., 2012). Due to high electrical conductivity (EC), sodium adsorption ratio (SAR) and residual sodium carbonate (RSC), the raw effluent has been established unfit for irrigation purposes (Murtaza et al., 2012). The concentration of Pb and other heavy metals in raw sewage (Iqbal et al., 2011) were found greater than permissible heavy metals limits for irrigation water (Iqbal et al., 2015). In case of corn plants, accumulation of Pb in shoots increased in response to high NaCl salinity compared to those grown in low salinity soil (Izzo et al., 1991). Moreover, the Pb uptake and transport in plants depend on the type of soil and plant species. Kabata-Pendias and Pendias (2001) reported that in salt-affected soils, heavy metals uptake by plants might raise or decline conditional to the type of plant,

salinity/sodicity, magnitude of metal ion and other environmental circumstances.

A number of strategies can be used to decrease the soil-plant transfer of Pb to produce crops with the lowest possible Pb uptake especially in edible parts. To decrease the solubility and bioavailability of Pb, chemical immobilization is also one of the promising techniques. The right selection of an amendment at a particular place is subjected to its comparative efficacy as judged from the enhancement of soil characteristics and plant growth, physiological processes, its availability, cost, management and application complications, and period vital to react in the soil (Ghafoor et al., 2004).

Gypsum is used in agriculture as a fertilizer as well as a soil amendment because of having essential plant nutrients (Ca and S) and an inexpensive amendment for reclamation of saline-sodic soils/waters, and for enhancing crop growth and yields (Murtaza et al., 2009). Gypsum dissolution releases its component ions (Ca^{2+} and SO$_4^{2-}$) into soil solution (Lottermoser, 2007) and consequently, new reactions take place. Sulfates rapidly counter with Pb which results in the formation of greatly insoluble anglesite like minerals (Garrido et al., 2005). Antosiewicz (2005) reported that Ca-regulated Pb deposition in cell walls of plants. It was found that phosphate which is accessible to the roots of crop plants might also be available to toxic heavy metals, conclusively insoluble metal phosphates form. The RP as a primary P source found very operative in diminishing Pb in the soil solution, transferable fraction of contaminated soils (Ma et al., 1995). The soluble fertilizers like DAP not only decrease the metal toxicity in plants but also as an excellent source of nutrients as well to increase the biomass production, which indirectly decreased the metal toxicity (Khan and Jones, 2008, Rehman et al. 2015; Arshad et al., 2016).

Keeping in view the present high costs of fertilizers and amendments detailed investigations are required for the minimum uptake, translocation of Pb and chlorophyll contents of rice as affected by applied amendments with their most appropriate amounts, both in normal and salt-affected Pb-spiked soils.

Materials and Methods

The present pot trial was conducted in wirehouse (sides being open and only having iron wire screens with no control over temperature and humidity) at Institute of Soil and Environmental Sciences,

University of Agriculture Faisalabad (UAF). The soil was taken from 0-20 cm depth from the Farms of UAF. Soil was air-dried, minced with wooden roller, passed through 2 mm sieve, systematically mixed and stored. The soil samples were analyzed for physico-chemical properties. The characters of soil used for present study are described in Table 1.

There were ten treatments arranged in completely randomized design each with three replications. The treatments used were as T_1 = Control (without applied Pb), T_2 = 100 mg Pb kg^{-1} soil, T_3 = 100 mg Pb kg^{-1} soil + 2.5 g gyp kg^{-1} soil, T_4 = 100 mg Pb kg^{-1} soil + 5 g gyp kg^{-1} soil, T_5 = 100 mg Pb kg^{-1} soil + 7.5 g gyp kg^{-1} soil, T_6 = 100 mg Pb kg^{-1} soil + 1.5 g RP kg^{-1} soil, T_7 = 100 mg Pb kg^{-1} soil + 3 g RP kg^{-1} soil, T_8 = 100 mg Pb kg^{-1} soil + 4.5 g RP kg^{-1} soil, T_9 = 100 mg Pb kg^{-1} soil + 115 mg DAP kg^{-1} soil, T_{10} = 100 mg Pb kg^{-1} soil + 130 mg DAP kg^{-1} soil.

Similar treatments were investigated in spiked salt-affected soil with EC: SAR = 6 dS m^{-1} : 22 (mmol L^{-1})$^{1/2}$. The required amounts of salts were applied by calculating through quadratic equation i.e., NaCl = 0.36, Na_2SO_4 = 0.53, $CaCl_2$ = 0.14 and $MgSO_4$ = 0.04, g kg^{-1} soil (Muhammed and Ghafoor, 1992, Iqbal et al., 2015). Both the normal and salt-affected soils were spiked at 100 mg Pb kg^{-1} soil using $Pb(NO_3)_2$ salt. White glazed ceramic pots containing 12 kg processed soil per pot (total 60 pots), following prescribed treatments layout were used.

Healthy seeds of rice were taken from Rice Research Institute, Kala Shah Kaku, Sheikhupura and then grown in polythene lined trays containing sand. Yoshida nutrient solution was applied to germinate the rice nursery (Yoshida et al., 1976). The twenty-eight days old rice seedlings were transplanted with three seedlings per hill and five hills per pot (Iqbal et al., 2015). Rice crop was fertilized at 80-50-38.5-7.5 mg NPKZn kg^{-1} soil using urea, DAP, sulfate of potash (SOP) and $ZnSO_4.7H_2O$, respectively. The pots were submerged with pumped groundwater (Table 2) upto 2-3 cm throughout the crop growth.

Later, sixty days of rice nursery transplantation, total chlorophyll content (TCC) of rice shoots in expressions of SPAD (Special Products Analysis Division, a division of Minolta) value was determined via a portable SPAD-502 meter (Minolta, Osaka, Japan). It is a low-cost mode to quantify plant photosynthetic capacity than expensive chlorophyll fluorescence (Munns et al., 2006). The TCC were determined from the leaf tip to the leaf base and then averaged following Saqib et al. (2012).

At harvest, the data about total biomass, plant height, paddy and straw yields was recorded. The concentration of Pb was determined from straw, paddy and post-experiment soil samples via atomic absorption spectrophotometer.

The translocation factor (TF) of Pb in rice was determined by using the ratio of paddy-Pb concentration to straw-Pb concentration (Majid et al., 2012). The Pb uptake by rice straw or paddy was calculated via Pb concentration in rice straw or paddy × straw dry matter or paddy yield / 1000 (Hadi and Bano, 2010).

The statistical analysis was performed via analysis of variance technique (ANOVA), and the least significant difference (LSD) test was functioned to evaluate the effectiveness of the treatments (Steel et al., 1997) at 5% significance level using "Statistix 8.1" statistical computer-based software package.

Results

Total chlorophyll contents

In the current experiment, the treatments, soil-type and their interaction significantly (p $\leq$ 0.05, Table 7) affected the total chlorophyll contents (TCC) of rice. The treatment effectiveness on mean TCC of rice were in the decreasing order of $T_5 > T_4 > T_3 > T_1 > T_{10} > T_9 > T_8 > T_7 > T_6 > T_2$ (Table 3). The mean TCC of rice were found higher in normal soil than salt-affected soil.

In control, TCC were 57.7 and 34.9 SPAD-value, in normal and salt-affected soils, respectively. At T_2, the TCC were 37.4 and 29.5 SPAD-value, respectively. Amendments and their increasing amounts gradually improved TCC in both normal and salt-affected Pb-spiked soils.

In normal Pb-spiked soil, application of gyp resulted in significant enhancement of TCC and these were found as 47.8, 48.5 and 52.6 SPAD-value with T_3, T_4 and T_5, respectively. Therefore, TCC were increased by 29.6, 27.7 and 40.7 % at T_3, T_4 and T_5, respectively over T_2. Moreover, the TCC were greater by 5.6, 10.5 and 15.9 % at T_6, T_7 and T_8, respectively over T_2. However, the TCC were enhanced by 17.5 and 22.6 % at T_9 and T_{10} respectively, over T_2.

In salt-affected Pb-spiked soil, TCC were found as 49.1, 51.1, and 54.5 SPAD-value with T_3, T_4 and T_5, respectively. Therefore, TCC were increased by 66.5, 73.3 and 85.0 % at T_3, T_4 and T_5, respectively over T_2. Furthermore, the TCC were greater by 16.0, 22.8 and 30.2 % at T_6, T_7 and T_8, respectively over T_2.

Moreover, the TCC were enhanced by 41.0 and 50.0 % at T_9 and T_{10}, respectively over T_2.

Pb uptake by rice straw

In the present study, a significant (p ≤ 0.05, Table 7) effect of treatments, soil-type and their interactive effects was found for Pb uptake by rice straw. The treatment effectiveness on mean Pb uptake by rice straw was in the decreasing order of $T_2 > T_6 > T_7 > T_8 > T_9 > T_{10} > T_3 > T_4 > T_5 > T_1$ (Table 4). The mean Pb uptake by rice straw was found higher in salt-affected soil than normal soil.

In control, Pb uptake by rice straw was 0.320 and 0.242 mg pot^{-1} DM in normal and salt-affected soils respectively. At T_2, Pb uptake by rice straw was 1.002 and 0.791 mg pot^{-1} DM in normal and salt-affected soils respectively. The gyp, RP and DAP with their increasing amounts gradually decreased Pb uptake by rice straw in both normal and salt-affected Pb-spiked soils.

In normal Pb-spiked soil, applied gyp resulted in significant decreased of Pb uptake by rice straw, and it was found as 0.565, 0.471 and 0.393 mg pot^{-1} DM with T_3, T_4 and T_5 respectively. Therefore, Pb uptake was decreased by 43.5, 53.0 and 60.7 % at T3, T_4 and T_5 respectively over T_2. However, the Pb uptake by rice straw was reduced by 13.2, 15.2 and 23.1 % with T_6, T_7 and T_8 respectively over T_2. Moreover, the Pb uptake by rice straw was declined by 29.1 and 42.4 % with T_9 and T_{10}, respectively over T_2.

In salt-affected Pb-spiked soil, Pb uptake by rice straw was found as 0.725, 0.624 and 0.585 mg pot^{-1} DM with T_3, T_4 and T_5, respectively. Therefore, Pb uptake was decreased by 8.3, 21.1 and 26.0 % at T_3, T_4 and T_5 respectively over T_2. Furthermore, the Pb uptake by rice straw was reduced by 1.0, 1.2 and 1.5 % at T_6, T_7 and T_8 respectively over T_2. However, the Pb uptake by rice straw was declined by 1.8 and 8.3 % at T_9 and T_{10} respectively over T_2.

Pb uptake by rice paddy

A significant (p ≤ 0.05, Table 7) effect of treatments was noted for Pb uptake by rice paddy. The treatment effectiveness on mean Pb uptake by rice paddy was in the decreasing order of $T_2 > T_7 > T_6 > T_8 > T_9 > T_{10} > T_3 > T_4 > T_5 > T_1$ (Table 5). The mean Pb uptake by rice paddy was found higher in salt-affected soil than normal soil.

In control, Pb uptake by rice paddy was 0.009 and 0.011 mg pot^{-1} DM in both normal and salt-affected soils respectively. At T_2, Pb uptake by rice paddy was

0.095 and 0.137 mg pot^{-1} DM in normal and salt-affected soils respectively. Gyp, RP and DAP with their increasing amounts gradually decreased Pb uptake by rice paddy in both normal and salt-affected Pb-spiked soils.

In normal Pb-spiked soil, applied gyp resulted in significant decreased of Pb uptake by rice paddy and it was found as 0.085, 0.057 and 0.041 mg pot^{-1} DM with T_3, T_4 and T_5, respectively. Therefore, Pb uptake was decreased by 10.4, 40.5 and 56.2 % at T_3, T_4 and T_5 respectively over T_2. Moreover, the Pb uptake by rice paddy was declined by 8.0, 8.3 and 14.6 % with T_6, T_7 and T_8 respectively over T_2. However, the Pb uptake by rice paddy was reduced by 1.8 and 11.8 % with T_9 and T_{10} respectively over T_2.

In salt-affected Pb-spiked soil, Pb uptake by rice paddy was found as 0.076, 0.056 and 0.036 mg pot^{-1} DM with T_3, T_4 and T_5 respectively. Therefore, Pb uptake decreased by 44.5, 59.1 and 73.4 % at T_3, T_4 and T_5 respectively over T_2. Furthermore, the Pb uptake by rice paddy was declined by 15.5, 14.8 and 19.7 % at T_6, T_7 and T_8 respectively over T_2. Likewise, the Pb uptake by rice paddy was reduced by 27.0 and 35.2 % at T_9 and T_{10} respectively over T_2.

Translocation factor of Pb

For Pb translocation factor (TF) from rice shoot to paddy, a significant (p ≤ 0.05, Table 7) effect of treatments, soil-type and their interaction was found. The treatment effectiveness on mean TF was in the decreasing order of $T_2 > T_6 > T_7 > T_8 > T_9 > T_{10} > T_3 > T_4 > T_5 > T_1$ (Table 6). The mean Pb TF was found higher in salt-affected soil than normal soil.

In control, Pb TF was 0.050 and 0.057 in normal and salt-affected soils respectively. At T_2, TF was 0.238 and 0.244 in normal and salt-affected soils respectively. The gyp, RP and DAP with their increasing amounts gradually decreased Pb TF in normal and salt-affected Pb-spiked soils.

In normal Pb-spiked soil, applied gyp resulted in significant decreased of Pb TF and it was found as 0.157, 0.120 and 0.061 with T_3, T_4 and T_5, respectively. Therefore, Pb TF was decreased by 34.0, 49.4 and 74.3 % at T_3, T_4 and T_5 respectively over T2. Moreover, the Pb TF was declined by 9.6, 16.8 and 19.7 % with T_6, T_7 and T_8 respectively over T_2. Nevertheless, the Pb TF was reduced by 22.6 and 24.9 % with T_9 and T_{10} respectively over T_2.

In salt-affected Pb-spiked soil, Pb TF was found as 0.162, 0.124 and 0.090 with T_3, T_4 and T_5 respectively. Therefore, Pb TF was decreased by 33.6, 49.3 and 63.1

% at T_3, T_4 and T_5 respectively over T_2. Likewise, the Pb TF was declined by 8.7, 17.2 and 19.6 % at T_6, T_7 and T_8 respectively over T_2. Similarly, the Pb TF was reduced by 20.9 and 22.9 % at T_9 and T_{10} respectively over T_2.

Table 1: Physico-chemical characteristics of soil used for pot trial

Parameter	Value
Textural class	Sandy loam
Sand (%)	69.20
Silt (%)	14.50
Clay (%)	16.30
pH_s	7.66
EC_e (dS m^{-1})	[a]1.11 (6)
TSS (mmol$_c$ L^{-1})	111
SAR (mmol L^{-1})$^{1/2}$	[a]3.28 (22)
Saturation percentage (%)	29.36
CEC (cmol$_c$ kg^{-1})	5.40
Organic matter (%)	0.83
CaCO$_3$ (%)	1.74
AB-DTPA extractable Pb (mg kg^{-1})	2.95
Total Pb (mg kg^{-1})	18.90

[a]Initial EC_e, SAR of the soil while values in parentheses represent artificially made saline-sodic soil as described by Muhammed and Ghafoor (1992)

Table 2: Composition of pumped ground water used for irrigation

Parameter	Value
pH	7.75
EC (dS m^{-1})	0.67
TSS (mmol$_c$ L^{-1})	6.70
CO$_3^{2-}$	Absent
HCO$_3^-$ (mmol$_c$ L^{-1})	3.1
Cl$^-$ (mmol$_c$ L^{-1})	2.60
SO$_4^{2-}$ (mmol$_c$ L^{-1})	0.10
Ca^{2+} + Mg^{2+} (mmol$_c$ L^{-1})	4.73
Na$^+$ (mmol$_c$ L^{-1})	1.97
RSC	Nil
SAR (mmol L^{-1})$^{1/2}$	1.28
Pb (mg L^{-1})	Traces

Table 3: Effect of applied amendments on total chlorophyll contents (TCC, SPAD-value) of rice

Treatment	Normal Soil	Salt-Affected Soil	Mean
T_1 = Control	57.7 a	34.9 mn	46.3 C
T_2 = 100 mg Pb kg^{-1} soil	37.4 klm	29.5 o	33.4 H
T_3 = 100 mg Pb kg^{-1} soil + 2.5 g gyp kg^{-1} soil	47.8 ef (27.7)	49.1 de (66.5)	48.4 B (47.1)
T_4 = 100 mg Pb kg^{-1} soil + 5 g gyp kg^{-1} soil	48.5 def (29.6)	51.1 cd (73.3)	49.8 B (51.4)
T_5 = 100 mg Pb kg^{-1} soil + 7.5 g gyp kg^{-1} soil	52.6 bc (40.7)	54.5 b (85.0)	53.6 A (62.8)
T_6 = 100 mg Pb kg^{-1} soil + 1.5 g RP kg^{-1} soil	39.5 jk (5.6)	34.2 n (16.0)	36.8 G (10.8)
T_7 = 100 mg Pb kg^{-1} soil + 3 g RP kg^{-1} soil	41.3 ij (10.5)	36.2 lmn (22.8)	38.8 F (16.6)
T_8 = 100 mg Pb kg^{-1} soil + 4.5 g RP kg^{-1} soil	43.4 ghi (15.9)	38.4 kl (30.2)	40.9 E (23.1)
T_9 = 100 mg Pb kg^{-1} soil + 115 mg DAP kg^{-1} soil	44.0 ghi (17.5)	41.6 hij (41.0)	42.8 D (29.3)
T_{10} = 100 mg Pb kg^{-1} soil + 130 mg DAP kg^{-1} soil	45.9 fg (22.6)	44.2 gh (50.0)	45.0 C (36.3)
LSD	2.8		1.7
Mean	45.8 A	41.4 B	
LSD	0.47		

Values in parenthesis are percent increase (+) or decrease (-) over that 100 mg Pb kg^{-1} soil (T_2) treatment. Means sharing dissimilar letter in a row or in a column are statistically significant (p ≤ 0.05, n = 3).
Small letters represent comparison among interaction means and capital letters are used for overall mean.

Table 4: Effect of applied amendments on Pb uptake (mg pot^{-1} DM) by rice straw

Treatment	Normal Soil	Salt-Affected Soil	Mean
T$_1$ = Control	0.320 jk	0.242 k	0.281 F
T$_2$ = 100 mg Pb kg^{-1} soil	1.002 a	0.791 bcd	0.896 A
T$_3$ = 100 mg Pb kg^{-1} soil + 2.5 g gyp kg^{-1} soil	0.565 gh (-43.5)	0.725 cde (-8.3)	0.645 D (-25.9)
T$_4$ = 100 mg Pb kg^{-1} soil + 5 g gyp kg^{-1} soil	0.471 hi (-53.0)	0.624 efg (-21.1)	0.547 E (-37.0)
T$_5$ = 100 mg Pb kg^{-1} soil + 7.5 g gyp kg^{-1} soil	0.393 ij (-60.7)	0.585 fgh (-26.0)	0.489 E (-43.4)
T$_6$ = 100 mg Pb kg^{-1} soil + 1.5 g RP kg^{-1} soil	0.869 b (-13.2)	0.783 bcd (-1.0)	0.826 AB (-7.1)
T$_7$ = 100 mg Pb kg^{-1} soil + 3 g RP kg^{-1} soil	0.849 bc (-15.2)	0.781 bcd (-1.2)	0.815 AB (-8.2)
T$_8$ = 100 mg Pb kg^{-1} soil + 4.5 g RP kg^{-1} soil	0.770 bcd (-23.1)	0.779 bcd (-1.5)	0.774 BC (-12.3)
T$_9$ = 100 mg Pb kg^{-1} soil + 115 mg DAP kg^{-1} soil	0.710 def (-29.1)	0.776 bcd (-1.8)	0.743 C (-15.5)
T$_{10}$ = 100 mg Pb kg^{-1} soil + 130 mg DAP kg^{-1} soil	0.576 gh (-42.4)	0.721 de (-8.3)	0.648 D (-25.6)
LSD	0.13		0.08
Mean	0.650 B	0.68 A	
LSD	0.02		

Table 5: Effect of applied amendments on Pb uptake (mg pot^{-1} DM) by rice paddy

Treatment	Normal Soil	Salt-Affected Soil	Mean
T$_1$ = Control	0.009 f	0.011 f	0.011 E
T$_2$ = 100 mg Pb kg^{-1} soil	0.095 abc (86.2)	0.137 a (108.8)	0.116 A (97.5)
T$_3$ = 100 mg Pb kg^{-1} soil + 2.5 g gyp kg^{-1} soil	0.085 bcde (-10.4)	0.076 bcde (-44.5)	0.081 BC (-27.5)
T$_4$ = 100 mg Pb kg^{-1} soil + 5 g gyp kg^{-1} soil	0.057 cde (-40.5)	0.056 cdef (-59.1)	0.057 CD (-49.8)
T$_5$ = 100 mg Pb kg^{-1} soil + 7.5 g gyp kg^{-1} soil	0.041 def (-56.2)	0.036 ef (-73.4)	0.039 DE (-64.8)
T$_6$ = 100 mg Pb kg^{-1} soil + 1.5 g RP kg^{-1} soil	0.088 bcd (-8.0)	0.116 ab (-15.5)	0.101 AB (-11.8)
T$_7$ = 100 mg Pb kg^{-1} soil + 3 g RP kg^{-1} soil	0.087 abcd (-8.3)	0.117 ab (-14.8)	0.102 AB (-11.6)
T$_8$ = 100 mg Pb kg^{-1} soil + 4.5 g RP kg^{-1} soil	0.081 bcde (-14.6)	0.111 ab (-19.7)	0.097 AB (-14.4)
T$_9$ = 100 mg Pb kg^{-1} soil + 115 mg DAP kg^{-1} soil	0.093 abc (-1.8)	0.100 abc (-27.0)	0.096 AB (-17.2)
T$_{10}$ = 100 mg Pb kg^{-1} soil + 130 mg DAP kg^{-1} soil	0.084 bcde (-11.8)	0.088 abcd (-35.2)	0.086 A-C (-23.5)
LSD	0.05		0.03
Mean	0.072 B	0.085 A	
LSD	0.0008		

Values in parenthesis are percent increase (+) or decrease (-) over that 100 mg Pb kg^{-1} soil (T$_2$) treatment. Means sharing dissimilar letter in a row or in a column are statistically significant (p ≤ 0.05, n = 3).
Small letters represent comparison among interaction means and capital letters are used for overall mean.

Table 6: Effect of applied amendments on translocation factor of Pb from rice shoot to paddy

Treatment	Normal Soil	Salt-Affected Soil	Mean
T_1 = Control	0.050 k	0.057 k	0.054 I
T_2 = 100 mg Pb kg^{-1} soil	0.238 ab	0.244 a	0.241 A
T_3 = 100 mg Pb kg^{-1} soil + 2.5 g gyp kg^{-1} soil	0.157 h (-34.0)	0.162 k (-33.6)	0.160 F (-33.8)
T_4 = 100 mg Pb kg^{-1} soil + 5 g gyp kg^{-1} soil	0.120 i (-49.4)	0.124 i (-49.3)	0.122 G (-49.3)
T_5 = 100 mg Pb kg^{-1} soil + 7.5 g gyp kg^{-1} soil	0.061 k (-74.3)	0.090 j (-63.1)	0.076 H (-68.7)
T_6 = 100 mg Pb kg^{-1} soil + 1.5 g RP kg^{-1} soil	0.215 cd (-9.6)	0.223 bc (-8.7)	0.219 B (-9.2)
T_7 = 100 mg Pb kg^{-1} soil + 3 g RP kg^{-1} soil	0.198 ef (-16.8)	0.202 de (-17.2)	0.200 C (-17.0)
T_8 = 100 mg Pb kg^{-1} soil + 4.5 g RP kg^{-1} soil	0.191 ef (-19.7)	0.196 ef (-19.6)	0.194 CD (-19.7)
T_9 = 100 mg Pb kg^{-1} soil + 115 mg DAP kg^{-1} soil	0.184 fg (-22.6)	0.193 ef (-20.9)	0.189 DE (-21.8)
T_{10} = 100 mg Pb kg^{-1} soil + 130 mg DAP kg^{-1} soil	0.179 g (-24.9)	0.188 fg (-22.9)	0.183 E (-23.9)
LSD	0.02		0.0009
Mean	0.159 B	0.168 A	
LSD	0.0002		

Values in parenthesis are percent increase (+) or decrease (-) over that 100 mg Pb kg^{-1} soil (T_2) treatment. Means sharing dissimilar letter in a row or in a column are statistically significant (p ≤ 0.05, n = 3).
Small letters represent comparison among interaction means and capital letters are used for overall mean.

Table 7: Mean squares of various rice traits as influenced by soil applied amendments in normal and salt-affected Pb-spiked soil conditions.

SOV	df	Total chlorophyll contents (TCC)	Pb uptake by rice straw	Pb uptake by rice paddy	Translocation factor (TF) of Pb from rice shoot to paddy
Soils	1	295.0**	0.01**	0.002**	0.001**
Treatments	9	234.1**	0.21**	0.006**	0.022**
Soils× Treatments	9	81.4**	0.02**	0.0005**	0.00008**
Error	40	0.8	0.001	0.0002	0.00002

NS = Non-significant (P > 0.05); * = Significant (P ≤ 0.05); ** = Highly significant

Discussion

Chlorophyll content is frequently determined in crop plants owing to evaluate the effect of abiotic and biotic stresses, as deviations in pigment content are interconnected to visual symptoms of plant disorders and photosynthetic efficiency (Purnama et al., 2015). In the present study, the total chlorophyll contents (TCC) of rice was decreased by both Pb and salinity stress (Table 3). The reduction in TCC of rice was more pronounced in salt-affected soil than normal Pb-spiked soil. The application of amendments significantly (p ≤ 0.05, Table 7) countered the toxic effects of Pb on TCC of rice both in normal and salt-affected soils. The application of gypsum was more effective followed by DAP in alleviating the harmful effects of Pb and salinity on TCC.

Previously, the Pb-influenced decrease in chlorophyll contents of rice (Li et al., 2012) and wheat (Bhatti et al., 2013) were also reported. Sharma and Dubey (2005) described more degradation of chlorophyll owing to increased chlorophyllase activity in Pb-treated plants. Ernst (1998) recognized that the inhibition of chlorophyll synthesis persuaded by Pb stress was often manifested as chlorosis. Similarly, Ewais (1997) reported that Pb decreased growth and chlorophyll contents in three weed *Chenopodium ambrosioides*, *Digitaria sanguinolis* and *Cyperus difformis*. The Pb inhibited synthesis of chlorophyll by affecting the uptake of essential nutrients by plants (Iqbal et al., 2017a). The Pb induced reduction in TCC can be due to the inhibition of the enzymes activities responsible for the synthesis of chlorophyll. The Pb stress hinders the growth of the plant and even causes the death of plant by disturbing the uptake of Mg and Fe, and thereby decreasing the photosynthesis via degradation of chlorophyll (Pourrut et al., 2011).

A significant effect of salinity/sodicity on chlorophyll concentration in tested rice cultivars being higher decrease in salt sensitive cultivars than tolerant once were reported by Khan and Abdullah (2003). Ali et al. (2004) reported that the synthesis of photosynthetic pigments was severely affected by soil salinity. The decrease in chlorophyll concentrations was attributed to the inhibitory effects of the accumulated ions of different salts on the synthesis of the diverse chlorophyll elements. In chloroplast, the salts disturb the ability of the forces interconnected the multifarious pigment protein liquid. Ali et al. (2004) described that membrane-bound chloroplast stability is reliant on its stability which cause a reduction in chlorophyll contents in higher salinity due to occasional endure integral. In the current study, the existence of salinity further intensified the harmful effects of Pb stress on chlorophyll contents of rice.

In present trial, the improvement in TCC by the application of amendments was due to decrease in Pb uptake and translocation in rice. Since the uptake of Pb in rice was highly reduced by the applied gyp in both normal and salt-affected Pb-spiked soils, hence the detrimental effects of Pb on TCC were ameliorated. The application of DAP also resulted in reduced Pb uptake and translocation in rice plants, resultantly the toxic effects of Pb on TCC were also decreased in normal and salt-affected Pb-spiked soils. However, RP was proved less effective in improving physiological functions such as TCC of rice plants.

The Pb uptake by rice shoots (Table 4) and paddy (Table 5) and translocation (Table 6) of Pb in rice plants grown in Pb-spiked soils was significantly ($p \leq$ 0.05, Table 7) higher than non-spiked (normal and salt-affected) soils. The application of gypsum was more useful to reduce Pb uptake and translocation. The DAP was ranked second effective treatment and its application resulted in decreased the uptake and translocation of Pb from rice shoot to paddy in both normal and salt-affected Pb-spiked soils. The Pb uptake in rice straw and paddy, and translocation

from shoot to paddy was not effectively reduced by RP. The chemical immobilization of Pb in soil, with the addition of amendments, reduced the mobility and phyto-availability of Pb, which resulted in reduced uptake of Pb by rice straw and paddy. The formation of less soluble anglesite mineral, i.e., $PbSO_4$, with the addition of gypsum in Pb-spiked soil had been earlier reported (Illera et al., 2004; Garrido et al., 2005; Iqbal et al., 2017b). The use of gypsum under reduced conditions was expected to transform Pb into less soluble PbS (Illera et al., 2004; Hashimoto et al., 2011). Similarly, application of phosphate amendments was also proved very effective for chemical immobilization of Pb (Basta and McGowen, 2004). The DAP increased the potential for the formation of Pb-pyromorphite which reduce phyto-availability of Pb (Khan and Jones, 2008). Chen et al. (2006) found that Pb uptake by wheat was considerably restrained by $Ca(H_2PO_4)$ fertilizer and there was a- significant negative correlation between the level of P and the uptake of Pb. The translocation of Pb from root to shoot was also reduced because P application reduced the soluble and exchangeable fractions of Pb in soil. Therefore, the application of gyp, DAP and RP can provide an efficient way to decrease Pb availability, uptake and translocation in plants grown on metal spiked soils.

In the present experiment, application of gypsum reduced the uptake of Pb by rice plants in Pb-spiked normal and salt-affected soil. The other beneficial effects of gypsum can be due to the provision of soluble Ca in soil solution which reduced the uptake of Pb (Antosiewicz, 2005). The concentration of Pb in the shoot was enhanced with a decrease in Ca in growth medium, which indicate increased root to shoot transport with decreased level of Ca. The enhanced uptake and translocation of Pb were presumed to be the result of Ca-dependent Pb transport system in plasma membrane (Sunkar et al., 2000). It was assumed that the Ca pathway was used by Pb ions to cross the membranes owing to the great affinity of Pb to Ca binding sites in biotic configurations (Vijverberg et al., 1994). The diverse Ca-channels were also found porous to a range of monovalent and divalent cations. Thus, at a greater concentration, Ca favorably engaged these sites and restricted the invasion of other cations (White, 2000; Sanders et al., 2002). According to Ernst (1998), Ca immobilized Pb in the plant roots by forming Pb precipitates in the cell walls. The form and size of Pb holding precipitates were found to be dependent on the Ca level in the medium, and large deposits were formed at high Ca (Antosiewicz, 2005). Calcium might also play role in cell signaling and Ca-mediated signaling that led to a partial block of metal uptake (Antosiewicz and Hennig, 2004). The contribution of Ca in the signal transduction succeeding a numeral abiotic stimulus was also recognized (Sanders et al., 2002). Calcium might also involve in the signal transduction against heavy metal stress and thereby reduce metal uptake. In salt-affected soils, the beneficial effects of Ca in reducing Pb uptake due to the role of Ca, to maintain cell membrane integrity, thereby reduced the uptake of Na and Pb. Soluble Ca reduce the binding of Na to the cell wall and plasma membrane (Rengel, 1992) and improved the integrity and functions of plasma membrane (Lauchli, 1990). Therefore, the application of gypsum was superior because it not only reduced Pb bioavailability by immobilizing it but also reduced the Pb transport across the cell membrane.

Conclusion

The chlorophyll contents of rice were significantly (p ≤ 0.05) decreased both in normal and salt-affected Pb-spiked soils than in non-spiked soils. In salt-affected Pb-spiked soil, decrease in chlorophyll contents, and increase in uptake and translocation of Pb were more prominent than that found in normal Pb-spiked soil. The decrease in chlorophyll contents of rice was counteracted by the application of gyp, RP and DAP with their variable amounts. With increasing amounts of applied amendments, the Pb uptake and translocation in rice were decreased gradually. Thus, application of 7.5 g gyp kg^{-1} soil was proved the most efficient in improving chlorophyll contents in rice as well as reducing Pb uptake and translocation in rice grown in both normal and salt-affected Pb-spiked soils. However, the present results need to be confirmed in field trial and economic feasibility must be worked out.

Acknowledgements

The first author is highly thankful and admires the Higher Education Commission of Pakistan for providing funds during present Ph.D. research work.

References

Abd-Elrahman SH, Mostafa MAM, Taha TA, Elsharawy MAO and Eid MA, 2012. Effect of different amendments on soil chemical characteristics, grain yield and elemental content of wheat plants grown on salt-affected soil irrigated with low quality water. Ann. Agric. Sci. 57: 175–182.

Ahmad HR, 2007. Metal ion pollution potential of raw waste effluent of the Faisalabad city: Impact assessment on soils, plants and shallow ground water. Ph.D. Thesis, Inst. Soil Environ. Sci. Univ. Agric. Faisalabad, Pakistan.

Ali Y, Aslam Z, Ashraf MY and Tahir GR, 2004. Effect of salinity on chlorophyll concentration, leaf area, yield and yield components of rice genotypes grown under saline environment. Int. J. Environ. Sci. Technol. 1: 221–225.

GoP, 2015. Agricultural Statistics of Pakistan. Government of Pakistan, Ministry of Food, Agriculture and Livestock (Economic Wing), Islamabad, Pakistan.

Antosiewicz DM and Hennig J, 2004. Overexpression of LCT1 in tobacco enhances the protective action of calcium against cadmium toxicity. Environ. Pollut. 129: 37–245.

Antosiewicz DM, 2005. Study of calcium-dependent lead-tolerance on plants differing in their level of Ca-deficiency tolerance. Environ. Pollut. 134: 23–34.

Arshad M, Ali S, Noman A, Ali Q, Rizwan M, Farid M and Irshad MK, 2016. Phosphorus amendment decreased cadmium (Cd) uptake and ameliorates chlorophyll contents, gas exchange attributes, antioxidants and mineral nutrients in wheat (Triticum aestivum L.) under Cd stress. Arch. Agron. Soil Sci. 62: 533–546.

Basta NT and McGowan SL, 2004. Evaluation of chemical immobilization treatments for reducing heavy metal transport in a smelter-contaminated soil. Environ. Pollut. 127: 73–82.

Bhatti KH, Anwar S, Nawaz K, Hussain K, Siddiqi EH, Sharif RU, Talat A and Khalid A, 2013. Effect of heavy metal lead (Pb) stress of different concentration on wheat (Triticum aestivum L.). Middle-East J. Sci. Res. 14: 148–154.

Chen S, Tie-heng S, Li-na S, Qi-Xing Z and Lei C, 2006. Influences of phosphate nutritional level on the phytoavailability and speciation distribution of cadmium and lead in soil. J. Environ. Sci. 18: 1247–1253.

Ernst WHO, 1998. Effects of heavy metals in plants at the cellular and organismic level, ecotoxicology. pp. 587-620. In: G. Schuurmann and B. Markert (eds.), Ecological Fundamentals, Chemical Exposure and Biological Effects. Wiley and Sons, Heidelberg, Germany.

Ewais EA, 1997. Effects of cadmium, nickel and lead on growth, chlorophyll content and proteins of weeds. Biol. Plantaum. 39: 403–410.

Garrido F, Illera V and Garclai-Gonzalez MT, 2005. Effect of the addition of gypsum- and lime-rich industrial by-products on Cd, Cu and Pb availability and leachability in metal-spiked acid soils. Appl. Geochem. 20: 397–408.

Ghafoor A, Qadir M and Murtaza G, 2004. Salt-affected soils: Principles of management. Allied Book Centre, Urdu Bazar, Lahore, Pakistan.

Gnanamanickam SS, 2009. Rice and its importance to human life. In: Biological Control of Rice Diseases. pp. 1–11. S.S. Gnanamanickam (eds.), Springer Press, The Netherlands.

Hadi F and Bano A, 2010. Effect of diazotrophs (rhizobium and azatebactor) on growth of maize (Zea mays L.) and accumulation of lead (Pb) in different plant parts. P. J. Bot. 42: 4363–4370.

Hashimoto Y, Yamaguchi N, Takaoka M and Shiota K, 2011. EXAFS speciation and phytoavailability of Pb in a contaminated soil amended with compost and gypsum. Sci. Total Environ. 409: 1001–1007.

Illera V, Garrido F, Serrano S and Garcia-Gonzalez MT, 2004. Immobilization of the heavy metals Cd, Cu and Pb in an acid soil amended with gypsum- and lime-rich industrial by-products. Eu. J. Soil Sci. 55: 24–145.

Iqbal MA, Chaudhry MN, Zaib S, Imran M, Ali K and Iqbal A, 2011. Accumulation of heavy metals (Ni, Cu, Cd, Cr, Pb) in agricultural soils and spring seasonal plants, irrigated by industrial waste water. J. Environ. Tech. Manage. 2: 1–9.

Iqbal MM, Murtaza G, Saqib ZA and Ahmad R, 2015. Growth and physiological responses of two rice varieties to applied lead in normal and salt-affected soils. Int. J. Agric. Biol. 17: 901–910.

Iqbal MM, Murtaza G, Naz T, Niazi NK, Shakar M, Wattoo FM, Farooq O, Ali M, Rehman MU, Afzal I, Mehdi SM and Mahmood A, 2017a. Effects of lead salts on growth, chlorophyll contents and tissue concentration of rice genotypes. Int. J. Agric. Biol. 19: 69–76.

Iqbal MM, Murtaza G, Naz T, Akhtar J, Afzal M, Meers E and Laing GD, 2017b. Amendments affect Pb mobility and modulated chemo-speciation under different moisture regimes in

normal and salt-affected Pb-contaminated soils. Int. J. Environ. Sci. Tech. 14: 113–122.

Izzo R, Izzo FN and Quartacci MF, 1991. Growth and mineral absorption in maize seedlings as Affected by increasing NaCl concentrations. J. Plant Nutr. 14: 687–699.

Britto A JD, Roshan SS, Gracelin S DH, 2011. Effect of lead on malondialdehyde, superoxide dismutase, proline activity and chlorophyll content in Capsicum annum. Biores. Bull. 5: 357–362.

Kabata-Pendias A and Pendias H, 2001. Trace elements in soils and plants, 3rd ed. CRC Press, Boca Raton, FL, USA.

Khan MJ and Jones DL, 2008. Effect of composts, lime and diammonium phosphate on the phytoavailability of heavy metals in a copper mine tailing soil. Pedosphere. 19: 631–641.

Khan, M.A and Z. Abdullah. 2003. Salinity-sodicity induced changes in reproductive physiology of rice (Oryza sativa) under dense soil conditions. Environ. Exp. Bot. 49: 145–157.

Lauchli A, 1990. Calcium, salinity and the plasma membrane, In: R.T. Leonard and P.K. Hepler (eds.), Calcium in Plant Growth and Development, The Am. Soc. Plant physiologists Symposium Series. 4: 26–35.

Li X, Bu N, Li Y, Ma L, Xin S and Zhang L, 2012. Growth, photosynthesis and antioxidant responses of endophyte infected and non-infected rice under lead stress conditions. J. Hazard. Mat. 213–214: 55–61.

Lottermoser B, 2007. Mine wastes: Characterization, treatment and environmental impacts. 2nd ed. Springer-Verlag, Berlin Heilderberg, Germany.

Ma QY, Logan TJ and Traina SJ, 1995. Lead immobilization from aqueous solutions and contaminated soils using phosphate rocks. Environ. Sci. Technol. 29: 1118–1126.

Maas EV and Grattan SR, 1999. Crop yields as affected by salinity. Agron. Monogr. 38: 55–108.

Majid NM, Islam MM and Enanee N, 2012. Heavy metal uptake and translocation by Semuloh (Fagopyrum dibotrys) from sawdust sludge contaminated soil. Bulgar. J. Agric. Sci. 18: 912–923.

Meenakshi C, Jetly UK, Khan MA, Zutshi S and Fatma T, 2006. Effect of heavy metal stress on proline, malondialdehyde and superoxide dismutase activity in the cyanobacterium Spirulina plantensis-S5. Ecotox. Environ. Saf.

Muhammed S and Ghafoor A, 1992. Irrigation water salinity/sodicity analysis lab. Manual of soil salinity research method. IWASRI, Lahore.

Munns R, James RA, Lauchli A, 2006. Approaches to increasing the salt tolerance of wheat and other cereals. J. Exp. Bot. 57: 1025–1043.

Murtaza G, Ghafoor A, Owens G, Qadir M and Kahlon UZ, 2009. Environmental and economic benefits of saline-sodic soil reclamation using low-quality water and soil amendments in conjuction with a rice-wheat cropping system. J. Agron. Crop Sci. 195: 124–136.

Murtaza G, Ghafoor A and Qadir M, 2008. Accumulation and implications of cadmium, cobalt and manganese in soils and vegetables irrigated with city effluent. J. Sci. Food Agric. 88: 100–107.

Murtaza G, Ghafoor A, Rehman MZ, Sabir M and Naeem A, 2012. Phytodiversity for metals in plants grown in urban agricultural lands irrigated with untreated city effluent. Commun. Soil Sci. Plant Anal. 43: 1181–1201.

Pourrut B, Shahid M, Dumat C, Winterton P and Pinelli E, 2011. Lead uptake, toxicity, and detoxification in plants. Rev. Environ. Contam. Toxicol. 213:113–136.

Purnama PR, Soedarti T and Purnobasuki H, 2015. The effects of lead [Pb(NO$_3$)$_2$] on the growth and chlorophyll content of Sea Grass [*Thalassia hemprichii* (ehrenb.) Aschers.] Ex situ. VEGETOS. 28: 09–15.

Rehman MZ, Rizwan M, Ghafoor A, Naeem A, Ali S, Sabir M, and Qayyum MF, 2015. Effect of inorganic amendments for in situ stabilization of cadmium in contaminated soil and its phyto-availability to wheat and rice under rotation. Environ. Sci. Pollut. Res. 22: 16897–16906.

Rengel Z, 1992. The role of calcium in salt toxicity. Plant Cell Environ. 15: 625–632.

Rizwan M, Ali S, Adrees M, Rizvi H, Rehman MZ, Hannan F, Qayyum MF, Hafeez F and OK YS, 2016. Cadmium stress in rice: Toxic effects, tolerance mechanisms and management: A critical review. Environ. Sci. Pollut. Res. 23: 17859–17879.

Sanders D, Pelloux J, Brownlee C and Harper JF, 2002. Calcium at the crossroad of signaling. Plant Cell. 14: 401–417.

Saqib ZA, Akhtar J, Haq MA, Ahmad I and Bakhat HF, 2012. Rationality of using various physiological and yield related traits in determining salt tolerance in wheat. Afr. J. Biotechnol. 11: 3558–3568.

Sharma P and Dubey S, 2005. Lead toxicity in plants. Braz. J. Plant Physiol. 17: 35–52.

Steel RGD, Torrie JH and Dickey DA, 1997. Principles and procedures of statistics. 172-177. A biometrical approach. (3rd ed.), McGraw Hill book Co., Inc. New york, NY, USA.

Sunkar R, Kaplan B, Bouche N, Arazi T, Dolev D, Talke IN, Maathius JM, Sanders D, Bouchez D, and Fromm H, 2000. Expression of a truncated tobacco NtCBP4 channel in transgenic plants and disruption of the homologous Arabidopsis CNGC1 gene confer Pb^{2+} tolerance. Plant J. 24: 533–542.

Vijverberg HPM, Oortgiesen M, Leinders T and van Kleef RGDM, 1994. Metal interactions with voltage- and receptor-activated ion channels. Environ. Health Perspect. 102: 153–158.

White PJ, 2000. Calcium channels in higher plants. Biochimica et Biophysics Acta. 1465: 171–189.

Yoshida S, Forno DA, Cock JH and Gomez KA, 1976. Laboratory manual for physiological studies of rice, pp. 62. 3rd ed. IRRI, Philippines.

4

Relationship between nutrient concentration in saffron corms and saffron yield in perennial fields of South Khorasan province

Mobina Maktabdaran[1]*, Mohammad Hassan Sayyari Zohan[2], Majid Jami Alahmadi[3],
Golam Reza Zamani[3]

[1]MSc Student, Department of Ecology, Faculty of Agriculture, University of Birjand, Iran
[2]Associate Professor, Department of Soil Science and Engineering, Faculty of Agriculture, University of Birjand, Iran
[3]Associate Professor, Department of Agronomy and Plant Breeding, Faculty of Agriculture, University of Birjand, Iran

Abstract

Saffron (*Crocus sativus* L.) is one of the most important exports products in Iran that proper concentration of nutrients is particularly important in the development and production. Since saffron is reproduced by corm, so always been considered production of replacement corms through appropriate nutrition; and concentration of element in corm in specific stage of growth and development has high correlation with plant yield. Therefore, this study was conducted to determine nutrients concentration in corm of saffron in perennial fields of Qaen and Nehbandan in South Khorasan in 2015. The information of two regions (Qaen and Nehbandan) were collected from 3, 5 and 7 years-old fields. Then, three fields with at least 500 m^2 under cultivation selected for each age of 3, 5 and 7 years fields and 3 plots from each field and one corm sample from each plot were selected. The concentration of phosphorus (P), potassium (K), nitrogen (N) and iron in corm of saffron were analyzed based on standard laboratory methods. Results revealed that there were significant differences in N, P, K and iron concentrations of saffron corm (P≤0.01), but the effects were distinct for corm number in different weight groups. The present study showed that in saffron fields, N, P, K and iron concentration in corm were the most effective parameters for saffron yield increment, which regression correlations showed yield increment compeer to the change in these indices. Investigation of nutrients concentration of the corm on saffron yield shows the important role of such elements in yield. Therefore, nutrient concentration in saffron corms affect the plant nutrition, promoting growth and yield of crop. Generally, corm selection with optimum weight for sowing and proper use of nutrients, especially nitrogen, phosphorus, potassium and iron, as well as their correct balance in soil, can be effective in yield increment and stability of soil fertility.

Keywords: Corm, Phosphorous, Potassium, Iron, Yield of saffron

*Corresponding author email:
mobina.maktabdaran@birjand.ac.ir

Introduction

Saffron (*Crocus sativus* L.) belongs to Iridaceae family, which is xerophyte, herbaceous and perennial (Husaini and Ashraf, 2010; Gresta et al., 2009). In Iran, saffron is being noteworthy in various aspects such as high water efficiency, creating jobs and non-oil export extension (Hosseini et al., 2004). Saffron is male sterile and its reproduction is performed through the corm (Kafi et al., 2002). Due to reproduction of saffron is only possible through corm propagation, it is necessary to increase the production of this plant in order to expand its cultivation. Today, in addition to the production of flowers and dry matter yield of saffron, special attention is paid to the production of saffron corms. It has been proven in different studies that the use of big, healthy and strong corms can increase the saffron yield (Kafi et al., 2002; Hassanzadeh Aval et al., 2013). Therefore, most of the physical and nutritional characteristics of corms (such as size and nutrient concentration) are evaluated as a benchmark.

Plant nourishment with nitrogen is important in terms of both quantity and quality of the product (Shen et al., 1994). Nitrogen (N) plays a key role in crop productivity. Indeed, N is involved in the functioning of meristem tic tissues, in photosynthesis, and in the determination of the protein content of harvested organs (Bertheloot et al., 2008).

Because of phosphorus importance role in vegetative growth and flower induction in saffron (Chaji et al., 2013; Naghdi Badi et al., 2011), the more phosphorus that a plant takes up, the more stability will be achieved during saffron perennial life cycle. In general, it can be said that suitable nourishment of saffron in the first year produce daughter corms, richer in phosphorous and higher reservoirs, which will act as seeds for the next year. Corm reservoirs increment will have a positive effect on yield and vegetative organs in the following year, and the phosphorus-rich corms will probably increase assimilates and production of bigger daughter corms in next years (Chaji et al., 2013). Amir Ghasemi (2001) also stated that phosphorus is involved in the completion of saffron corm reservoirs.

Potassium is one of the most important nutrients after nitrogen, which in addition of important physiological functions, has a special place in improving the quality of agricultural products that made it known as quality element (Malakoti and Homai, 2004; Marschner, 1995). Sadeghi et al. (1989) studied the effect of NPK fertilizers and manure on leaf production and the average weight of saffron corm, and concluded that the highest effect was obtained from NPK, NP and manure, respectively.

Medicinal herbs require adequate amounts of micronutrients to grow and produce extract (Leilah, 1988; Sarmadnia and Koocheki, 1992), and among micronutrients, iron is the most necessary element (Kafi et al., 2008). It was reported that iron is one of the most important micronutrient elements for plants (Ksouri et al., 2007). Micronutrient elements are essential for plant growth and development, but their

application rates are lower than macronutrients such as nitrogen, phosphorus and potassium (Nateghi et al., 2015). Baghai and Maleki Farahani (2014) reported that the application of iron fertilizer has affected flower fresh weight, so that the application of 5 and 10 kg of both two iron fertilizers types, increased flower fresh weight.

According to all mentioned above, the current study was aimed to determine nutrients concentration in corm of saffron and relationship between nutrient concentration in saffron corms and saffron yield in perennial fields of South Khorasan.

Material and Methods

The study was carried out in two Nehbandan and Qaen regions and their geographical locations are presented in Table 1. This research was carried out in a factorial arrangement based on completely randomized design, with three replications in Nehbandan and Qaen regions, in 2015 in 3, 5 and 7 years old fields. The first factor was the location and the second factor was field's age. Firstly, based on the available information, received from Agriculture Jihad Organization of Nehbandan and Qaen Province and the opinion of relevant experts and considering area under saffron cultivation and awareness about farmers' ability, 3, 5 and 7 years old fields were identified. Then, three fields with at least 500 m^2 under cultivation selected for each age of 3, 5 and 7 years fields and 3 plots from each field and one corm sample from each plot were selected. 54 corm samples were collected from two regions about 10 day after the first irrigation. In order to determine nitrogen, phosphorus, potassium and iron concentrations in saffron corms, one corm sample was randomly harvested in each plot from 0-30 cm depth and 20×20 cm^2 dimensions with saffron shoot; then transferred to Soil Laboratory of Faculty of Agriculture in University of Birjand and the corms were separated from the soil. Soil physical and chemical characteristics of the studied regions are presented in Table 2. Phosphorus concentration was determined by spectrophotometer at 660 nm wavelength, and nitrogen was determined by Kjeldahl method in saffron corms (Soil and Plant Analysis Council, 1999). Potassium concentration of corms were performed by flame photometer (Thomas, 1982). Iron concentration was determined using acid digestion method and atomic absorption (Page et al., 1982). To estimate saffron yield in studied fields, it was stipulated that 3 plots in each field, to be harvested and weighed daily. The average yield of 10 days was collected and weighed, that totally was generalized to hectare. Due to the fact that farmers in these regions have been cultivating saffron using traditional and indigenous methods, thus there is an interact in terms of agronomic and nutritional management with each other, so all 3 selected fields from each age, have almost the same management. Accordingly, the basis for selection of the fields was the uniformity of field management in each two regions. At the end, the correlation coefficients of measured factors were investigated with saffron yield. Data analysis was performed using SAS software. Means comparison was performed by LSD test at a significant level of 5%.

Results and Discussion

The ANOVA of the effect of region and fields age on the nutrients concentration in saffron corms and saffron dry matter yield are shown in Table 3. As it is obvious, there was a significant difference between three age groups of 3, 5 and 7 years old fields in terms of nutrients concentration in saffron corms and saffron dry matter yield. There was a significant difference between Nehbandan and Qaen regions in terms of nutrients concentration in saffron corms and saffron dry matter yield. Among different nutrients concentrations, only nitrogen concentration was not affected by the interaction between region and field's age (Table 3).

Relationship between corm nitrogen concentration and dry matter yield

The study on the relationship between corm nitrogen content and dry matter yield at Nehbandan and Qaen regions in three age groups of 3, 5 and 7 years old fields showed a significant relationship. Fitting the linear regression function between saffron yield and nitrogen content in 3, 5 and 7 years fields of two studied regions showed that increasing in nitrogen percentage caused an increase in yield, while the five years old field in Qaen had the highest yield (Fig. 1). Correlation coefficients of fitted equations also showed a significant linear relationship between nitrogen percentage and the yield. However, the slope of the yield curve in comparison to the percentage of nitrogen was higher than the rest, in five years old field in Nehbandan. Higher organic matter in five years old field of Qaen has led to nitrogen uptake via corm, resulting higher yield. Nitrogen, which is one of the

most important elements which causes yield increment of saffron flowers and corms (Chaji et al., 2013), is known as a mobile element in the plant (Bertheloot et al., 2008) and can be transport during plant growth, from vegetative organs to underground parts, especially at the end of each season (Ourry et al., 1988; Masclaux-Daubresse et al., 2010). The increasing process of yield compared to nitrogen concentration of corm in both Nehbandan and Qaen regions shows that the five year fields of these two regions illustrate better reaction to nitrogen concentration and would obtain higher yield by corm nitrogen increasing. The yield response to low nitrogen content in five years old fields of Nehbandan and Qaen was more than those of three and seven; therefore, these fields had higher yields at lower nitrogen concentrations. Nitrogen is an essential and fundamental element for crops. Thus, nitrogen compounds affect dry matter yield, which has a direct relationship with photosynthesis and nitrogen uptake, and increasing of these compounds will increase total nitrogen in plants. One of the important requirements in agronomic planning is the evaluation of various plant nutrition systems in order to achieve higher yield and desirable quality, especially in case of medicinal herbs (Rezaenejad and Afyuni, 2001; Balkcom and Monks, 2007). The organic matter presence in Nehbandan and Qaen regions which was available due to manure addition to the soils of these fields by farmers has led to nitrogen uptake via plant corm, resulting yield increase, which is consistent with the results of Brussard (1997). He reported that organic matter addition into the soil increases soil nutrient content including nitrogen and its absorbance ability by crop, increasing nitrogen balance and the efficiency of phosphorus uptake.

Relationship between corm phosphorus concentration and dry matter yield

The relationship between phosphorus percentage of saffron corms in Nehbandan and Qaen fields with dry matter yield in three age groups of 3, 5 and 7 years old indicates that increase in corm phosphorus percentage increased yield (Fig. 2). Correlation coefficients of explained equations for all years indicate a significant relationship between them. In the five years fields of Qaen, phosphorus concentration of corm was higher than other field's ages, of course had the highest yield. This can be attributed to the geographical location and soil fertility of Qaen region, which higher elevation and colder climate has led to soil fertility. Also, the lowest soil pH was found in Qaen's five year fields, which increased phosphorus uptake and availability. In a study by Mohammady-Aria et al., (2010) in association with soil pH and phosphorus solubility in laboratory conditions, it was reported that soil pH changes had a negative and significant correlation with soluble phosphorus, that soil pH decreasing, will increases the amount of absorbable phosphorus in the soil and thus the plant will be able to absorb more phosphorus from the soil.

Coefficients of fitted equations for fields with different ages in Nehbandan and Qaen regions indicate that X coefficient of the equation (fitted line slope) in Qaen five year fields is greater than that of Nehbandan five year fields. In other ages of fields, the three year fields of Nehbandan have given the highest X coefficient. Among the nutrients, phosphorus plays a special role in reproductive phase of crops (White and Veneklaas, 2012) and in addition of saffron yield improvement, it can affect the growth of daughters' corms in saffron (Naghdi Badi et al., 2011). The balanced nutrients availability based on proper fertilizers management, is one of the most effective factors in sustainability of saffron production, especially in arid and semi-arid regions (Amiri, 2008; Koocheki et al., 2009), so that, up to 80% of flower yield changes in saffron are affected by variables that govern the soil especially, the amount of organic matter (Nehvi et al., 2010). The organic matter presence in Nehbandan and Qaen regions due to manure addition to the soils of these fields by farmers has led to phosphorus uptake by saffron corm, resulting yield increase, which is consistent with the results of Koocheki et al. (2014). They found that organic fertilizers application and mother corms sowing with more than 8 gr weighing could have play more effective role in yield improving and also phosphorus uptake of saffron per unit area. According to the results of this experiment, it seems that more phosphorus uptake by daughter corms is one of the reasons for higher production of saffron daughter corms as a result of organic fertilizers application.

Relationship between corm potassium concentration and dry matter yield

Studying the relationship between potassium of saffron corm and dry matter yield in 3, 5 and 7 year fields (Figure 3) shows a positive linear relationship between corm potassium content and yield. The explanation coefficients were significant for all years ($P \leq 0.01$). The fitted correlation evaluation between corm potassium concentration and saffron yield

indicates that corm potassium concentration is lower in three and seven year fields compared to five year fields. Considering the greater curve slope in the five year fields of Qaen, it is concluded that due to higher potassium concentrations in the five year fields of Qaen, dry matter yield of saffron was more than the rest. This is probably due to higher soluble potassium content of soil in Qaen's five year fields, which has resulted in more potassium uptake by saffron corm in the county. Potassium is one of the essential elements for plant growth, and if not be sufficient in the soil, plant growth will greatly decrease, resulting yield reduction. The results of this study are consistent with the findings of Zabihi and Feizi (2014). Considering the multiyear saffron cultivation of saffron, they examined the effect of one time potassium application at sowing in a 4 year period. The results showed that there is a little need for potassium in saffron cultivation, but providing such a low amount of potassium can have a considerable and significant effect on saffron yield and should not neglect potassium fertilizer application in the nutrition management of saffron. Organic matter presence in Nehbandan and Qaen regions which was available due to manure addition to the soils of these fields by farmers, has led to potassium uptake via plant corm, resulting yield increase. Some studies (Munshi, 1994; Shahande and Mousavi, 1988; Negbi, 1999) have shown that there is a positive and high correlation between soil organic matter and saffron yield. Increasing yield by organic matter application is probably caused by nutrients supply, especially phosphorus, potassium, nitrogen, and the improvement of soil physical properties. Improving soil structure or increasing cation exchange capacity of the soil reduces potassium, calcium and magnesium leaching, which has a positive effect on saffron yield.

Relationship between corm iron concentration and dry matter yield

The relationship between iron concentrations of saffron and dry matter yield of Nehbandan and Qaen regions in each groups of 3, 5 and 7 year fields, indicate that there is a positive linear relationship between corm's iron concentration and saffron yield (Fig. 4). The explanation coefficients of the equations for all years indicate a significant relationship between them. Since iron is one of the most important nutrient for plants, its deficiency leads to leaf yellowing, changing iron and other metal elements concentration in plant tissues, which are closely related to crops yield. It can be concluded that increasing of iron concentration in saffron corm resulted in dry matter yield increment. The importance of iron in plant nutrition has been emphasized considering the strategies that plants use to absorb this element under iron stress conditions (Baghai and Maleki Farahani, 2014). Micro elements are used in plants in small amounts. But they have important effects. In case of deficiency, these elements can sometimes act as a limiting factor for growth and uptake of other nutrients, and this issue makes it necessary to pay more attention to their application (Malakoti, 2000).

The relationship between yield and corm number
The ANOVA of the effect of region and fields age on saffron corms number are shown in Table 4. As it can be seen, the number of saffron corms was significantly affected by field's age. There is a significant difference between Nehbandan and Qaen regions in terms of corm number of 4-8 gr and 8-12 gr groups. Among the different groups of corm, only corm number of 8-12 gr was affected by the interaction between the region and fields age (Table 4).

The ANOVA showed that total corm number per m^2 was significantly affected by fields age (Table 4). Generally, the studied regions did not differ in total number corm number, but field's age increased total corm number per unit area. Total corm number was only significant in three year fields. The highest corm number observed in 7 and 5 year fields, respectively. Although there were no significant difference between 5 and 7 year fields, but total corm number in 7 year fields increased 2.95%. The lowest amount of corm number was allocated to three year fields, so that total corm number per m^2 decreased 60.20% compared to 5 year fields (Fig. 5). The total corm number of saffron corms in both Nehbandan and Qaen regions was not statistically significant (Table 4). Fields with 3 and 7 years age had more corm number in 2 gr and 2-4 gr groups. In these fields, corm number was high in terms of quantity, while these corms were not been recommended for cultivation (Sadeghi, 1996; Kafi et al., 2002). One of the reasons for low yielding in traditional Iranian fields is the use of small corms of old fields as seed corm (Hemmati-Kakhki and Hosseini, 2003). Comparison of the average corm number of saffron fields showed that increasing the

age of the saffron fields increased the reproduction of corms and consequently the total number of corms per m^2 (Fig. 5). However, in the seventh year, the highestcorm number was produced, but small corms did not produce economical yield. Small corms due to intense competition with each other have less nutritional reserve and weak vegetative growth resulting yield reduction (Ramezani, 2000). Fields with a life span of five years because of higher number of large corm had better condition. According to other researches, corm size and density are two major factors for yield increment. In this study, saffron fields have the highest number of corms with 8-12 gr weight in fifth year, which has a significant effect on saffron yield, and practically farmers obtain highest yield in third to fifth years from their fields (Mollafilabi, 2012; Sadeghi, 2012). It is recommended that farmers choose their corms form young or three year fields for cultivation and gaining maximum yield.

Among the measured nutrients in saffron corm, dry matter yield had a significant correlation with nitrogen (r = 0.311), phosphorus (r = 0.561), potassium (r =0.545) and iron (r = 0.433) concentrations (Table 5). It can be concluded that the increase in yield is associated with the increase in concentrations of nitrogen, phosphorus, potassium and iron in saffron corms. Proper use of nutrients especially nitrogen, phosphorus and potassium and its correct balance in the soil can be effective in yield increment and soil fertility stability (Sacco et al., 2003; Bassanino et al., 2011).

Among the different sizes of corms in this study, corms with 8-12 gr weight had a positive and significant correlation with nutrient concentrations of nitrogen, phosphorus, potassium and iron of saffron corm in both regions (Table 5). This correlation shows that the highest concentrations of nutrients are belong to 8-12 gr corms. Also, the number of corms with 8-12 gr weight, had a positive and significant correlation with saffron yield (r = 0.690) in both regions. Also corms with 4-8 gr weight, had a positive and significant correlation with saffron yield (r = 0.533). Regarding the correlation with the increase in corm number of 8-12 gr, it can be said that the increase in yield in five year fields is associated with the increase in corm number with 8-12 gr weight.

Table1: Geographical characteristics of Qaen and Nehbandan South Khorasan

Region	Longitude (E)		Latitude (N)		Altitude (m)	Total annual precipitation (mm)
	Degrees	Minutes	Degrees	Minutes		
Nehbandan	60	30	31	33	1196	128.6
	58	32	30	28		
Qaen	58	38	15	33	1440	180
	60	56	12	34		

Table 2- Physical and chemical properties of soil (0-30 cm depth)

	Field age	P (mg/kg soil)	K (mg/kg soil)	Na (meq/l)	EC_e (dS/m)	pH	SAR	O.M (%)	Sp (%)	Soil texture
Nehbandan	3	28.0	240.8	19.24	2.6	7.90	6.89	0.8	35	Sandy loam
	5	35.5	350.6	15.31	2.0	7.91	5.18	1.1	36.33	loam
	7	21.4	280.6	12.43	2.2	7.76	3.72	0.8	36.11	Sandy loam
Qaen	3	18.9	392.2	24.49	3.6	7.9	8.74	0.9	34.29	loam
	5	29.5	380.8	46.86	6.1	7.72	9.60	1.4	42.11	loam
	7	19.8	391.4	58.84	8.9	7.71	12.77	1.2	36	loam

Table 3- Variation analysis (Sum of square) of the effects of field age and region on the yield of dry matter saffron and some of nutrients concentration of saffron

Source of variation	Degree of freedom	Sum of square				
		Corm nitrogen concentration (%)	Corm phosphorus concentration (%)	Corm potassium concentration (%)	Corm iron concentration (mg Kg^{-1})	yield of dry matter saffron (Kg h^{-1})
Region	1	3.833^{**}	0.0029^{**}	0.267^{**}	382.71^{**}	11.926^{**}
Field age	2	1.290^{**}	0.0025^{**}	0.196^{**}	47.45^{**}	82.988^{**}
Region ×Field age	2	0.598^{ns}	0.0009^{*}	0.025^{*}	32.25^{**}	18.882^{**}
Error	48	0.251	0.0002	0.008	5.07	0.613
%(cv)	-	32.58	19.18	13.28	32.68	24.46

ns, * and ** are non significant, significant at 5 and 1% probability levels, respectively.

A) 3 Years

y = 2.3656x - 0.2031
R² = 0.8328**

y = 0.3776x + 1.6824
R² = 0.9361

B) 5 Years

y = 2.5896x + 4.3981
R² = 0.9404**

y = 3.2773x + 1.0762
R² = 0.9572

C) 7 Years

y = 0.8873x + 1.0341
R² = 0.9245**

y = 0.9351x + 0.098
R² = 0.9963

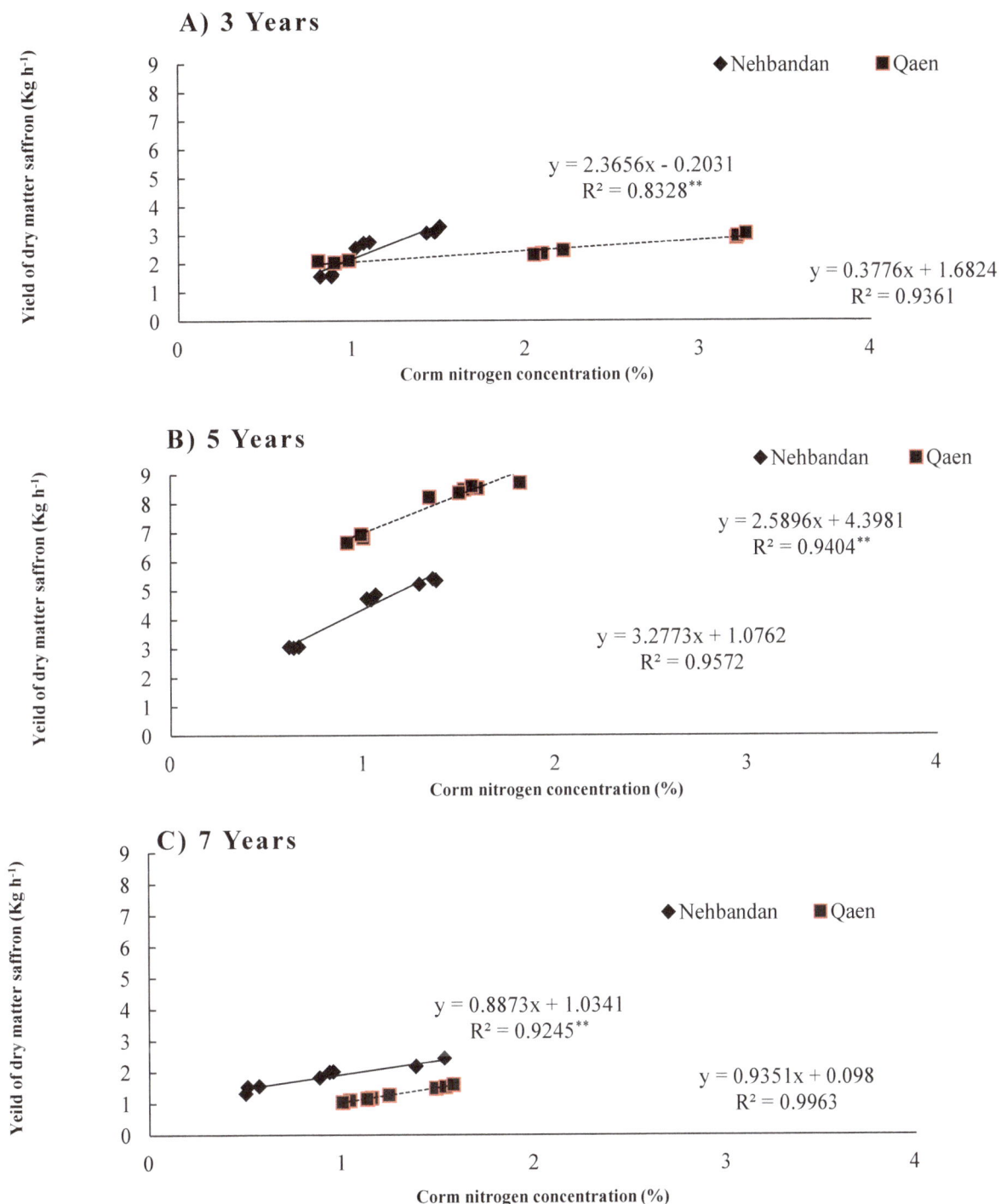

Figure 1: Relationship between nitrogen concentration of saffron corms in Nehbandan and Qaen fields with dry matter yield in three age groups of 1) Tree years, 2) Five years and 3) Seven years

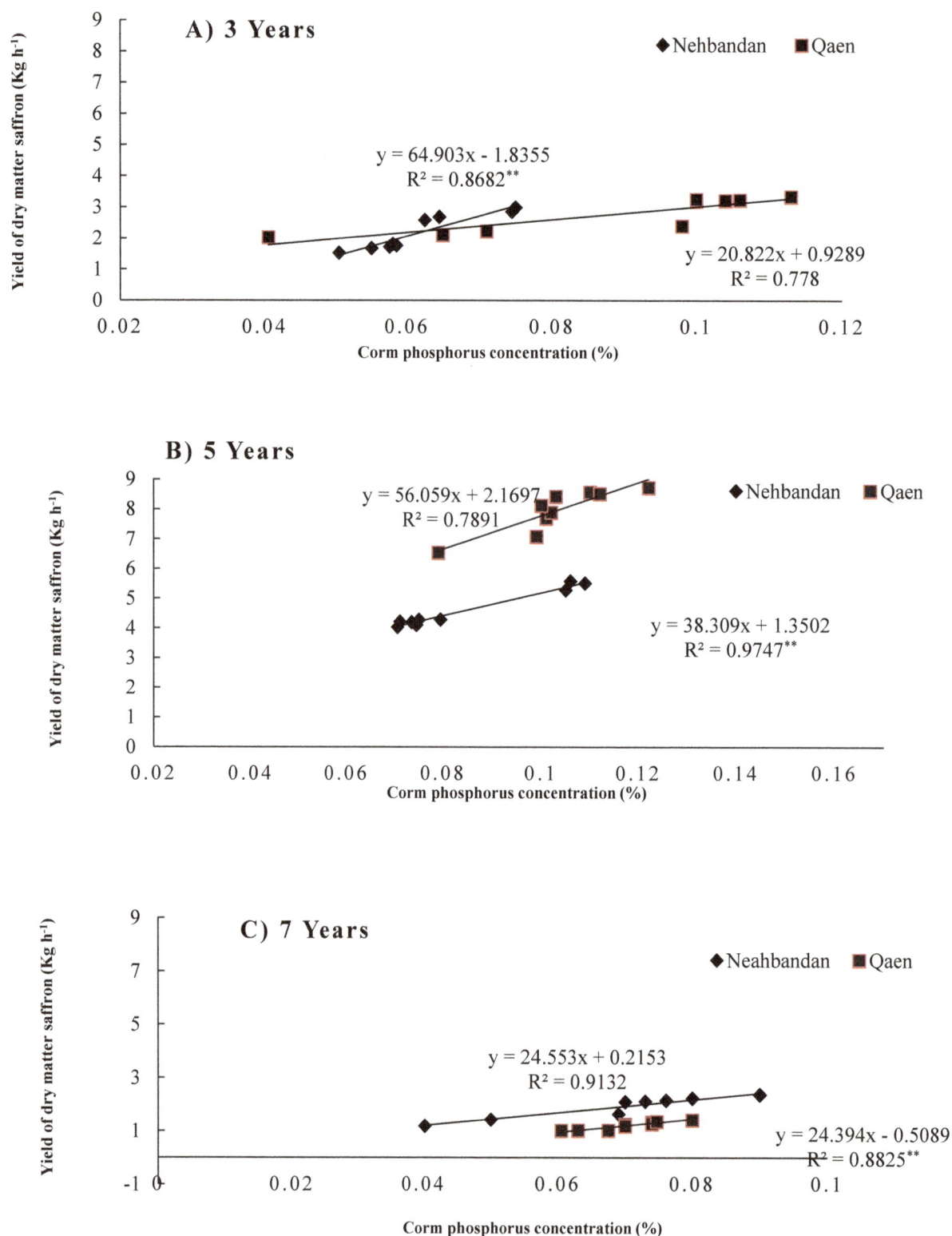

Figure 2: Relationship between phosphorus concentration of saffron corms in Nehbandan and Qaen fields with dry matter yield in three age groups of 1) Tree years, 2) Five years and 3) Seven years

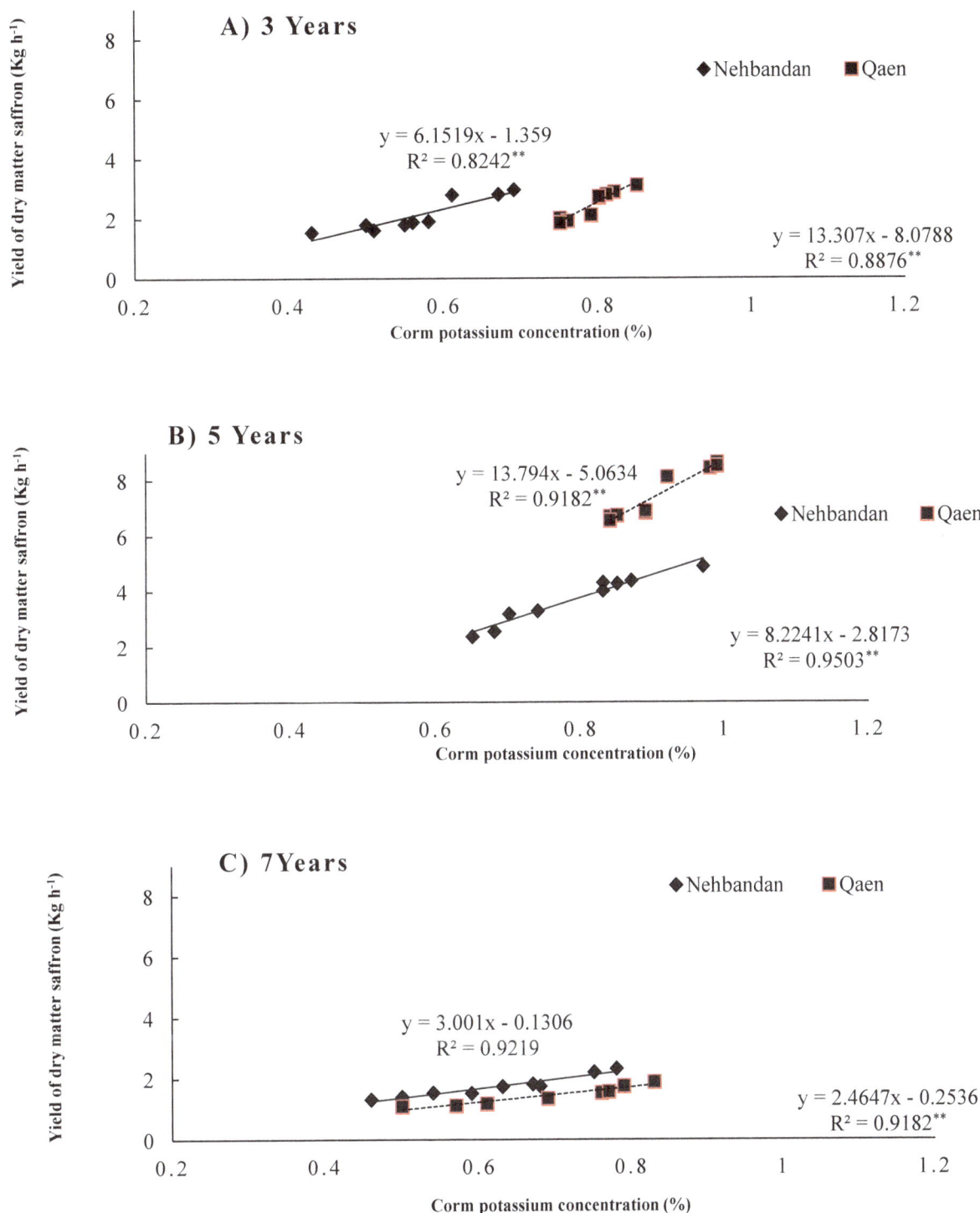

A) 3 Years

$y = 6.1519x - 1.359$
$R^2 = 0.8242^{**}$

$y = 13.307x - 8.0788$
$R^2 = 0.8876^{**}$

◆ Nehbandan   ■ Qaen

Yield of dry matter saffron (Kg h⁻¹)

Corm potassium concentration (%)

B) 5 Years

$y = 13.794x - 5.0634$
$R^2 = 0.9182^{**}$

$y = 8.2241x - 2.8173$
$R^2 = 0.9503^{**}$

◆ Nehbandan   ■ Qaen

Yield of dry matter saffron (Kg h⁻¹)

Corm potassium concentration (%)

C) 7 Years

◆ Nehbandan   ■ Qaen

$y = 3.001x - 0.1306$
$R^2 = 0.9219$

$y = 2.4647x - 0.2536$
$R^2 = 0.9182^{**}$

Yield of dry matter saffron (Kg h⁻¹)

Corm potassium concentration (%)

Figure 3: Relationship between potassium concentration of saffron corms in Nehbandan and Qaen fields with dry matter yield in three age groups of 1) Tree years, 2) Five years and 3) Seven years

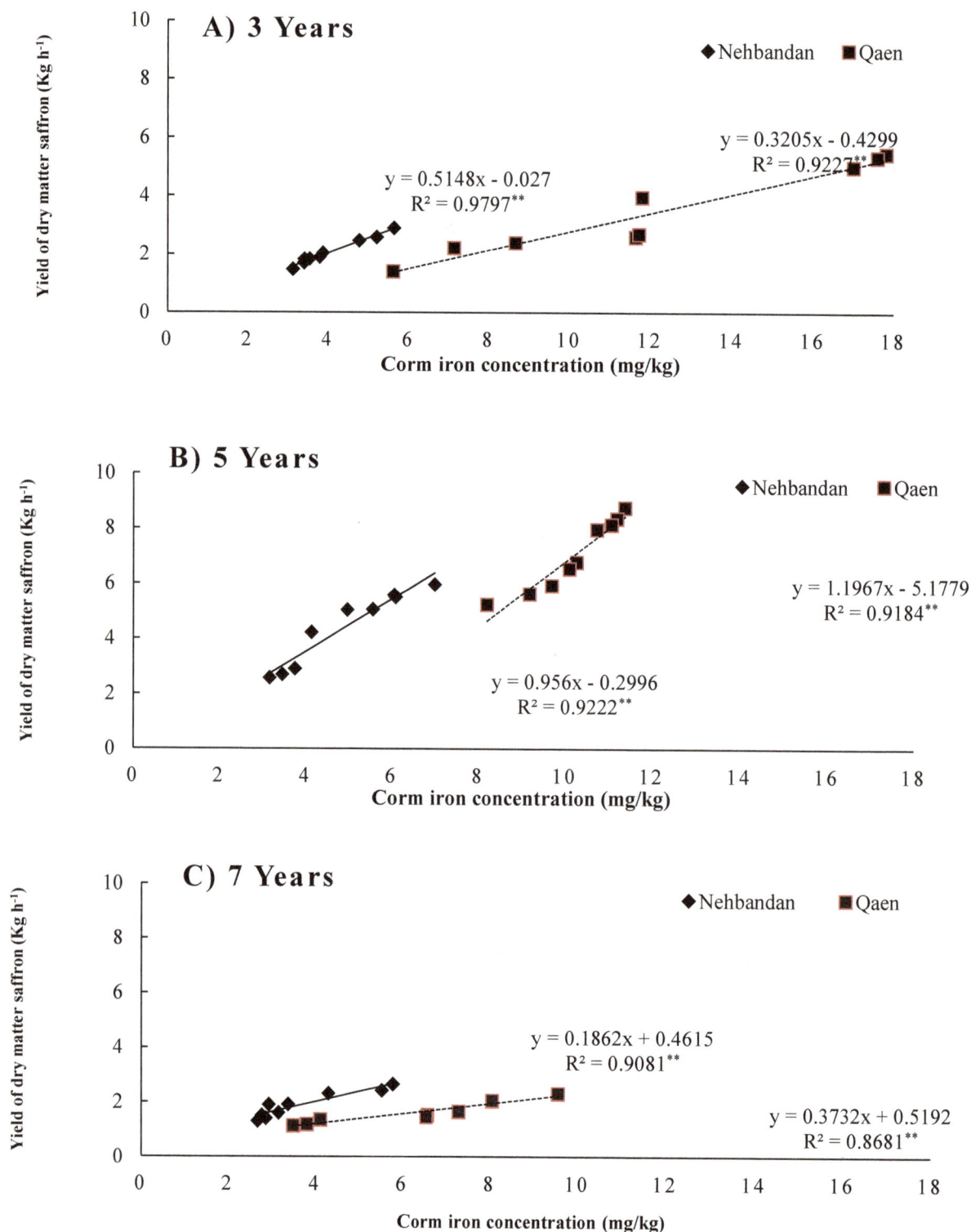

Figure 4: Relationship between iron concentration of saffron corms in Nehbandan and Qaen fields with dry matter yield in three age groups of 1) Tree years, 2) Five years and 3) Seven years

Table 4- Variation analysis (Sum of square) of the effects of field age and region on the number of saffron corms

Source of variation	Degree of freedom	The number of saffron corms				
		Under 2 g	2-4 g	4-8 g	8-12 g	Total
Region	1	12300.46^{ns}	6122.70^{ns}	16712.96^{*}	1956.02^{*}	244.91^{ns}
Field age	2	286072.69^{**}	119867.13^{**}	47604.17^{**}	10983.80^{**}	1030467.1^{**}
Region ×Field age	2	15908.80^{ns}	9061.57^{ns}	8761.57^{ns}	1956.025^{**}	74436.57^{ns}
Error	48	24041.09	15924.54	3075.81	384.84	58402.89

ns, * and ** are non significant, significant at 5 and 1% probability levels, respectively.

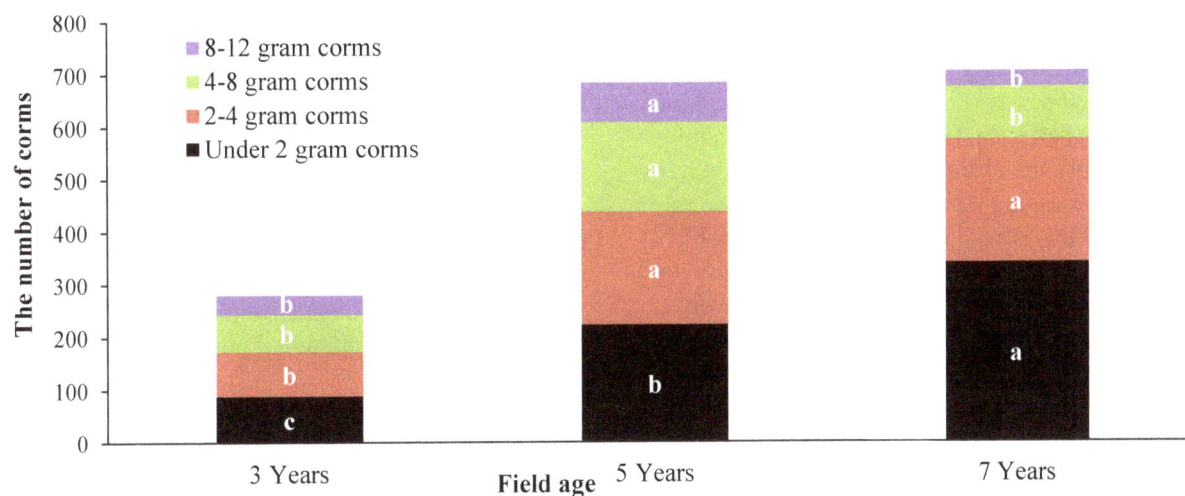

Figure 5: The effects of field age on the number of total saffron corms in perennial fields of Qaen and Nehbandan

Table 5: Pearson Correlation Coefficients of some nutrients concentration and number of saffron corms

	Corm nitrogen concentration (%)	Corm phosphorus concentration (%)	Corm potassium concentration (%)	Corm iron concentration (mg/Kg)	The number of saffron corms					yield of dry matter saffron (Kg h^{-1})
					Under 2 g	2-4 g	4-8 g	8-12 g	Total	
Corm nitrogen concentration (%)	1									
Corm phosphorus concentration (%)	0.227	1								
Corm potassium concentration (%)	0.284^{*}	0.560^{**}	1							
Corm iron concentration (mg/Kg)	0.642^{**}	0.481^{**}	0.461^{**}	1						
The number of Under 2 g corms	-0.89	-0.164	-0.280^{*}	-0.249	1					
The number of 2-4 g corms	0.60	0.006	0.133	-0.187	0.424^{**}	1				
The number of 4-8 g corms	0.457^{**}	0.416^{**}	0.492^{**}	0.316^{**}	0.141	0.216	1			
The number of 8-12 g corms	0.443^{**}	0.491^{**}	0.579^{**}	0.399^{**}	0.082	0.139	0.507^{**}	1		
yield of dry matter saffron (Kg h-1)	0.311^{**}	0.561^{**}	0.545^{**}	0.433^{**}	0.193	0.024	0.533^{**}	0.690^{*}	0.167	1

Conclusion

The results of this study showed that saffron yield was positively and significantly correlated with nutrient concentrations of nitrogen, phosphorus, potassium and iron in saffron corms. It can be said that increase in the yield is related to nutrients content increment in saffron corms, and can be attributed to animal manure application in these regions, which increased nitrogen, phosphorus and potassium uptake via saffron corm. The nutritional availability can increase yield due to the impact on crop growth processes. Its nature of perennial growth pattern in field conditions is effective in increasing nutrients absorption over time. In other words, with increasing life cycle of saffron, the number of daughter corms increases in the soil, which in proportion to this increase, the plant's ability to absorb nutrients such as nitrogen, phosphorus, potassium and iron also increase. Investigation of nutrients concentration of the corm on saffron yield shows the important role of such elements in yield. So that nitrogen, phosphorus, potassium and iron concentrations of the corm had a positive effect on saffron yield, among the investigated indices. Therefore, corm selection with optimum weight for sowing and proper use of nutrients, especially nitrogen, phosphorus, potassium and iron, as well as their correct balance in soil, can be effective in yield increment and stability of soil fertility. Generally it can be concluded that climatic condition, genotypic or management can stimulate plant growth and absorbing more elements from the soil.

References

Amir Ghasemi T, 2001. Saffron: Red Gold of Iran. Nashre Ayandegan Publication, Tehran, Iran. 112 p.

Amiri ME, 2008. Impact of animal manures and chemical fertilizers on yield components of (*Crocus sativus* L.). J. Soil Water Conserv. 25: 196-206.

Baghai N and Maleki Farahani S, 2014. Comparison of Nano and micro Chelated iron fertilizers on quantitative yield and assimilates allocation of saffron (*Crocus sativus* L.). J. Saffron Res. 1(2): 156-169.

Balkcom KS and Monks CD, 2007. Nitrogen plant growth regulator rates on Cotton yield and fiber quality. Cotton Res. Exten. Rep. 32: 21-22.

Bassanino M, Sacco D, Zavattaro L and Grignani C, 2011. Nutrient balance as a sustainability indicator of different agro-environments in Italy. Ecological Indicators. 11: 715-723.

Bertheloot J, Martre P and Andrieu B, 2008. Dynamics of Light and Nitrogen Distribution during Grain Filling within Wheat Canopy. J. Plant Physiol. 148: 1707-1720.

Brussard L and Ferrera-Cenato R, 1997. Soil ecology in sustainable arrigultural systems. New York: Lewis publishers, U.S.A. 168 p.

Chaji N, Khorassani R, Astaraei AR and Lakzian A, 2013. Effect of phosphorous and nitrogen on vegetative growth and production of daughter corms of saffron. J. Saffron Res. 1 (1): 112.

Gresta F, Avola G, Lombardo GM and Ruberto G, 2009. Analysis of flowering, stigmas yield and qualitative traits of saffron (*Crocus sativus* L.) as affected by environmental conditions. Sci. Hortic. 119: 320-324.

Hassanzadeh Aval F, Rezvani moghaddam P, Bannayan aval M and Khorasani R, 2013. Effects of maternal corm weight and different levels of cow manure on corm and flower yield of saffron (*Crocus sativus* L.). J. Saffron Agr. And Tech. 1 (1): 22-39.

Hemmati-Kakhki A and Hosseini M, 2003. A review of 15 years research on Saffron in Khorasan research and technology institute. Ferdowsi University of Mashhad Press. Iran. 134 p.

Hosseini M, Sadeghiand B and Aghamiri SA, 2004. Influence of foliar fertilization on yield of saffron (*Crocus sativus* L.). In: Proceedings of the 1st International Symposium on Saffron Biology and Biotechnology. Acta Hortic. 650: 207-209.

Husaini A and Ashraf N, 2010. Understanding saffron biology using bioinformatics tools. Glob. Sci. Books. 4: 31-34.

Kafi M, Lahoti M, Zand A, Sharifi HR and Goldani M, 2008. Plant physiology (Vol.1). Publications University of Mashhad (Translation), Mashhad, Iran, 456 p

Kafi M, Rashed M, Koocheki A and Mollafilabi A, 2002. Saffron: Production and Processing. Ferdowsi University of Mashhad Press. 276p.

Koocheki A, Najibnia S and Lalehgani B, 2009. Evaluation of saffron yield (*Crocus sativus* L.) in intercropping with cereals, pulses and medicinal plants. Iranian J. Field Crop Res. 7(1): 173-182.

Koocheki A, Seyyedi SM, Azizi H and Shahriyari R, 2014. The effects of mother corm size, organic fertilizers and micronutrient foliar application on corm yield and phosphorus uptake of saffron (*Crocus sativus* L.). J. Saffron Agr. Tech. 2(1): 3-16.

Ksouri R, Debez A, Mahmoudi H, Ouerghi Z, Gharsalli M and Lachaal M, 2007. Genotypic variability within Tunisian grapevine varieties (*Vitis vinifera* L.) facing bicarbonate-induced iron deficiency. Plant Physiol. Biochem. 45: 315-322.

Leilah AA, Badawi MA, EL-Moursy SA and Attia AN, 1988. Response of soybean plants to foliar application of zinc and different levels of nitrogen. J. Sci. Food Agric. 13: 556-563.

Malakoti MJ and Homai M, 2004. Dry Land Soil Fertility, Problems and Solutions. Tarbiat Modaress University Publications, Iran, Second Edition. 518 p.

Malakoti MJ, 2000. Comprehensive Methods of Diagnosis and Necessity of Efficient Chemical Fertilizers. Tarbiat Modarres University press, Iran. 158 p.

Marschner H, 1995. Mineral nutrition of higher plants. 2nd ed. Academy Press, New York, USA.

Masclaux-Daubresse C, Daniel-Vedele F, Dechorgnat J, Chardon F, Gaufichon L and Suzuki A, 2010. Nitrogen uptake, assimilation and remobilization in plants: challenges for sustainable and productive agriculture. Ann. Bot. J. 105: 1141-1157.

Mohammady-Aria M, Lakzzian A, Haghnia GH and Berengi AR, 2010. Effect of Thiobacillus, Sulfur and vermicompost on the water-soluble phosphorus of hard rock phosphate. Bioresour. Technol. 101: 551-554.

Mollafilabi A, 2012. Effect of extensive range of corm weights on yield components and flowering characters of saffron (*Crocus sativus* L.) under greenhouse conditions. In 4th International Saffron Symposium. India, 22-25 October, Kashmir, India.

Munshi AM, 1994. Effect of N and K on the floral yield and corm production in saffron under rain fed condition. Indian J. Arcanut Spices. 18: 24-44.

Naghdi Badi HA, Omidi H, Golzad A, Torabi H and Fotookian MH, 2011. Change in crocin, safranal and picrocrocin content and agronomical characters of saffron (*Crocus sativus* L.) under biological and chemical of phosphorous fertilizers. J. Med. Plants. 4 (40): 58-68.

Nateghi SH, Pirzad AR and Darvishzadeh R, 2015. The impact of micronutrient fertilizers, iron and

zinc on yield and yield component of anise. Sci. Hortic. 29 (1): 37-46.

Negbi M, 1999. Saffron cultivation: past, present and future prospects. In: Negbi M. (ed), Saffron (*Crocus sativus* L.). Harwood Academic Publishers, Australia, pp. 1-17.

Nehvi FA, Lone AA, Khan MA and Maghdoomi MI, 2010. Comparative study on effect of nutrient management on growth and yield of saffron under temperate conditions of keshmir. Acta Horticulturae: Third International Symposium on Saffron: Forthcoming Challenges in Cultivation, Research and Economics. 850: 165-170.

Ourry A, Boucaud J and Salette J, 1988. Nitrogen mobilization from stubble and roots during regrowth of defoliated perennial ryegrass. J. Exp. Bot. 39: 803-809.

Page AL, Miller RH and Keeney DR, 1982. Methods of Soil Analysis: Part II-Chemical and Microbiological Properties. Second Edition. Agronomy Society of America. Madison, WI.

Ramezani A, 2000. Evaluation of the effect of corm weight on saffron yield in Neyshabur condition. MSc Thesis of Faculty of Agriculture. Tarbiat Modarres University, Tehran, Iran.

Rezaenejad Y and Afyuni M, 2001. Effect of Organic Matter on Soil Chemical Properties and Corn Yield and Elemental Uptake. J. Water Soil Sci. 4 (4):19-29.

Sacco D, Bassanino M and Grignani C, 2003. Developing a regional agronomic information system for estimating nutrient balances at a larger scale. Eur. J. Agron. 20: 199-210.

Sadeghi B, 1996. Effects of corm size on flower production in saffron. Annual report, Scientific and Industrial Research Organization of Khorasan, Mashhad, Iran.

Sadeghi B, 2012. Effect of corm weight on Saffron flowering. Proceedings of the 4th International Saffron Symposium. Iran, 4-5 November, Keshmir, Iran. pp. 49-53.

Sadeghi B, Razavi M and Mollafilabi A, 1989. Effect of chemical fertilizers and animal manure on the leaf and corm production of saffron. Khorasan Agriculture Research Center, Mashhad, Iran.

Sarmadnia GH and Koocheki A, 1992. Physiological aspects of dryland farming (by Gupta, U.S.). Jihad Daneshgahi Mashhad Press. 424 p.

Shahande H and Mousavi M, 1988. Evaluation of physical and chemical properties of soils in relation to water and saffron in Gonabad. Research

Projects, Scientific Research and Industrial Organizations of Khorasan, Iran.

Shen D, Shen Q, Liang Y and Liu Y, 1994. Effect of nitrogen on the growth and photosynthetic activity of salt stressed barley. J. Plant Nutr. 17: 187-199.

Soil and Plant Analysis Council, 1999. Handbook on reference methods for soil analysis. Council on Soil Testing and Plant Analysis, CRC Press, Boca Raton, FL.

Thomas GW, 1982. Exchangeable Cations, pp. 159-165. In: A.L. Page R.H. Miller and D.R. Keeney (eds.), Methods of Soil Analysis. Part 2. Chemical and Microbiological Properties. Second Edition. Agron. Monogr. 9. ASA and SSSA, Madison, WI.

White PJ and Veneklaas EJ, 2012. Nature and nurture: the importance of seed phosphorus content. Soil Sci. Plant Nutr. 357: 1–8.

Zabihi HR and Feizi H, 2014. Saffron response to the rate of two kinds of potassium fertilizers. J. Saffron Agr. Tech. 2(3): 191-198.

5

The effect of different levels of amino acid and zinc on the quality and quantity of Berseem (*Trifolium alexandrinum*)

Muhammad Zakirullah[1]*, Sumayya Innayat[1], Tariq Jan[1], Muhammad Arif[2], Muhammad Ali[1], Mehboob Alam[3]

[1]Agricultural Research Institute Tarnab, Peshawar, Khyber Pakhtunkhwa, Pakistan
[2]Directorate of Outreach, Agricultural Research, Khyber Pakhtunkhwa, Pakistan
[3]Department of Horticulture, The University of Agriculture, Peshawar, Khyber Pakhtunkhwa, Pakistan

Abstract

The unavailability of green fodder throughout the year and low quality fodder are some of the main constraints that contribute in low yield of livestock. To overcome the restraint in quality and quantity, an experiment was designed at Agriculture Research Institute, Tarnab - Peshawar to investigate the effect of different levels of amino acid (aspartic acid) and zinc (zinc sulphate) on the quality and quantity of berseem. The experiment was laid out in randomized complete block design with split plot arrangement having three replications. Different levels of amino acid were applied @1000, 2000 and 3000 ml ha^{-1} to main plot while Zinc was applied @14 and 28 kg ha^{-1} to sub plot. Maximum stem height (73.77 cm), branches per stem (20.39), highest percentage of crude protein (17.67 %), crude fiber (33.59 %), dry mater (21.75 %) and green fodder yield (27.84 t ha^{-1}) was recorded in the plots that received the amino acid @ 3000 mlha^{-1}, while the plots that received amino acid at lowest rate i.e. 1000 ml ha^{-1} yielded the lowest stem height (60.22 cm), minimum number of branches per stem (12.72), low percentage of crude protein (15.85 %), crude fiber (31.21 %), dry matter (19.38 %) and lowest green fodder yield (23.72 t ha^{-1}). Similarly, zinc applied at higher rate of 28 kg ha^{-1} boosted the stem height (70.89 cm), number of branches (16.62), crude protein (17.86 %), crude fiber (35.01 %), dry matter (21.53 %) and green fodder yield (28.91 t ha^{-1}) compared to zinc applied @ 14 kg ha^{-1}. It is therefore, recommended that while growing berseem amino acid @ 3000 ml ha^{-1} and Zinc @ 28 kg ha^{-1} should be applied in order to get good yield and a quality crop.

*Corresponding author email:
mzakirtarakzai@gmail.com

Keywords: Amino acid, Zinc, Berseem, Quality and quantity

Introduction

For maintaining the normal health and reproduction of livestock it is essential to feed them with green fodder and forage of high quality (Roy and Khandaker, 2010). In Pakistan, 90% of the livestock diet consists of poor quality of roughages and yield of these forages are also less than its diet requirement. As a result the livestock is under fed and their health is poorly maintained.

Furthermore, this problem is becoming more severe as the population of cattle are increasing and most of the land is diverted to the cereals grain production for human consumption (Roy and Khandaker, 2010). Therefore cultivation of good quality forages with high yield is necessary to alleviate the shortage of fodder and forage for feeding livestock in Pakistan. Berseem is one of the fast growing and high quality leguminous forage that is fed to the animals as green chopped forage. In Pakistan, the yield and quality of berseem is low due to many environmental factors and certain management constrains. Among these, zinc deficiency is considered as one of the major nutritional constraint (Asif et al., 2013). Zinc plays a vital role in plant growth process and enhance metabolism. It is also considered to be involved in the formation of auxin and chlorophyll pigment which is essential for carbohydrate metabolism, enzyme formation and proper root development (Asif et al., 2013). This micronutrient is also required for nitrogen metabolism and is considered important element in the process of photosynthesis. In soil, deficiency of zinc may lead to poor quality and low yield of fodder crop (Rathore et al., 2015). Zinc is an important nutrient; therefore, its proper application may mitigate zinc deficiency in animals. Zinc interacts with other macro and micronutrients in crops and tends to increase the production. In pearl millet it was reported that application of zinc along with nitrogen increased the dry matter yield of a crop (Kumar et al., 1985). Furthermore, it was also observed that application of zinc enhances the fresh matter production of rice (Malik et al., 2011). Zinc also plays an important role in improving the quality of the crop and in mung bean it is observed that zinc treatment tends to increase the protein content (Krishna, 1995). Similarly, in sweet potatoes application of zinc increases the crude fiber content of a crop (Khairi et al., 2016).

Similarly, crops yield is also improved by the application of amino acids. Amino acid is used to increase the overall production and quality of a crop. It plays fundamental role in the synthesis of photo assimilates and can directly or indirectly influence the physiological activities of a crop (Liu and Lee, 2012). Amino acid applied to the soil is believed to improve the soil micro flora, which further facilitate the availability of nutrients to the crops. It is also not known that application of amino acid along with zinc can bring some changes in the morphological and quality characters of crops. Therefore the present study was designed to study the effect of different level of zinc and amino acid on the yield and nutritive value of berseem. Objective was to investigate the optimum level of amino acid and zinc for improving the growth, yield and quality of berseem.

Material and Methods

The present study was conducted at Agriculture Research Institute, Tarnab-Peshawar. The experiment was conducted to know the impact of different levels of amino acid and zinc on the quality and yield of berseem during rabi season 2016. Experiment was arranged in randomized complete block design with split plot arrangement having three replications. Zinc sulphate was used as a source of zinc whereas aspartic acid was used as a source for amino acid. Zinc sulfate was applied as soil application @ 14 kg ha^{-1} and 28 kg ha^{-1} while amino acid was applied as foliar application@ 1000, 2000 and 3000 ml ha^{-1}. All other agronomic practices were kept constant to all the treatments of the experiment.

Crop morphological data
Plant growth parameters including stem height (cm) and number of branches plant^{-1} were collected from five randomly selected plants in each treatment. The stem height was recorded from base of the plant to the top using measuring tape and the average stem height of the plant was calculated using eq. 1. Similarly, numbers of branches per plant were counted on the selected plants and their average was worked out with the help of eq. 2.
$\sum$ Values of stem height ÷ total number of selected stem............... Eq. 1
$\sum$ Number of branches ÷ total number of selected plants .... Eq. 2

Crude protein (%)
Took 30 ml of concentrated sulphuric acid (H_2SO_4), which was added to 1.0 g of finely powdered dried plant material. Then 5g of digested mixture were added to the sample and material was digested in the digestion chamber at 400^{0}C for 2-3 hours. After digestion the material was cooled at the room temperature and the volume of the sample was made 250ml with the addition of distilled water. From this sample 10 ml of aliquot was taken and distillation was done in the kjeldhal apparatus. Nitrogen was evolved from this distillation in the form of ammonia, which was collected in a receiver containing 2% boric acid. This sample was then titrated against 0.1N H_2SO_4until the golden colour appeared. At that point the volume of the acid was recorded. Nitrogen reading was multiplied with conversion factor 06.25 to get crude protein percentage of berseem.

Crude fiber (%)
For determination of crude fiber about 1.0 g of dried plant sample was taken and was wetted with 1.25% H_2SO_4. Distilled water was added to the beaker in order

to make up the volume up to 200ml. Material was then boiled on the flame for 30 min and was filtered and washed. To the filtrate 1.25% NaOH and distilled water was added and the volume was made up to 200ml. Again the sample was heated for 30 minute, filtered and the filtrate was washed again. The filtrate was then taken in a pre-weighed crucible and placed in oven for drying at 105^{0}C for 24 hours. After 24 hours the dry weight of the sample (W1) was recorded. The sample was then placed in the muffle furnace at 600^{0}C until the grey or white ash was obtained. Sample was then cooled at the room temperature and the weight of the ash (W2) was recorded. From the above recorded data the crude fiber percentage was calculated by using the formula.

Crude fiber (%) = W1 – W2 (g) x 100 ÷ Sample weight

Dry matter (%)
For dry matter estimation, weight of the oven dried aluminum container was recorded. Then 10 g of the green fodder was weighed and was placed in oven at 105^{0}C for 24 hours. Percentage of dry matter was calculated by using the following formula;

Dry Matter (%) = Weight of dry sample (g) x 100 ÷ Weight of green sample (g)
Crude protein, crude fiber and dry matter yield was estimated according to the procedures recommended by AOAC, (1990).

Green fodder yield (t ha^{-1})
Each plot in the experiment was harvested separately. Green fodder of the harvested plots was weighed on the balance so as to determine the total green fodder yield per plot. Data was recorded on kg per plot, which was then converted to tons per hectare.

Statistical analysis
The recorded data was statistically analyzed according to Steel and Torrie (1982) using RCB design with split plot arrangements. To determine treatments mean difference, least significant difference (LSD) was used at 5 % level of significance (P ≤ 0.05).

Results and Discussion

Stem height (cm)
Data pertaining to the stem height treated with different levels of ZnSO$_4$ and amino acids is presented in Table 01. The data revealed that stem height increased with increasing the level of amino acid and ZnSO$_4$. Maximum plant height was obtained with the application of highest level of amino acid and ZnSO$_4$and vice versa. Similar results were reported in mung bean by Samreen et al. (2017) who found out that application of ZnSO$_4$ increased the stem length of mung bean as compared to control. Furthermore, our results are also in agreement with the findings of Alam and Shereen (2002) who studied the effect of various levels of zinc and phosphorous on wheat and observed that the stem length of wheat was increased vs control in all the treatments. On the other hand application of amino acid in the form of aspartic acid increased the plant height of rice seedling, when applied in foliar form (Rizwan et al., 2017).The gain in stem height due to zinc and amino acid might be due to the zinc appetizing effects on crop enabling it to absorbed maximum available plants nutrients while the amino acid helped the crop in making photosynthates rapidly during photosynthesis.

Branches per stem
The results as reflected in table 02 revealed that both amino acid and zinc significantly affected number of branches per stem while the interaction of amino acid and zinc on branches per stem were found non-significant. The results of increased branches per stem with increased levels of amino acid and zinc are aligned with the findings of Sahito et al. (2014) who reported the increase in number of branches per plant of mustard with increase in level of zinc. The increased in number of branches per stem is due to the foliar application of amino acids which specifically useful in providing the readymade building blocks for protein synthesis without going through the cycle of amino acids synthesis within the plant.

Crude protein (%)
Table 03 shows the mean crude protein content of berseem treated with zinc sulphate and amino acid. The data exposed that crude protein content was positively correlated with zinc and amino acid treatment.
These results are in agreement with the findings of Hisamitsu et al. (2001). Similarly, Krishna also found increase in the protein content of mung bean treated with zinc. This increase might be due to the fact that zinc is involved in a catalytic and structural component of protein and enzyme which is essential in normal growth and development of a crop (Broadley et al., 2007). Similarly, Mohan also reported

significant increment in the protein content of maize crop treated with zinc sulphate at the rate of 30kg ha^{-1}.Similarly the results of amino acid on increasing the protein content of berseem clover are supported with the findings of Abd Allah et al. (2015)who reported increment in protein content when amino acid was applied in the form of glutamic acid.

Crude fiber (%)

Data regarding crude fiber (Table 04) discovered that application of different levels of $ZnSO_4$ and amino acid enhanced the percentage of crude fiber. Maximum crude fiber (33.59%) of berseem was recorded when amino acid was applied at highest rate of 3000 ml ha^{-1} while minimum crude fiber (31.21%) was observed with the application of amino acid @ 1000 ml ha^{-1}.The same pattern results were observed for zinc application as increase in zinc levels has increased the crude fiber content of berseem clover. All these outcome are in line with the findings of Khairi et al. (2016) who find out that increased the dose of zinc increased the crude fiber content of sweet potatoes. The increase in crude fiber of berseem with the increase in amino acid and zinc application might be due to the appetizing effect of zinc in crop and the foliar application of amino acid resulted in the increase in chlorophyll content which expedite the rate of photosynthesis, boosted up the carbohydrates formation which ultimately contributed in the increase crude fiber content of berseem clover.

Dry matter (%)

It is evident (Table 05) that application of zinc and amino acid increases the dry matter production of berseem. Increasing the levels of zinc and amino acids has positive effect on the dry matter yield. Maximum dry matter yield was observed when zinc was applied at the rate of 28 kg ha^{-1} and amino acids at the rate of 3000 ml ha^{-1}. Malik et al. (2011) reported that application of zinc increases the average root and shoot dry matter yield of rice. Our results are also supported by the findings of Kumar et al. (1985) who also observed increase in the dry matter production of pearl millet fertilized with high dose of zinc. Similar results were also discussed by Kumar et al. (2016) where application of zinc sulphate enhances the dry matter content of maize over the control. Abd Allah et al. (2015) reported increase in the dry matter content of rice when treated with high level of amino acid.

Table 1: Stem height (cm) of berseem as affected by different levels of amino acids and zinc sulphate.

Amino Acid Levels	Zinc Levels		Mean
	14 (kg ha^{-1})	28 (kg ha^{-1})	
1000 (ml ha^{-1})	56.10	64.33	60.22b
2000 (ml ha^{-1})	57.11	70.11	63.61b
3000 (ml ha^{-1})	71.44	76.11	73.77a
Mean	61.55b	70.89a	

LSD value for amino acid and zinc sulphate levels at (P≤0.05) = 10.96 & 5.17

Table 2: Number of branches stem^{-1}of berseem as affected by different levels of amino acids and zinc sulphate.

Amino Acid Levels	Zinc Levels		Mean
	14 (kg ha^{-1})	28 (kg ha^{-1})	
1000 (ml ha^{-1})	13.11	12.33	12.72c
2000 (ml ha^{-1})	15.22	16.99	16.11b
3000 (ml ha^{-1})	20.22	20.55	20.39a
Mean	16.18a	16.62a	

LSD value for amino acid and zinc sulphate levels at (P≤0.05) = 1.46 & 1.53

Table 3: Crude Protein (%) of berseem as affected by different levels of amino acids and zinc sulphate.

Amino Acid Levels	Zinc Levels		Mean
	14 (kg ha^{-1})	28 (kg ha^{-1})	
1000 (ml ha^{-1})	14.42	17.28	15.85c
2000 (ml ha^{-1})	15.77	17.75	16.76b
3000 (ml ha^{-1})	16.76	18.57	17.67a
Mean	16.65a	17.86b	

LSD value for amino acid and zinc sulphate levels at (P≤0.05) = 0.85 & 0.74

Table 4: Crude fiber (%) of berseem as affected by different levels of amino acids and zinc sulphate.

Amino Acid Levels	Zinc Levels		Mean
	14 (kg ha^{-1})	28 (kg ha^{-1})	
1000 (ml ha^{-1})	28.30	34.13	31.21b
2000 (ml ha^{-1})	30.14	34.70	32.42ab
3000 (ml ha^{-1})	30.98	36.20	33.59a
Mean	29.81b	35.01a	

LSD value for amino acid and zinc sulphate levels at (P≤0.05) = 2.12 & 2.08

Table 5: Dry matter (%) of berseem as affected by different levels of amino acids and zinc sulphate.

Amino Acid Levels	Zinc Levels		Mean
	14 (kg ha^{-1})	28 (kg ha^{-1})	
1000 (ml ha^{-1})	18.03	20.74	19.38b
2000 (ml ha^{-1})	19.60	21.14	20.37b
3000 (ml ha^{-1})	20.79	22.71	21.75a
Mean	19.47b	21.53a	

LSD value for amino acid and zinc sulphate levels at (P≤0.05) = 1.09 & 1.44

Table 6. Green Fodder yield (t ha^{-1}) of berseem as affected by different levels of amino acids and zinc sulphate.

Amino Acid Levels	Zinc Levels		Mean
	14 (kg ha^{-})	28 (kg ha^{-1})	
1000 (ml ha^{-1})	20.93	26.52	23.72c
2000 (ml ha^{-1})	22.15	28.89	25.52b
3000 (ml ha^{-1})	24.37	31.32	27.84a
Mean	22.48b	28.91a	

LSD value for amino acid and zinc sulphate levels at (P≤0.05) = 1.21 & 1.44

Green fodder yield (t ha^{-1})

Various levels of amino acid and zinc positively affected the green fodder yield as shown in (table 06). It was observed that increment in application of zinc correspondingly increased the growth and green fodder yield which are supported by the statements of (Kumar and Bohra, 2014) who observed increased yield of corn with increasing the level of zinc. These results are also in line with the findings of Kumar et al. (2016) where zinc soil and foliar application both led to an increase in green fodder yield of the maize crop. Likewise, the linear increase in fodder yield was also noted with the increase of foliar application of aspartic acid. These results are according to the findings of (Abd Allah et al., 2015) who observed the increase in fresh weight of rice tiller and roots with the application of amino acid. The increase in green fodder yield of berseem with increase in amino acid and zinc might be due to the zinc appetizing effects and most of the crop likes to have amino acid as source of nitrogen. Amino acid being a building block of protein, plays a vital role in enhancing the photo synthetic cells division, readily observed and easily converted in to photo assimilates without consuming much energy by the crop.

Conclusion

It is concluded from the study that amino acids and zinc can be applied at the rate of 3000 ml ha^{-1} and 28 kg ha^{-1} respectively to enhance the quality and yield of berseem grown under the agro ecological conditions of Peshawar.

References

Allah A, El-Bassiouny M, Bakry H and Sadak B, 2015. Effect of arbuscular mycorrhiza and glutamic acid on growth, yield, some chemical and nutritional quality of wheat plant grown in newly reclaimed sandy soil. Res. J. Pharm. Biol. Chem. Sci. 6(3):1038-1054.

Alam S and ShereenA, 2002. Effect of different levels of zinc and phosphorus on growth and chlorophyll content of wheat. Asian J. Plant Sci. 1: 364-366.

AOAC, 1990. Official methods of analysis of Association of Official Analytical Chemists International. 17th Ed., Washington, USA.

Ashok K, Bisht B, Manish K and Lalit K, 2010. Effects of Ni and Zn on growth of Vigna mungo, Vigna radiata and Glycine max. Int. J. Pharm. Bio. Sci. 1(2):1083-1090.

Asif M, Saleem MF, Anjum SA, Wahid M and Bilal MF, 2013. Effect of nitrogen and zinc sulphate on growth and yield of maize (*Zea mays*). J. Agric. Res. 51(4): 455-460.

Broadley MR, White PJ, Hammond JP, Zelko and Lux A, 2007. Zinc in plants. New Phytologist.173(4): 677–702.

Hisamitsu T, Ryuichi O and Hidenobu Y, 2001. Effect of zinc concentration in the solution culture on the growth and content of chlorophyll, zinc and nitrogen in corn plants (*Zea mays* L.). J. Trop. Agric. 36: 58-66.

Khairi M, Nozilaudi M, Sarmila MA, Naqib S and Jahan S, 2016. Compost and zinc application enhanced production of sweet potatoes in sandy soil. Open Access J. Agric. Res. 1(2):000107

Krishna S, 1995. Effect of sulphur and zinc application on yield, S and Zn uptake and protein content of mung (green gram). Legume Res. 18: 89-92.

Kumar R and Bohra JS, 2014. Effect of NPKS and Zn application on growth, yield, economics and quality of baby corn. Arch. Agron. Soil Sci. 60: 1193-1206.

Kumar R, Rathore D, Meena B, Singh M, Kumar U and Meena V, 2016. Enhancing productivity and quality of fodder maize through soil and foliar zinc nutrition. Indian J. Agric. Res. 50(3): 259-263.

Kumar V, Ahlawat V and Antil R, 1985. Effect of nitrogen and zinc levels on dry matter yield and concentration and uptake of nitrogen and zinc in pearl millet. Soil Sci. 139: 351-356.

Liu XQ and Lee KS, 2012. Effect of mixed amino acids on crop growth. Agric. Sci. DOI: 10.5772/37461.

Malik NM, Chamon A, Mondol M, Elahi S and Afaiz S, 2011. Effects of different levels of zinc on growth and yield of red amaranth (Amaranthus sp.) and rice (*Oryza sativa*, Variety-BR49). J. Bangladesh Assoc. Young Researchers. 1: 79-91.

Mohan S, Singh M and Kumar R, 2015. Effect of nitrogen, phosphorus and zinc fertilization on yield and quality of kharif fodder-A review. Agric. Rev. 36:218-226.

Rathore DK, Kumar R, Singh M, Kumar P, Ttyagi N, Datt C, Meena B, Soni PG and Makrana G, 2015. Effect of phosphorus and zinc Application on nutritional characteristics of fodder cowpea (*Vigna unguiculata*). Indian J. Anim. Nutr. 32: 388-392.

RizwanM, Ali S, Aakbar MZ, Shakoor MB, Mahmood A, Ishaque W and Hussain A, 2017. Foliar application of aspartic acid lowers cadmium uptake and Cd-induced oxidative stress in rice under Cd stress. Envir. Sci. Poll. Res. 24(27):21938-21947..

Roy P and Khandaker Z, 2010. Effects of phosphorus fertilizer on yield and nutritional value of sorghum (*Sorghum bicolor*) fodder at three cuttings. Bangladesh J. Anim. Sci. 39:106-115.

Sagardoy R, Morales F, Lopez AF, Abadia A and Abadia J, 2009. Effects of zinc toxicity on sugar beet (*Beta vulgaris* L.) plants grown in hydroponics. Plant Biol. 11: 339-350.

Sahito HA, Solangi WA, Lanjar AG, Solangi AH and Khuhro SA, 2014. Effect of micronutrient (zinc) on growth and yield of mustard varieties. Asian J. Agric. Biol. 2: 105-113.

SamreenT, Shah HU, Ullah S and Javid M, 2017. Zinc effect on growth rate, chlorophyll, protein and mineral contents of hydroponically grown mungbeans plant (*Vigna radiata*). Arabian J. Chem. 10:1802-1807.

Steel R and Torrie J, 1982. Principles and procedures of statisties 2nd edition. McGraw, Hill Book, New York, USA.

Zakirullah M, Ali N, Jan T, Akakhil H and Ikramullah M, 2017. Effect of different nitrogen levels and cutting stages on crude protein, crude fiber, dry matter and green fodder yield of oat (*Avena sativa* L.). Pure App. Biol. 6: 448-453.

6

Determination of the critical period for weed control of sweet corn under tropical organic farming system

Marulak Simarmata*, Uswatun Nurjanah, Nanik Setyowati
Department of Agronomy, University of Bengkulu, Jalan W.R. Supratman Kandang Limun, Bengkulu 38371, Indonesia

Abstract

An understanding of the critical period of the crop for weed control (CPWC) is needed before making a decision on weed management. A field experiment to study a CPWC of sweet corn was carried out at the highland of Bengkulu Province, Indonesia from October 2015 to January 2016. The objective was to determine the CPWC of sweet corn under the tropical organic farming system. Weed infestations in the research plots including 14, 28, 42, 56, 70, and 84 days after planting (DAP) of weedy and weed-free periods were arranged in a completely randomized block design (CRBD) with three replications. Results showed that the plant height, leaf area, and yield of sweet corn descended and ascended due to the increase of weedy and weed-free periods, respectively. The biomass and yield losses due to weed competition during the growing season reached 49.5 and 54.7 %, respectively. The relative yield descended or ascended in logistic equation curves due to the increase of weedy or weed-free periods, respectively. Based on the acceptable yield loss (AYL) of 5 %, the CPWC of sweet corn under organic farming system was determined from 2 to 77 DAP, and with the AYL of 10 %, the CPWC was determined from 3 to 53 DAP.

*Corresponding author email:
marulak_simarmata@yahoo.com

Keywords: Sweet corn, CPWC, Tropical organic farming, Weed control

Introduction

Sweet corn (*Zea mays saccharata*) is a horticultural crop cultivated over the world. The consumption of sweet corn in Indonesia continues to increase because of its nutrition and naturally sweet taste, but production remains low. Efforts to optimize the yields of sweet corn were done intensively using many synthetic inputs such as fertilizers and pesticides (Johnson et al., 2010; Akintoye and Olaniyan, 2012). But, using of synthetic inputs intensively may cause adverse impacts to the environment due to the deterioration of soil and water quality (Mulvaney et al., 2009; Savci, 2012; Ruark et al., 2012). Therefore, the organic farming system, known as an environmentally friendly practice without using synthetic materials, becomes a wise choice not only to save soil environment but also to produce healthy food (Hue and Silva, 2000; Taguling, 2013). An organic farming system is trending over the world today, especially for horticultural crops (Ruark et al., 2012; Muktamar et al., 2017).

Organic farming can be interpreted as a crop production system based on biological recycling of nutrients. Soil fertility can be improved not only by recycling the organic material in-situ but also by using the organic materials from the outside of the farming areas, such as composted forage plants and animal manures (Taguiling, 2013). However, some organic materials such as cattle manure and forage composts may carry weed propagules (Barberi, 2002). The number of weed populations in planting area may increase because of the emerging weeds both from animal manure and from soil seed bank (Carr et al., 2013). Weeds will be very detrimental to the crops if they are not controlled by the right measures at the appropriate time. Since organic farming does not use any synthetic chemicals such as herbicides, the choice of weed control practices may be limited to cultural, mechanical, biological, or integrated control practices by combining those methods (Barberi, 2002). An appropriate weed management should be carried out when the presence of weeds is harmful to the crops through a significant reduction of yield known as the CPWC (Knezevic at al., 2002; Johnson et al., 2010). According to Zimdahl (2004), CPWC is the growth stages of the crop where weeds must be controlled to prevent the apparent losses of crop yields. An understanding of the CPWC is very useful to make timely decisions for specific weed control on each plant species (Knezevic and Datta, 2015).

Determination of the CPWC can be approached by investigation of the weedy and weed-free periods on a specific crop. The limit of the acceptable yield loss (AYL) in general varies from 5 to 10 % (Knezeviz and Datta, 2015). The yield or relative yield of weedy and weed-free trials are described in logistic equation curves, where the intercept of the weedy and weed-free curves at the AYL on x-axis can determine the maximum of weedy periods and the minimum of weed-free periods of crop growth stages to weed competition, respectively (Juraimi et al., 2009; Mekonnen et al., 2017).

Some researchers have reported the CPWC among crops worldwide that included corn (Evans et al., 2003; Gantoli et al., 2013), soybeans (Knezevic et al., 2003), rice (Juraimi et al., 2009; Chauhan and Johnson, 2011; Mekonnen, 2017), peanuts (Everman et al, 2008), and sweet corn (Williams II, 2006). The CPWC varies due to climates factors, environmental conditions, cropping system, and cultivation technology such as row spacing, planting date, and fertilizer application (Williams II, 2006; Juraimi et al., 2009; Chauhan and Johnson, 2011). A field experiment was carried out at the highland of tropical areas to determine the critical periods of sweet corn to weed competition under a closed organic farming system.

Material and Methods

The research was carried out at the Closed Agricultural Production System (CAPS) Research Station of the University of Bengkulu located at District of the Rejang Lebong, Bengkulu Province, Indonesia from October 2015 to January 2016. The field is positioned at 1020 36' 56" E, 30 27' 37" S, and the altitude of 1,054 meters above the sea level. The soil type is Andept and the soil texture is classified as sandy loam. The site was regularly cultivated for organic vegetable production since 2009.

Weed assessment on the research site was conducted in 3 blocks based on weeds stratification. Weeds were enumerated in 5 sampling plots for each block using a wooden square plot size of 0.5 m x 0.5 m each, following the methods of Simarmata et al. (2015). Data observed were density, frequency, and dry biomass weight of weed species in each sampling plot. The rank of dominant weed species was determined based on the values of summed dominance ratio (SDR) calculated from the average of relative density, frequency, and dry biomass weight as modified from Janiya and Moody (1989) (Eq. 1).

$$SDR = \frac{Dr + Fr + Br}{3} \qquad (1)$$

Where, SDR is summed dominance ratio; Dr is relative density calculated from density of one species divided by total density of all species; Fr is relative frequency calculated from frequency of one species divided by total of frequency of all species; and Br is relative biomass calculated from biomass of one species divided by total biomass of all species.

Land preparation was started by cutting the weeds and land was cultivated twice using hoes. The experimental site was formed for 36 plots (12 trials with 3 replications) with sizes of 3 m x 1.5 m each. Organic fertilizer was applied one week before planting using composted solid cow's manure of 10 ton ha^{-1}. The manures were mixed homogeneously within 20 cm depth of soil surface. Seeds of sweet corn var. Secada were planted in a hole of 3 cm depth with planting spaces of 75 cm x 25 cm.

Sweet corn plants were maintained regularly by watering and pest were controlled mechanically as needed. In addition to the manure fertilizers, liquid organic fertilizer (LOF) was sprayed on soil surface at 1 and 6 WAP at the rates of 5 ml m^{-2}. The LOF was produced at the University of Bengkulu, consisted of dairy cattle feces, dairy cattle urine, soil containing local microorganism, green leaves of *Tithonia diversifolia*, and solution of EM-4, diluted and fermented in water for 10, 10, 1, 2.5, 10 %, respectively (Muktamar et al., 2017). Weeds were controlled by physical control method (PCM) in accordance with weedy and weed-free periods (Table 1). The CPWC of sweet corn under tropical organic farming systems was evaluated by variations of weed infestations in the research plots including 14, 28, 42, 56, 70, and 84 DAP of weedy and weed-free periods.

The weed-free periods were maintained by manually removing weeds that appear in the specified weed-free period trial, likewise, the weedy periods were allowing weeds to grow within the specified period trial and after that period, the emerged weeds were manually removed from the plots. The experiment was arranged in a completely randomized block design (CRBD) with three replications.

Table 1. Periods of weed infestations on sweet corn under tropical organic farming system.

No.	Treatment	0-14 DAP	0-28 DAP	0-42 DAP	0-56 DAP	0-70 DAP	0-84 DAP
1.	Weed-free 14 DAP						
2.	Weed-free 28 DAP						
3.	Weed-free 42 DAP						
4.	Weed-free 56 DAP						
5.	Weed-free 70 DAP						
6.	Weed-free 84 DAP =Weedy 0 DAP						
7.	Weedy 14 DAP						
8.	Weedy 28 DAP						
9.	Weedy 42 DAP						
10.	Weedy 56 DAP						
11.	Weedy 70 DAP						
12.	Weedy 84 DAP =Weed-free 0 DAP						

DAP = Days After Planting   = Weedy   = Weed-free

Ten sample plants from each plot were harvested at 84 DAP and data recorded for plant height, leaf area, and yield of unhusked cob's weight, cob's length, cob's diameter, and biomass weight. Residual of weed biomass were collected from 2 sampling plots (sizes of 0.5 m x 0.5 m each) of weed-free period trials. Biomass of sweet corn and weeds were oven-dried at 70 C for 72 hours. Data were subjected to one-way analysis of variances (ANOVA) and further separated by Duncan's multiple range test (DMRT) at 5 % level. Data of the yield were converted to relative yield as a percent of control and further analyzed using non-linear regression model. The equation with the highest determination factor (R^2) was judged as the most appropriate model to determine critical periods of sweet corn (Williams II, 2006; Juraimi et al., 2009). The intercept of weedy and weed-free curves on the x-axis at 5 % and 10 % of acceptable yield loss (AYL) were chosen arbitrarily to determine the CPWC of sweet corn under the tropical organic farming system (Knezeviz and Datta, 2015).

Results and Discussion

Weed analysis

Weed vegetation in the experimental site was identified before and after the experiment. Initial analysis was conducted in 3 blocks based on visual views of weed stratifications. Weed assessment in block I (Fig. 1A) identified 11 species of weeds in which 4 species have SDR > 10 which were *Ageratum conyzoides, Mimosa invisa, Echinochloa colona, Stachytarpeta jamaicensis*, respectively with SDR of 26.0, 18.0, 14.6, and 13.2 %. In block II (Fig. 1B) there were 7 species of weeds in which 5 species have SDR > 10 which were *Stachytarpeta jamaicensis, Cyperus kyllingia, Echinochloa colona, Ageratum conyzoides, and Mimosa invisa* with SDR of 29.4, 16.8, 16.4, 15.1 and 14.7 %, respectively. In Block III (Fig. 1C) there were 8 weed species but only 3 species have SDR > 10 which were *Echinochloa colona, Ageratum conyzoides, Stachytarpeta jamaicensis* with SDR of 40.9, 25.6, and 10.7 %, respectively. Based on the differences in weed distributions in the three blocks before the experiment, the study was designed in a randomized block design (RBD) with 3 replications as blocks.

Weeds that grew at the end of the study as residuals of the treatment were only observed in the plots of weed-free periods. There were 9 species that grew at the harvested time but only 3 species had SDR > 10 which were *Euphorbia prunifolia, Ageratum conyzoides, Echinochloa colona* with SDR of 24.0, 22.7, and 15.6 %, respectively. Compared to the initial analysis, two new weed species were *Amaranthus spinosus, Borreria latifolia* with SDR of 4.4 and 3.7 %, while five weed species were absent were *Cyperus kyllingia, Phyllanthus niruri, Syndrella nodiflora, Spilanthes acmella*, and *Stachytarpeta jamaicensis* (Fig. 1D). The presence or absence of weed species indicated the shifting in weed vegetation due to manipulated microenvironment such as the different periods of weed infestation (Simarmata et al., 2015). Newly emerging weeds may be carried away from the manure or emerged from dormant seed or from the soil seed bank (Barberi, 2002). Weed residue harvested at the end of the study as the total weed biomass was only observed in weed-free trials because there was no weed residue in the weedy trials (William II et al., 2008). The heaviest weed residue in the weed-free trials reached 656 g m^{-2}. The weed residue decreased with longer periods of weed-free and there was no weed residue if the plot was free from weeds during the season (Fig. 2).

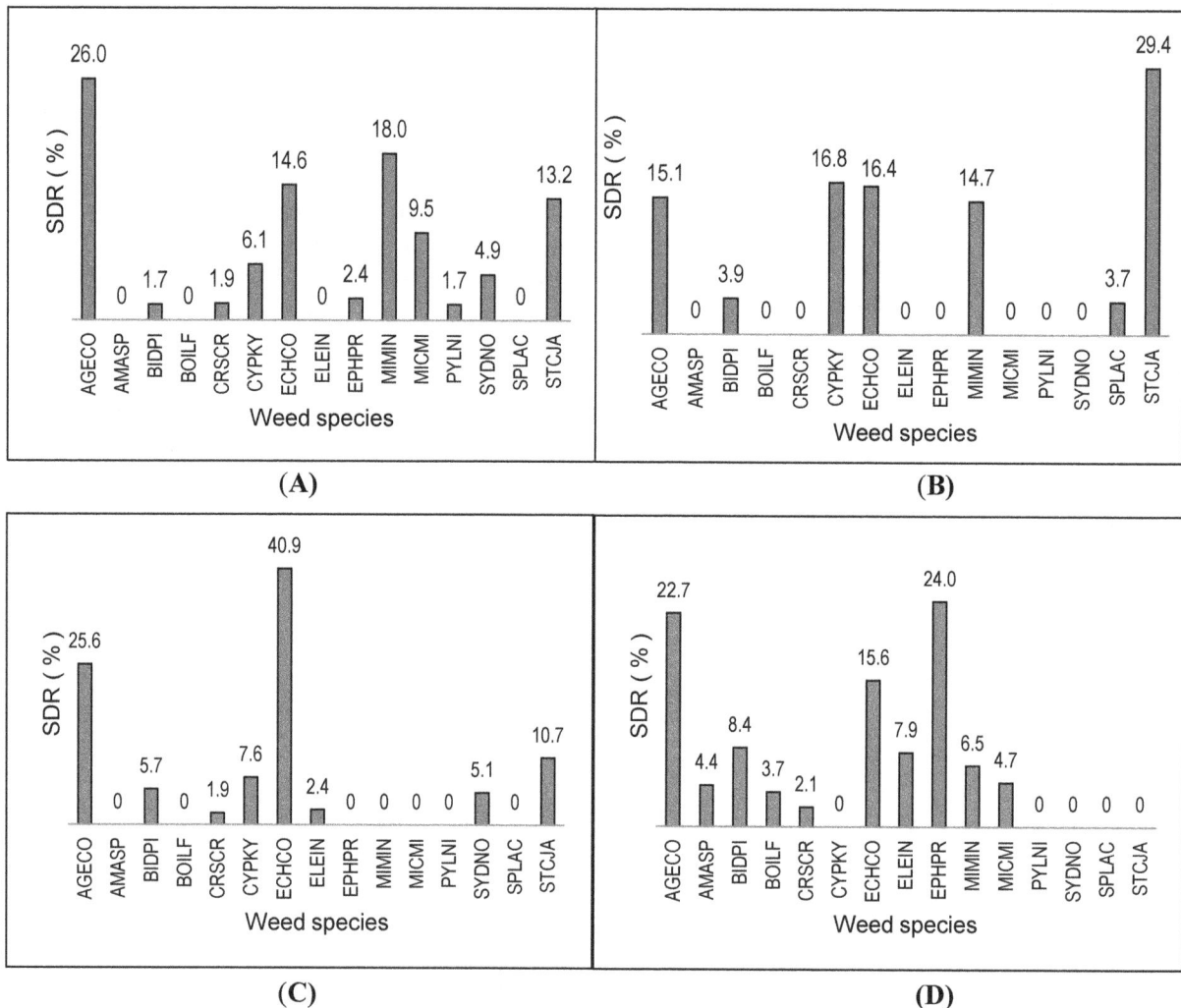

Fig. 1. Initial assessment of weed vegetation in research sites (A, B, C) , and final assessment at the end of the growing season (D) counted as summed dominance ratio (SDR); AGECO = *Ageratum conyzoides,* **AMASP =** *Amaranthus spinosus,* **BIDPI =** *Bidens pilosa,* **BOILF =** *Borreria latifolia,* **CRSCR =** *Crassocephalum crepidioides,* **CYPKY =** *Cyperus kyllingia,* **ECHCO =** *Echinocloa colona,* **ELEIN =** *Eleusine indica,* **EPHPR =** *Euphorbia prunifolia,* **MIMIN =** *Mimosa invisa,* **MICMI =** *Mikania micrantha,* **PYLNI =** *Phyllanthus niruri,* **SYDNO =** *Synedrella nodiflora,* **SPLAC =** *Spilanthes acmella,* **STCJA =** *Stachytarpheta jamaicensis.*

Growth, yield and biomass production

The sweet corn seeds germinated and grew at 99 %, so seedlings were thinned become one plant per planting hole. With low rain in October 2015, 35 mm in 5 days of rain categorized as a dry month, seedlings were watered every day. But in November, December 2015, and January 2016, the rainfalls were 355, 592, and 391 mm with rainy days of 23, 26, and 19 days, respectively. These rainfalls were optimum for sweet corn growth (Fig. 3). Overall, the crops grew well and

there was no evidence of diseases and insects in the experimental plots.

Data on the growth variables, yield, and plant biomass are presented in Table (2). The period of weed infestations significantly affected plant height, leaf area, yield, and biomass production, but no effect was found on the yield components of cob's diameter and length. The longer the weedy period, the lower the height and the less the leaf area of sweet corn was. If the plots were weedy during agrowing season, the plant height and leaf area were depressed to 161.9 cm and 567.5 cm^2 compared to weed-free in a season, the plant height and leaf area were 230.9 cm and 786.6 cm^2, respectively. On the other hand, the opposite was found on the weed-free trial, where the longer the weed-free period the higher the plant height and the more the leaf area. If the plots were free from weeds during a season the plant height increased from 161.6 to 230.9 cm and the leaf area increased from 567.6 to 786.6 cm^2. The responses of plant growth to weed infestations can be explained by the competition periods between crop and weeds to the life necessities such as nutrition, growing space, water, and CO$_2$ (Zimdahl, 2004). If weeds were suppressed by increasing the weed-free periods then the crop growth and yield increases (Williams II et al., 2008). The decrease and increase of the plant height and the leaf area on weedy and weed-free trials affected the yield and plant biomass production. The yield decreased when the weedy period increased and the yield increased when the weed-free periods increased (Table 2). If the plot was free from weeds during a season, the yield of unhusked cobs was 463.5 g plant^{-1}. But, when the plots were weedy during the season, the yield decreased to 209.9 g plant^{-1}. The decrease or increase of yield was not correlated to the yield components of the diameter and length of the cob, but it was suspected due to the size and the number of seeds. The more opportunities the crops free from weeds, the higher the growth and yield of crops that were harvested (Zimdahl, 2004; Williams II et al., 2008).

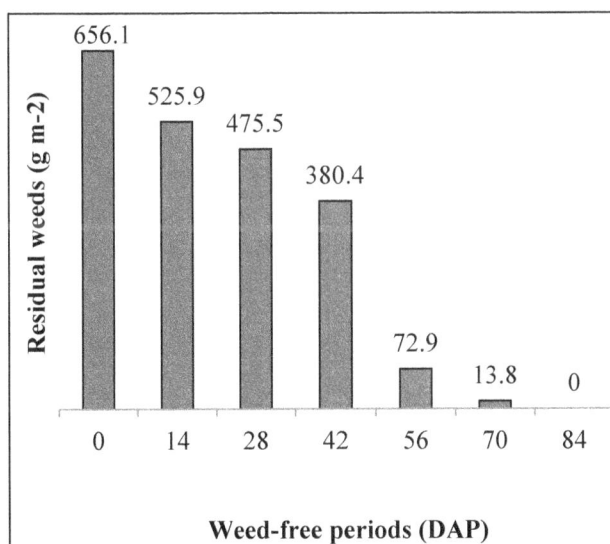

Fig. 2. Residual weeds of in sweet corn plots harvested at the end of experiment.

Biomass production of sweet corn also showed the same pattern with the growth parameters because biomass was the accumulation of the plant height and leaf areas of the plant. Biomass production decreased from 104.7 to 52.9 g plant^{-1} if the plots were weedy during the growing season and vice versa occurred if the plots were weed-free during the season. The biomass production was also an important variable of sweet corn because it can be utilized for industries such as bioethanol or for local needs as fresh ruminant food (Barros-Rioss et al., 2015).

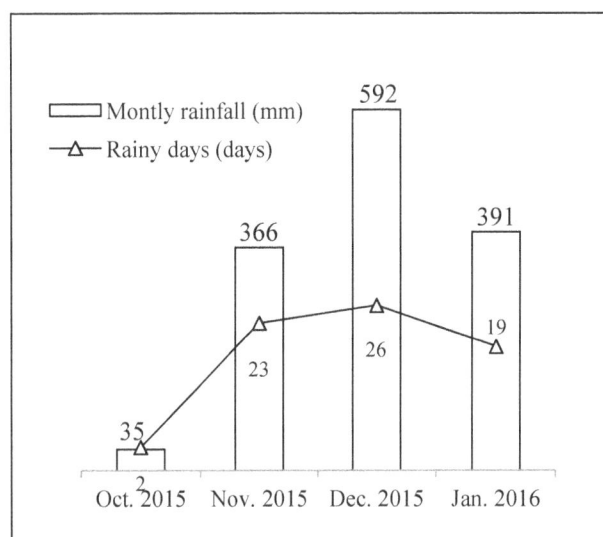

Fig. 3. Montly rainfall and number of rainy days in the research location from October 2015 to January 2016.

Determination of the CPWC

The yield and biomass relatives of sweet corn expressed in the percent of control are presented in Table 3. It appeared that the yield losses increased or decreased when the weedy or weed-free period increased, respectively. The highest yield loss of sweet corn grown under tropical organic farming reached 54.7 % if the weeds were not controlled during the season (84 days of weedy). Similarly, the highest biomass loss reached 49.5 % due to uncontrolled weeds during the season (84 days of weedy).

The logistic equation curves of relative yield of weedy and weed-free periods were used to determine the CPWC (Knezevic et al., 2002; Gantoli et al., 2013). The AYL due to adverse effects of weeds varies from 5 – 10 % (Knezevic and Datta, 2015). Determination of the critical period was judged by analysis of non-linear curves of the relative yields of weedy and weed-free treatments. In some publications, the critical periods of crops to weed control were fitted to Gompertz and logistic equation curves (Juraimi et al., 2009). In this study, a logistic equation was used to determine the CPWC with determination factors (R^2) reaching 94.9 and 85.8 % on the weedy and weed-free curves, respectively (Fig.4).

Based on the intercept of the curves on the x-axis with the AYL of 5 %, the maximum weedy period was 2 DAT, and the minimum of the weed-free period was 77 DAP. Thus, the CPWC of sweet corn under the tropical organic farming system with the AYL of 5 % was from 2 to 77 DAT (Table 4). If the AYL become 10 %, then the maximum weedy period was 3 DAT, and the minimum of the weed-free period was 53 DAP. Thus, the CPWC of sweet corn under the tropical organic farming system with the AYL of 10 % was from 3 to 53 DAT (Table 4).

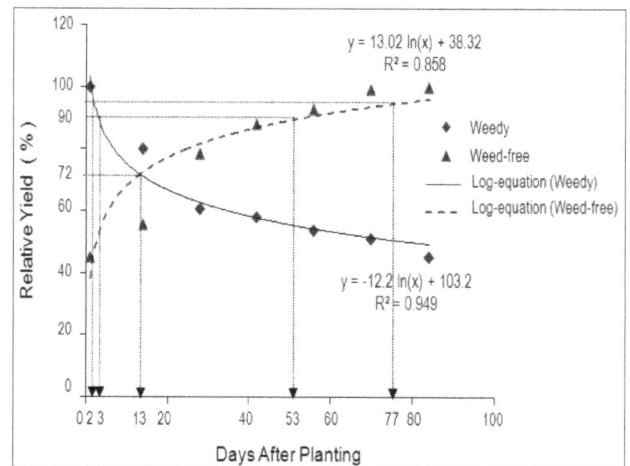

Fig. 4. The critical periods for weed control (CPWC) of sweet corn under tropical organic farming system.

Table 2. Effect of weed infestation on growth, yield, yield components, and biomass of sweet corn.

Periods of weed infestation (DAP)		Plant height (cm)	Leaf area (cm²)	Unhusked - cob (g plant⁻¹)	Cob- diameter (cm)	Cob- lenght (cm)	Oven-dried biomass (g plant⁻²)
Weedy	0	230.9 a	786.6 a	463.5 a	5.77	20.8	104.7 a
	0-14	233.2 a	775.8 a	371.8 b	5.88	20.8	93.0 ab
	0-28	189.1 ab	668.2 b	281.9 c	5.76	19.7	78.8 bcde
	0-42	190.4 ab	633.7 b	269.2 c	5.80	21.9	67.5 def
	0-56	196.9 ab	626.4 b	250.2 cd	5.54	19.2	60.9 ef
	0-70	189.9 ab	591.0 bc	237.2 cd	5.02	19.5	54.9 f
	0-84	161.9 b	567.5 c	209.9 d	5.00	21.5	52.9 f
Weed-free	0	161.9 b	567.5 c	209.7 d	5.00	21.5	52.9 f
	0-14	206.3 a	586.6 bc	258.4 bc	5.17	20.7	70.9 cdef
	0-28	215.2 a	742.8 a	364.6 b	5.13	21.4	83.9 bcd
	0-42	221.2 a	749.6 a	405.0 b	5.23	21.3	90.8 abc
	0-56	226.8 a	760.4 a	431.1 a	5.25	21.8	91.3 abc
	0-70	222.6 a	772.1 a	460.4 a	5.84	21.3	96.2 ab
	0-84	230.9 a	786.6 a	463.5 a	5.77	20.8	104.7 a
ANOVA (P<0.05)		*	*	*	NS	NS	*

* = Significant effect; NS = Non-significant effect; Numbers followed by the same letter in one column are not significantly different by DMRT (P<0.05).

Table 3. Effect of the weed infestations on yield and biomass losses of sweet corn.

Period of weed infestation (DAP)		Relative Yield (%)	Yield losses (%)	Relative biomass (%)	Biomass losses (%)
Weedy	0-14	80.2	19.8	88.8	11.2
	0-28	60.8	39.2	75.3	24.7
	0-42	58.1	41.9	64.5	35.5
	0-56	53.9	46.1	58.2	41.8
	0-70	51.2	48.8	52.4	47.6
	0-84	45.3	54.7	50.5	49.5
Weed-free	0	45.3	44.7	50.5	49.5
	0-14	55.8	44.2	67.7	32.3
	0-28	78.7	21.3	80.1	19.9
	0-42	88.2	11.8	86.7	13.3
	0-56	93.0	7.0	87.2	12.8
	0-70	99.3	0.7	91.9	8.1
	0-84	100	0.0	100	0

DAP = days after planting

Table 4. Determination of the CPWC of sweet corn based the AYL 5 and 10 % of weedy and weed-free curves.

Relative Yield (%)	Yield Loss (%)	Maximum weedyperiods (DAP)	Minimum weed-free periods (DAP)
95	5	2	77
90	10	3	53

CPWC = critical periods for weed control; AYL = acceptable yield loss; DAP = days after planting.

Conclusion

The periods of weed infestations influenced the growth and yield of sweet corn cultivated under tropical organic farming systems which ascended or descended with the non-linear curves of the logistic equations due to the increase of weedy or weed-free periods, respectively. Based on the AYL of 5 %, the CPWC of sweet corn under tropical organic farming was from 2 to 77 DAP and with the AYL of 10 %, the CPWC was from 3 to 53 DAP.

Acknowledgment

Appreciation is expressed to staffs of Agriculture Research Center, the University of Bengkulu for cooperation to facilitate field experiments. Special thanks are also presented to the students who helped the works at the laboratory and for field experiments.

References

Akintoye HA and Olaniyan AB, 2012. Yield of sweet corn in response to fertilizer sources. J. Agric. Sci. 1(5):110-116.

Barberi P, 2002. Weed management in organic agriculture: Are we addressing the right issue? Weed Res. 42(3):177-193.

Barros-Rios J, Romani A, Garrote G. and Ordas B, 2015. Biomass, sugar, and bioethanol potential of sweet corn. GCB Bioenergy 7:153-160.

Carr PM, Gramig, GG and Liebig MA, 2013. Impacts of organic zero tillage system on crops, weeds, and soil quality. Sustainability. 5:3172-3201.

Chauhan BS and Johnson DE, 2011. Row spacing and weed control timing affect yield of aerobic rice. Field Crop Res. 121(2):226-232.

Evans SP, Knezevic SZ, Lindquist JL, Shapiro CA and Blankenship EE, 2003. Nitrogen application influences the critical period for weed control in corn. Weed Sci. 51(3):408-417.

Everman WJ, Clewis SB, Thomas WE, Burke IC and Wilcut JW, 2008. Critical period of weed interference in peanut. Weed Tech. 22(1):63-67.

Gantoli G, Ayala VR and Gerhards R, 2013. Determination of the critical period for weed control in corn. Weed Tech. 27(1):63-71.

Hue NV and Silva JA, 2000. Organic Soil Amendments for Sustainable Agriculture: Organic Sources of Nitrogen, Phosphorous and Potassium. In: Silva, J.A. and Uchida, R. (eds). Plant Nutrient Management in Hawaii's Soils: Approaches for Tropical and Subtropical Agriculture. College of Tropical Agriculture and Human Resources, University of Hawaii at Manoa, Hawaii. pp. 133-144.

Janiya JD and Moody K, 1989. Weed populations in transplanted and wet-seeded rice as affected by weed control method. Tropical Pest. Manag. 35(1):8-11.

Johnson HJ, Colquhoun JB, Bussan AJ and Rittmeyer RA, 2010. Feasibility of organic weed management in sweet corn and snap bean for processing. Weed Tech. 24(4):544-550.

Juraimi AS, Najib MYM, Begum M, Anuar AR, Azmi M and Puteh A, 2009. Critical period of weed competition in direct seeded rice under saturated and flooded conditions. Pertanika J. Trop. Agric. Sci. 32(2):305-316.

Knezevic SZ, Evans SP and Blankenship EE, 2002. Critical period for weed control: the concept and data analysis. Weed Sci. 50(6):773-786.

Knezevic SZ, Evans SP and Mainz M, 2003. Row spacing influences the critical timing for weed removal in soybean (Glycine max). Weed Tech. 17(4):666-673.

Knezevic SZ and Datta A, 2015. The critical period for weed control: Revisiting data analysis. Weed Sci. Special Issue: 188-202.

Mekonnen G, Woldesenbet M and Yegezu E, 2017. Determination of critical period of weed-crop competition in rice (Oryza sativa L.) in Bench Maji and Kaffa Zone, South Western Ethiopia. J. Plant Sci. 5(3):90-98.

Muktamar Z, Sudjatmiko S, Chozin M, Setyowati N and Fahrurrozi, 2017. Sweet corn performance and its major nutrient uptake following application of vermicompost supplemented with liquid organic fertilizer. IJASEIT 7(5):602-608.

Mulvaney RL, Khan SA and Elsworth TR, 2009. Synthetic nitrogen fertilizers deplete soil nitrogen: A global dilemma for sustainable cereal production. J. Environ. Qual. 38(6):2295-2314.

Ruark M, Brundy L, Andraski T and Peterson A, 2012. Fifty years of continuous corn: Effects on soil fertility. Proc. Wisconsin Crop Manag. Conf. 51:127-132.

Savci S, 2012. An agricultural pollutant: Chemical fertilizer. Int. J. Environ. Sci. Dev. 3(1):77-80.

Simarmata M, Sitanggang CD and Djamilah D, 2015. Shifting weed compositions and biomass production in sweet corn field treated with organic composts and chemical weed controls. Agrivita J. Agric. Sci. 37(3):226-236.

Taguiling LG, 2013. Quality improvement of organic compost using green biomass. Eur. Sci. J. 9(36):319-341.

Williams II MM, 2006. Planting date influences critical period of weed control in sweet corn. Weed Sci. 54(3):928–933.

Williams II MM, Rabaey TL and Boerboom CM, 2008. Residual weeds of processing sweet corn in the North Central Region. Weed Tech. 22(4):646–653.

Zimdahl RL, 2004. Weed Crop Competition: A Review, 2nd edition. Blackwell, Oxford, UK. pp. 220.

7

The impact of organophosphorus pesticide on *Solanum melongena, Capsicum annum* and Soil

Bindu Singh*, Virendra Kumar Singh, Khalid Monowar Alam
Department of Environmental Science, Integral University, Lucknow, UP, India

Abstract

Eggplant (*Solanum melongena*) and green chilli (*Capsicum annum*) are an important vegetables crop grown throughout the year in the India. However, these vegetables crops suffer heavily from the ravages of various insect pests and disease, which reduce not only the yield but also the quality of the fruit. Malathion is part of the widely used insecticides all around the world. The present study emphasizes the effect of various concentrations and exposure periods of an organophosphorus pesticide. Malathion on two very important vegetative crops eggplant (*Solanum melongena*) green chilli (*Capsicum annum*). This study was performed at the department of environmental studies, Integral University, Lucknow (U.P.). The effect of Malathion on growth of the two vegetable crops was observed under greenhouse condition. Sampling of eggplant and green chilli were grown in assorted sets with 2, 4 & 6 ml/L foliar treatment of Malathion. Two sets of control plants were grown without Malathion treatment. They were harvested after 7 & 14 days. Experimental observation revealed that low concentration of Malathion had synergistic effect while higher level had an adverse effect on growth of plant and soil properties. This study infers that whereas lower levels are beneficial. Higher level of pesticides should be avoided and special care be taken to prevent their entry in the food chain.

Keywords: Malathion, Pesticides, Environmental pollution, Vegetative crops

Corresponding author email:
bindujnp917@gmail.com

Introduction

Fruit and vegetables are important components of the human diet since they provide various nutrients essential for our body. Their quality and yield are influenced by attacks of pests during their production and storage, leading to a huge loss. With revolutionary advancements in agricultural practices during the last few decades, use of pesticides has largely increased worldwide to protect crops from pest and diseases and increase agricultural yield for rapidly growing global population (Abhilash et al., 2009; Fenner et al., 2013), but their excess and indiscriminate use had adverse effects on growth, yield and quality of crops. Due to extensive use of pesticides across the world since 1950, most of the agricultural land is contaminated (Singh et al., 2012). Harmful effect of pesticide is part of the major factors responsible for reduced yield, especially for most sensitive horticultural crops, the vegetables. Reduced yield due to pests is common all over the world including India, where crop losses are nearly 40% under normal condition (Yuya, 2014). However, lack of awareness regarding pesticide use has led to environmental pollution, including contamination of soil and underground water (Nasrabadi et al., 2011). Currently, among the various groups of pesticides used all over the world, organophosphates from a major group and extend to more than 36% of the total world market (Kanekar et al., 2003). Organophosphates have gradually replaced organochlorine for pest control and higher crop yields. Chemically, organophosphates are esters of phosphoric acid and used to control a variety of insects, spiders, mites, aphids and pests which attack crops like tobacco, sugarcane, cotton, fruit, vegetables and ornamentals (Gafar et al., 2013).

Malathion [S-(1,2- dicarbethoxyethyl) -O, O-dimethyl dithio-phosphate], also known as carbophos, maldison and mercaptothion is a non-systemic, wide-spectrum organophosphorus pesticide used in agricultural settings (Srinivas et al. 2016). It is commonly used for the control of sucking and chewing insects of fruit and vegetables, mosquitoes, flies, household insects, animal parasites (ectoparasites). Malathion is largely used for public health and agricultural purposes (Singh et al., 2013)

Nevertheless, Malathion is a highly toxic compound and is indicated by the United State Environmental Protection Agency (USEPA) as toxicity class (Group 2A). It is considered to be carcinogenic to humans and animals. Its high-level exposure will affect nerve fibers and is neurotoxic in animals and affect immunity of higher vertebrates (Rai et al., 2016). Malathion is not supposed to be toxic to plants or aquatic algae because its mode of action targets only nervous systems (Qing et al., 2009). However, in humans, Malathion toxicity can be exerted via skin contact, ingestion, and inhalation exposure (Vasiliki et al., 2007). In South Africa, researchers found negative results of Malathion on Alfalfa, Maize and watermelon. Pesticide's effect on soil was investigated by many researchers (Aktar et al., 2009). Some study reported a decrease in some nutrients in soil, mainly heavy minerals and increase in phosphorus level due to the application of Malathion (Gafar et al., 2014). Ahmed et al. (2011) reported the negative effect of the pesticide on wheat growth. Srinivas et al., 2016 considered the effect of different concentrations of Malathion on protein and chlorophyll content of Green gram and Fenugreek. Similarly, Saleh et al., 2006 reported the comparative effect of Malathion and Mancozeb on growth parameters of Zea mays and Viciafaba under the presence or absence of mycorrhizal, inoculation.

In the present study, two widely used vegetable crops in India such as Eggplant (*Solanum melongena*) and green chilli (*Capsicum annum*) were selected. These are the principal vegetable crops grown in India. However, both of them are susceptible to pests and diseases. Farmers use higher amounts of pesticides than recommended doses because of ignorance, lack of training, experience, awareness etc. (Sabur et al., 2001). They also have the mistaken notion that over application of pesticides will enhance the plant growth and yield.

Hence, they utilize excess pesticides. On the basis of higher pesticides use. These vegetative plants were chosen for this study. They were treated with foliar application of Malathion under the controlled conditions in a greenhouse. The effort was to assess the effect of Malathion on plant growth and soil.

Material and Methods

Experiment layout

The experiment was conducted at the experimental field of Integral University (Lucknow, Uttar Pradesh 26.9585°N, 80.9992°E). Garden soil, with no previous insecticide history, was collected from the university campus, for pot culture experiment. Larger particles were removed from the soil for homogeneity. 2kg garden soil was filled in earthen pots. Eight sets of pots were taken in triplicate for each sampling day. Among this, six sets of vegetable crops (three sets each of eggplants and green chilli) were treated with three different concentrations of Malathion 2, 4, and 6 ml L-1. Remaining two sets were kept as control (pots with control eggplants and green chilli plants) without any treatments. The two vegetable crops are Eggplant (*Solanum melongena*) and green chilli (*Capsicum annum*) were grown in the nursery tray before being transplanted to the experimental pots. Different concentration of Malathion (EC-50%) was applied by foliar spray at the stage when leaves overspread. All pots were watered daily and growth variables such as root length, shoot length and growth rate were recorded periodically.

Growth analysis of plants

Plant samples were collected periodically and washed with tap water to remove the soil particles. Shoot length, root length, no. Of leaves and leaf area of collected sample was recorded.

Physico-chemical analysis of soil

The pH was measured in soil–water suspension using cyberscan 500 pH meters; electrical conductivity (EC) was measured using cyberscan 500 EC Meter. Soil potassium was measured using flame photometer (systronics 128). Total nitrogen was determined using the Kjeldahl method and available phosphorus was determined by the method of Olsen (Anderson et al., 1994).

Results

Growth performance of *Solanum melongena* and *Capsicum annum*

Effect of Malathion on growth of eggplant and green chilli is shown in Fig.1A and Fig.2A respectively, and their relative growth levels compared with the respective control plants are depicted in Fig. 1B and Fig. 2B. There was significant difference in growth of plants with low and higher concentration of Malathion and exposure periods.

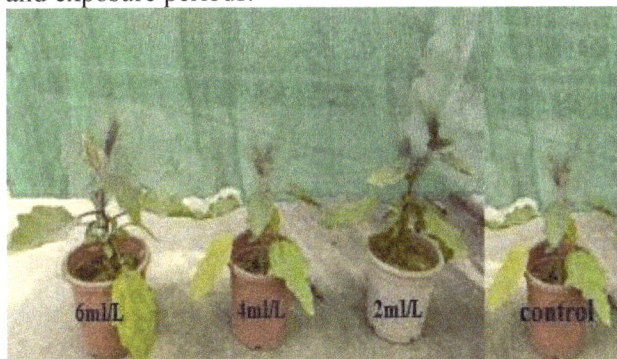

Figure 1-A: Growth response of eggplants grown at three different Malathion concentrations (ml L^{-1})

Figure - 2-A: Growth response of green chilli grown at three different Malathion concentrations (ml L^{-1})

Thus from Fig. 1A and 1B, we can see that, 7days after the application of pesticide, at 2 ml L^{-1} concentration, the growth parameters (root length, shoot length, no. of leaves and leaf area) of eggplant increased as compared to other concentrations (i.e. 4 and 6 ml L^{-1}) and control. At 4 ml L^{-1} concentration, the growth parameters increased as compared to control and then decreased at 6 ml L^{-1}. After 14 days, the root and shoot length were increased in 2 ml L^{-1} and decreased in control, 4 ml L^{-1} and 6 ml L^{-1}L. The overall changes in root and shoot length in response to various concentrations of Malathion treatments were in the order of 2 ml L^{-1}> control > 4 ml L^{-1}> 6 ml L^{-1}. Similarly, the leaf area was highest at 2 ml L^{-1} but lower at 6 ml L^{-1} concentration and but increased in 4 ml L^{-1} and control plant (2 ml L^{-1}> 4 ml L^{-1}> control > 6 ml L^{-1}). whereas at 6 ml L^{-1}, the number of leaves increases as compared to control and 4 ml L^{-1} concentration (2 ml L^{-1}> 6 ml L^{-1} > control > 4 ml L^{-1}). The visual symptom was seen in fig. 1A that the plant leaves of eggplant were yellowish in colour at 4 and 6 ml L^{-1} concentration and also in control plant. *Capsicum annum* has better growth response in comparison with that of *Solanum melongena*. There was no visual symptom seen in chilli plant except the reduction in plant height and total number of leaves. After 7th day the growth parameters increased at 2 ml L^{-1} concentration compared to other concentrations and control. On day 14th, the root length of chilli plant recorded better result in 2 ml L^{-1} concentration. Barr et al. (2006) studied the effect of Malathion on tomato. He found that there Malathion has little or no effect on plant height but strongly affected the root system. The root length and no. of leaves also increased at 6 ml L^{-1}. But the leaf area was decreased at 6ml L^{-1} concentration and increased at 2 ml L^{-1} and 4 ml L^{-1} concentration. This reveals that Malathion has significant effect on growth of both chilli and eggplant. Both the figures show that, during 7th day of exposure, all the growth parameters were positively affected by lower concentration (2 ml L^{-1}) of Malathion. However, with control plant and with higher levels (4 & 6 ml L^{-1}) of Malathion and increasing exposure days, growth parameters viz. root length, shoot length and foliage were declined. A study revealed a positive effect of Malathion on radish growth at lower concentration and negative effect at higher concentrations (Dennis et al., 1999). Gafar et al. (2014) also found positive effect of Malathion on carrot plant at lower levels and negative effect at higher doses. There is no significant difference in growth parameters with higher dose and control plants.

Physico-chemical properties of soil

Soil quality changes after the addition of Malathion in test plants at various exposure days is presented in Table - 1. The effect of Malathion addition to the soil of test plants was variable.

The total nitrogen content, extractable phosphorous and potassium were higher in soil before planting. Before the addition of Malathion, soil had pH 7.37 ± 0.3, electrical conductivity (EC) 1.34 ± 0.11, total nitrogen percentage 0.009 ± 00.3, available phosphorus 6.91±0.99 ppm and potassium 36.11 ± 0.99 ppm

Soil pH

In eggplant, the soil pH was highest in control (7.34 ± 0.4) and slightly reduced with increasing concentration of Malathion (2 ml L^{-1}, 4 ml L^{-1}, and 6 ml L^{-1}). After 14 days, the soil pH was maximum in control (7.31 ± 0.6), and minimum in 6 ml L^{-1}(7.14 ± 0.2) concentration. It was increase in 2 ml L^{-1}(7.26 ± 0.5) compared to 4 ml L^{-1}(7.22±0.8). In green chilli plant, soil pH was maximum in control experiment and was slightly reduced with increasing concentration of Malathion (2 ml L^{-1}, ml L^{-1}, and 6 ml L^{-1}) at both harvesting days (day 7 and day14).

Electrical conductivity

On day 7 and day 14, the electrical conductivity was minimum in control and maximum in 2 ml L^{-1} and slightly decreased with increasing concentrations for both plant species.

Total nitrogen

From table 1, after the application of pesticide at day 7, the total nitrogen was maximum in 2 ml L^{-1}(0.021 ± 0.05, 0.025 ± 0.08) and minimum in control (0.006 ± 0.01, 0.009 ± 00.3).

There was no significant difference in 4 ml L^{-1}(0.013 ± 0.004, 0.016 ± 0.03) and 6 ml L^{-1}(0.011 ± 0.01, 0.017 ± 0.12) for eggplant and green chilli respectively. At day 14, the same trend was found for both plants.

Available phosphorus

For eggplant, at day 7, the available phosphorus was higher in 2 ml L^{-1}(6.21±1.55) and reduced with 4 ml L^{-1} (6.12 ± 1.95), 6 ml L^{-1}(6.02 ± 0.93) and control (6.05 ± 0.86). On 14th day exposure, phosphorus was maximum in 2 ml L^{-1}(6.54 ± 1.38) whereas minimum in control plants (6.19 ± 0.69). But in the case of chilli plants, phosphorus was maximum in 2 ml L^{-1}(6.88 ± 0.93, 7.31 ± 0.99) and minimum in control (6.01±.731, 6.42 ± 0.64) for 7 and 14 day.

Potassium

Potassium was maximum in 2 ml L^{-1} (33.60 ± 1.08, 40.72 ± 1.48) and minimum in control (28.43 ± 2.24, 30.88 ± 2.03) with intermediate levels at 4ml/L (32.26 ± 2.01, 36.69 ± 2.00) and 6 ml L^{-1} (29.01 ± 1.82, 35.65 ± 3.39) concentration for day 7 and 14, respectively, in eggplant. However, in chilli, it was maximum in 2 ml L^{-1} (36.58 ± 3.24, 38.32 ± 3.54) whereas minimum in control (29.41±2.06, 29.52±1.47) but no significant difference in 4 ml L^{-1} (33.84 ± 1.09, 41.78 ± 5.24) and 6 ml L^{-1} (34.32 ± 1.76, 41.04 ± 0.62) was observed.

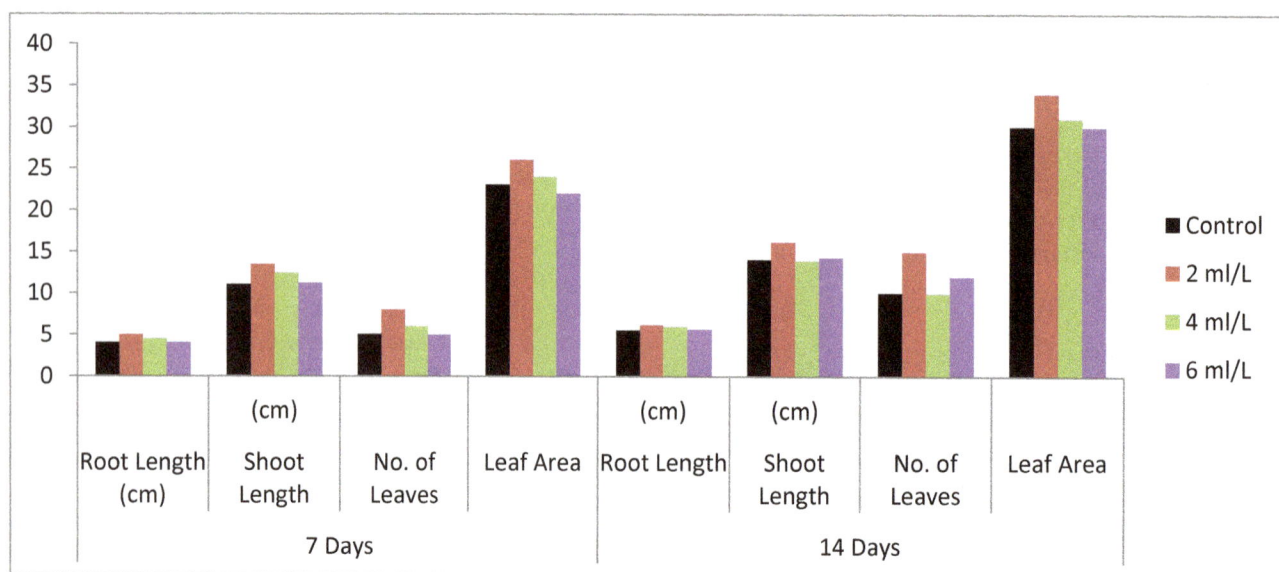

Figure - 1-B: Growth response of eggplant during Malathion exposure

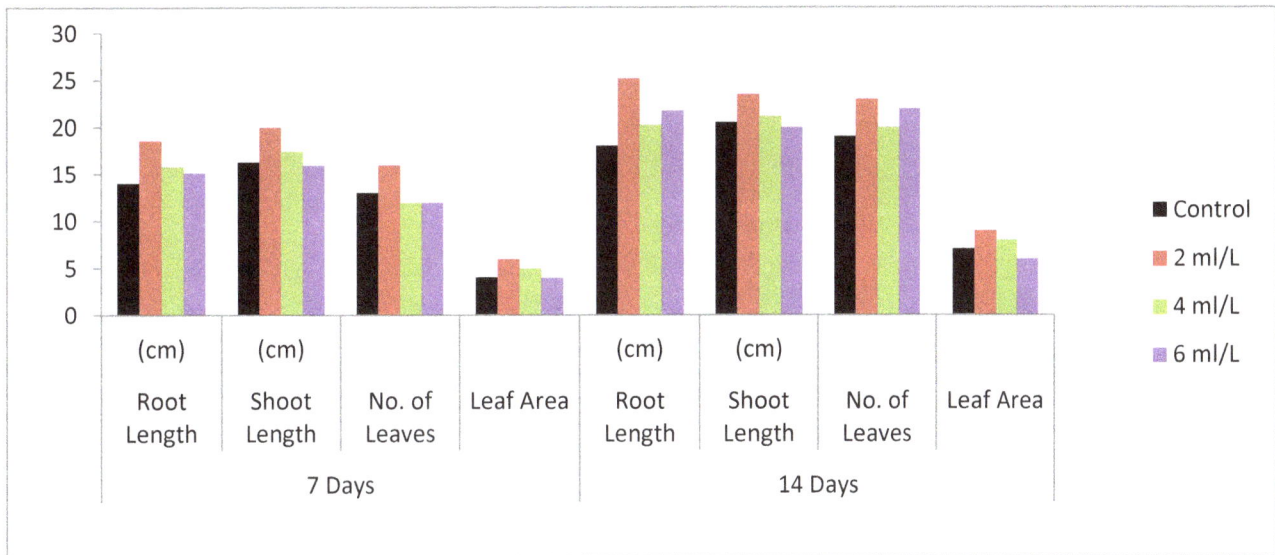

Figure - 2-B: Growth response of Chili Plant during Malathion exposure

Table – 1: Changes in soil quality after the addition of Malathion

Malathion concentration	Eggplant									
	Total Nitrogen (%)		Available Phosphorus (ppm)		Potassium (ppm)		pH		EC	
	7 days	14 days	7 days	14 days	7 days	14 days	7 days	14 days	7 days	14days
Control	0.006±0.01	0.012±0.003	6.05±0.86	6.19±0.69	28.43±2.24	30.88±2.03	7.34±0.4	7.31±0.6	0.92±0.1	0.95±0.05
2 ml L^{-1}	0.021±0.05	0.027±0.02	6.21±1.55	6.54±1.38	33.60±1.08	40.72±1.48	7.23±0.5	7.26±0.5	1.15±0.1	1.04±0.09
4 ml L^{-1}	0.013±0.004	0.022±0.005	6.12±1.95	6.29±0.99	32.26±2.01	36.69±2.00	7.23±0.2	7.22±0.8	1.09±0.8	1.00±0.3
6 ml L^{-1}	0.011±0.01	0.020±0.004	6.02±0.93	6.27±2.04	29.01±1.82	35.65±3.39	7.21±0.2	7.14±0.2	1.01±0.4	0.96±0.6
Green chilli Plant										
Control	0.009±00.3	0.017±0.11	6.01±1.73	6.42±0.64	29.41±2.06	29.52±1.47	7.32±0.09	7.33±0.5	0.96±0.07	0.99±0.02
2 ml L^{-1}	0.025±0.08	0.031±0.01	6.88±0.93	7.31±0.99	36.58±3.24	38.32±3.54	7.26±0.28	7.28±0.61	1.21±0.02	1.08±0.01
4 ml L^{-1}	0.016±0.03	0.025±0.003	6.43±0.88	6.98±1.39	33.84±1.09	41.78±5.24	7.24±0.03	7.27±0.01	1.16±0.04	1.13±0.08
6 ml L^{-1}	0.017±0.12	0.021±0.013	6.08±1.92	6.79±1.02	34.32±1.76	41.04±0.62	7.18±0.01	7.21±0.7	1.15±0.09	1.07±0.04

Discussion

The results of the present study have revealed that a higher dose of Malathion on test plants has affected the growth response of both eggplant and green chilli and also altered the soil properties. In general, low level of Malathion gave better results than higher doses and control plant. Gafar et al., 2014 also found a positive effect of Malathion on carrot plant at lower levels and negative effect at higher doses. There is no significant differences in growth parameters of higher dose and control plants.

pH and EC were not affected that much by the addition of different doses of Malathion but total nitrogen, available phosphorus and potassium were significantly affected with increasing concentration of Malathion (i.e. from 2 to 6 ml L-1) and exposure days. The physico-chemical properties of soil viz. total nitrogen, phosphorus and potassium gave a better result in

Capsicum annum compared to *Solanum melongena* that's why *Capsicum annum* has better growth response than *Solanum melongena*. Walia et al., 2018 studied the effect of Chlorpyrifos and Malathion on soil microbial population and enzyme activity and he found that the lower concentration of Malathion and chlorpyrifos is beneficial but higher concentration lead to reduction in microbial action in soil, which is also responsible for plant growth. Similar results were obtained by researchers who stated that Sevin pesticide reduce the activity of microorganisms which lead to the reduction of the absorption of some minerals, especially, in garden rocket plant (Elbashier et al., 2016). He also concluded that the stability of pesticides in soil depends on the nature of pesticides, especially their concentration, solubility and evolution in the air. It is reported that pesticides reduce absorption of some trace elements (Fe. Zu) and affect the viruses found at the root zone of the garden rocket plant which reduce their tolerance to some diseases (Mohanty et al., 2013).

Conclusion

The study explored the effect of Malathion on eggplant and green chilli plants in greenhouse condition. The experimental results revealed that the excessive amounts of Malathion on vegetative crop distorted the soil properties and also the growth response of eggplant and green chilli. It is clearly seen that the higher doses of Malathion are deleterious to plant growth and soil properties, but lower or recommended dose is helpful to enhance the growth of plants and soil properties compared to control. Therefore, the wise and judicious use of Malathion will not only help in killing pests, but also, improves the soil quality and vegetable production. However, more studies are required to elucidate the fate and behaviour of Malathion in different kind of vegetables and soil.

Acknowledgement

All authors contributed equally to this work. K.M.A and V.K.S design the work. B.S performed the experiment. The authors are thankful to the Integral University for providing facilities, critical suggestion regarding the improvement of the manuscript and special thanks to research community for providing MCNIU/R&D/2017/MCN-000192.

References

Abhilash PC, Jamil S and Singh N, 2009. Transgenic plants for enhanced biodegradation and phytoremediation of organic xenobiotics. Biotechnol. Adv. 27(4):474-488.

Ahmed N, Ahman I, Englund JE and Johansson E, 2011. Effect on radish pests by application of insecticides in a nearby spring oilseed rape field. J. Appl. Entomol. 135(2011):168-176.

Aktar MW, Sengupta D and Chowdhury A, 2009. Impact of pesticides use in agriculture: their benefits and hazards. Interdiscip. Toxicol. 2(1):1-12.

Anderson JM and Ingram JSI, 1994. Tropical soil biology and fertility: a handbook of methods Soil Science. 157(4):265.

Barr DBandAngerer J, 2006. Potential Uses of Biomonitoring Data: A Case Study Using the Organophosphorus Pesticides Chlorpyrifos and Malathion. Environ. Health Perspect. 114(11):1763-1769.

Chang M, Wang M, Kuo DTF and Shih Y, 2013. Sorption of selected aromatic compounds by vegetables. Ecol. Eng. 61(2):74-81.

Dennis GA and Lee PN, 1999. Phase 1 volunteer study to establish the degree of absorption and effect of cholinesterase activity of four head lice preparation containing Malathion. Clin. Drug Investig. 18(2):105-115.

Elbashier MMA, Shao X, Mohmmed A, Ali AAS and Osman BH, 2016. Effect of Pesticide Residues (Sevin) on Carrot (Daucus carota L.) and Free Nitrogen Fixers (Azotobacter spp). Agri. Sci. 7(2): 93-99.

Fenner K, Canonica S, Wackett LP and Elsner M, 2013. Evaluating pesticide degradation in the environment: blind spots and emerging opportunities. Sci. 341(6147):752-758.

Gafar MO and Dagash YM(b), 2011. The effect Sevin pesticdes on garlic growth. Res. J. Agric. Biol. Sci. 7(3):332-334.

Gafar MO, Elhag AZ, Warrag MOA and Yagi ME, 2014. The Reseidual Effect of Malathion (Organophosphate) and Sevin (Carbamate) application on Soil and Carrot (Daucus carota L.) Growth. J. Agric. Environ. Sci. 3(1):203-207.

Gafar MO, Yagi MI, Elhag AZ and Musa SA, 2013. The effect of Malathion and Sevin pesticides application on soil and garden rocket. Univer. J. App. Sci. 1(3):82-85.

Gafar, MO and Dagash YM(a), 2011. The effect of Malathion pesticides on garlic growth. Research J. Agric. Biol. Sci. 7(3):332-334.

Geed SR, Rai BN, Kureel MK, Shukla AK and Singh RS, 2016. Biodegradation of Malathion and evaluation of kinetic parameters using three bacterial species. Resour. Efficient Technol. 2(1):S3-S11.

Mohanty MK, Behera BK, Jena SK, Srikanth S, Mogane C, Samal S and Behera AA. 2013. Knowledge attitude and practice of pesticide use among agricultural workers in Puducherry, South India. J. Forensic Leg. Med. 20(8):1028-1031.

Nasrabadi M, Ghayal N and Dhumal KN, 2011. Effect of Chlorpyrifos and Malathion on Stress and Osmolyte Parameters in Tomato and Brinjal. Int. J. Pharm. Biosci. 2(2):778-787.

Qing G, Xia L, Li T and Yang B, 2009. Simultaneous determination of 26 pesticide residues in Chinese medicinal materials using solid-phase extraction and GC-ECD method. Chin. J. Nat. Med. 7(3):210-216.

Saleh M and Saleh Al-G, 2006. Influence of Malathion and Mancozeb on Mycorrhizal Colonization and Growth of Zea mays and Vicia faba. World J. Agric. Sci. 2(3):303-310.

Singh B, Kaur J and Singh K, 2013. Bioremediation of malathion in soil by mixed Bacillus culture. Adv. Biosci. Biotechnol. 4(5):674-678.

Singh B, Kaur J and Singh K, 2012. Transformation of malathion by Lysinibacillus sp. isolated from soil. Biotechnol. Lett. 34(5):863-867.

Singh BK and Walker A, 2006. Microbial degradation of organophosphorus compounds. FEMS Microbiol. Rev. 30(3):428-471.

Srinivas SK and Damani J, 2016. Effect of Malathion on the Concentration of Primary Metabolites from Vigna radiate and Trigonella foenum-graecum. Bull. Environ. Pharmacol. Life Sci. 5(7):34-38.

Vasiliki IV, Vasilios AS and Triantafyllos AA, 2007. Determination of the pesticides considered as endocrine-disrupting compounds (EDCs) by solid phase extraction followed by gas chromatography with electron capture and mass spectrometric detection. J. Sep. Sci. 30(12):1936-1946.

Walia A, Sumal K and Kumari S, 2018. Effect of Chlorpyrifos and Malathion on Soil Microbial Population and Enzyme Activity. Acta Scientific Microbiol. 1(4):14-22.

Xie S, Liu JX, Li L and Qiao CL, 2009. Biodegradation of malathion by Acinetobacter johnsonii MA19 and optimization of cometabolism substrates. J. Environ. Sci. 21(1):76-82.

Yuya AI, 2014. Studies on some farmers practices and combinations of malathion and neem seed powder management options on stored sorghum and maize insect pests at Bako, West Shoa, Ethiopia. Asian J. Agri. Biol. 2(1):67-79.

Kanekar PP, Bhadbhade BJ, Deshpande NM and Sarnaik SS, 2004. Biodegradation of Organophosphorus Pesticides. Proc. Indian Nat'l. Sci. Acad. 70(1):57-70.

Sabur SA and Molla AR, 2001. Pesticide Use, Its Impact On Crop Production And Evaluation Of IPM Technologies In Bangladesh. Bangladesh J. Agric. Econom. 24(1&2):21-38.

8

Effectiveness of compost and gypsum for amelioration of saline sodic soil in rice wheat cropping system

Muhammad Anwar Zaka[1], Khalil Ahmed[1]*, Hafeezullah Rafa[1], Muhammad Sarfraz[1], Helge Schmeisk[2]
[1]Soil Salinity Research Institute, Pindi Bhattian, Punjab, Pakistan
[2]Faculty of Organic Agriculture Sciences, University of Kassel, Germany

Abstract

A lot of crop residues, kitchen wastes and tree leaves are wasted annually. These materials can be composted and used for improvement of soil health. The possibility of using compost in reclamation of salt affected soil was studied with the treatments *i.e.* control (no amendment), gypsum @ 100 % GR, compost 20 t ha^{-1}, gypsum 50 @ % GR+ compost @ 20 t ha^{-1}, gypsum 50 @ % GR+ compost @ 10 t ha^{-1}, gypsum 25 @ % GR+ compost @ 20 t ha^{-1} and gypsum @ 25 % GR+ compost @ 10 t ha^{-1} in rice-wheat rotation at farmer field in Haveli Karimdad, Pindi Bhattian district Hafizabad, Punjab, Pakistan. The selected field was prepared and leveled. The design of the experiment was randomized complete block (RCBD) with four replications having the plot size of 8m x 6m. The prepared compost was applied and incorporated according to the treatments plan thirty days before transplanting of rice. Uniform cultural practices were applied to all the treatments. Rice-wheat crop rotation was used. The data for paddy yield of rice and wheat grain were recorded at maturity. The results showed that gypsum application @ 50% GR + compost @ 20 t ha^{-1} remained statistically at par with gypsum application @ 100% GR for producing biomass, paddy and wheat grain yield. However other treatments remained inferior but significantly better than control. The pH$_s$ and SAR were decreased significantly after harvesting of 2nd rice crop in two treatments *i.e.* gypsum application @ 100% GR and gypsum @ 50% GR+ compost @ 20 t ha^{-1}. The EC$_e$ was reduced to less than 4 dS m^{-1} in all the treatments except control after 1st rice crop. The physical properties of soil such as bulk density, porosity and hydraulic conductivity were also improved with passage of time. Results of current study suggested that salt affected soil can rehabilitate to their original potential if gypsum is applied at the full rate (100% GR) alone or decrease its quantity 25 or 50% by combining it with compost at rate of 20 or 10 t ha^{-1}.

Keywords: Saline sodic soil, Gypsum, Compost, Reclamation, Rice-wheat rotation

Corresponding author email:
khalilahmeduaf@gmail.com

Introduction

The amelioration of sodic or saline sodic soils is a very important for obtaining reasonable yield from salt affected soils. Such soils with low fertility and poor physical properties, adversely affect the growth and yield of most crops (Grattan and Grieve 1999). The worldwide occurrence of such soils urge the need for cost-effective, efficient and environmentally acceptable management practices (Abbas et al., 2016). More than half of the rice cultivated area (~ 1.0 mha) in the Punjab, Pakistan is subjected to salinity, causing 30-70% paddy yield reduction (FAO, 2011; NFDC, 2012)

Application of organic matter is considered as one of effective strategy for reclamation of salt affected soils and increasing crop growth (Ahmad et al., 2014).

Tejada et al. (2006) in a long-term study of five year reported steady removal of salt and Na^+ with noticeable increased in plant growth and soil porosity with the addition of compost. In a study conducted by Boateng et al. (2006) poultry manure and compost stimulated the removal of Na^+ from root zone, reduced EC, improved soil aggregate stability and water-holding capacity. Integrated use of gypsum and organic material has been successfully used to improve the physical and chemical properties of salt affected soils and the effect was more pronounced as compared to sole application of gypsum (Vance et al., 1998; Wright et al., 2007).

Use of gypsum with organic material like water hyacinth compost and rice straw compost showed that integrated use of these treatments was more efficient in decreasing the EC_e, pH, SAR, and ESP of clay saline-sodic soils as compared to their individual use. Rice straw compost was more effective in diminishing EC, pH, SAR and ESP than water hyacinth compost (Mikanova et al., 2012; Shaaban et al., 2013; Abdel-Fattah 2012). The most efficient and economical methods for reclamation of salt affected soil is addition of Ca^{2+} source which change the ionic composition of soil solution and replace the Na from exchange site which is leached down out of soil profile (Ghafoor et al., 2008).

Physical characteristic of salt affected soil *e.g.* water permeability, porosity, void ratio, bulk density, were significantly improved with chemical amendments and FYM @10 t ha^{-1} subsequently rice and wheat yields was also increased (Hussain et al., 2001). Similarly other organic materials *e.g.* wheat straw, rice husk, rice straw and chopped grass also has positive effect on chemical and physical properties of saline sodic soil (Ould Ahmed et al., 2010; Akhtar et al., 2014; Lakhdar, et al., 2009). Soil organic matter increases cation exchange capacity (CEC), promotes granulation and is responsible for up to 90% adsorbing power of soil (Diacono and Montemurro, 2015). During decomposition of organic matter, mineral nutrients such as Ca, Mg and K are released and become available for plants (Awaad et al., 2009; Mahmood et al., 2015). Compost is not only good alternative of organic matter in soil but also play very vital role for amelioration of saline sodic oils. Organic acid released during decomposition can improve the physical properties of such soils, which have been deteriorated to such an extent that passage of water and air become extremely difficult. The water stands on the surface of soils for weeks long. The plants when grown under such conditions ultimately die due to suffocation/deficiency of air for root respiration. So, compost can be a good organic amendment for reclamation of salt affected soils. The present study was conducted with the following objectives:

1-To determine the feasibility of compost as a reclaiming agent for saline sodic soil.

2- Monitoring the gradual improvement in soil health.

Material and Methods

The research work was conducted in the farmer's field at Havaily Karim dad, Pindi Bhattian (Punjab), for two consecutive years in rice-wheat cropping system to devise the effective technology for reclamation of salt affected soils with compost, gypsum and their different combinations. A saline sodic field was selected and composite soil samples were collected from upper (0-15 cm) and lower (15-30 cm) soil depth before starting the experiment. Soil samples were air dried, passed through 2 mm sieve and analyzed for physio- chemical parameters (Table-1). The selected field was prepared and leveled. The design of the experiment was randomized complete block (RCBD) with four replications having the plot size of 8m x 6m. Compost was prepared using residues of crops (rice and wheat) and wastes (tree leaves, grasses and kitchen waste).They were piled in the pit (How to build a compost www.exsands. com/ Gardening/ how to buildac).The treatments tested were control (no amendment), gypsum @ 100 % GR, compost 20 t ha⁻¹, gypsum 50 @ % GR+ compost @ 20 t ha⁻¹, gypsum 50 @ % GR+ compost @ 10 t ha⁻¹, gypsum 25 @ % GR+ compost @ 20 t ha⁻¹ and gypsum @ 25 % GR+ compost @ 10 t ha⁻¹.

The prepared compost was applied and incorporated according to the treatments 30 days before transplanting of rice. Field was irrigated with canal water (EC_{iw} = 0.23 dS m⁻¹, RSC= nil and SAR = 0.14 (mmol L⁻¹)^{1/2} as and when needed. Gypsum was applied in their respective treatments and leaching was provided for 15 days before transplanting of rice. The soil samples were collected at 0-15 and 15-30 cm depths before sowing of each crop for analysis of pH, EC_e, SAR, bulk density, hydraulic conductivity and % pore space according to U.S. Salinity Laboratory Staff (1954) and bulk density (d_b) by (Klute,1986).

Table-1 Soil analysis of the experimental site

Determinations	Units	0-15 cm	15-30 cm
pH_s	-	9.3	9.3
EC_e	dSm^{-1}	8.52	7.80
SAR	$(m\ mol\ L\text{-}1)^{½}$	96.2	104.8
Sand	%	62	-
silt	%	20	-
Clay	%	18	-
Textural Class	-	Sandy Clay Loam	
Hydraulic Conductivity	$cm\ hr^{-1}$	0.037	
Bulk Density	$M\ gm^{-3}$	1.72	-
Gypsum Requirement	$Tons\ acre^{-1}$	3.7	-

Table-2 Chemical composition of compost

Determinations	Units	
pH	-	7.93
Ece	dSm^{-1}	6.83
Organic-C	%	30.56
Total-N	%	2.63
C/N ratio	-	11.62
P	Ppm	12.6
K	Ppm	9.7

Rice and wheat crops were grown. Recommended dose of fertilizer (110-90-70 NPK Kg ha^{-1}) was used to grow rice crop. The concentration of nitrogen, phosphorus and potash of compost was also considered in calculations of fertilizers. Half of the recommended nitrogen (N) and full dose of P (P$_2$O$_5$) and K (K$_2$0) were applied at transplanting while the remaining half dose of N was applied 30 days after transplanting (DAT). Macheiti weedicide was applied seven days after transplanting of rice seedling to control the weed growth. Padan insecticide was applied 45 DAT to control the attack of stem borer. Sundaphos insecticide was sprayed to cover the risk of rice leaf roller and ensure the good yield of rice. Zinc sulphate was applied to avoid the deficiency of Zn. At maturity crop was harvested and paddy yield data were recorded.

After harvesting of rice, wheat seed was treated with Benlate to avoid the effect of fungal disease and wheat was sown with single row drill in the same layout plan in all the treatments without addition of any amendment. Fertilizer was added to wheat according to recommended dose (120-90-70 NPK Kg ha^{-1}). Single super phosphate, potassium sulphate and urea were applied as a source of P, K, and N. Full dose of P, K and half of the recommended dose of N was applied at sowing while remaining N was applied at first irrigation. Pumasuper and Isoprotone weedicides were applied 30 days after sowing of wheat. At maturity, paddy and grain yield of wheat data were recorded. Subsequent 2nd rice and wheat crop were also sown in same layout with above mentioned all agronomical and/cultural practices. After harvest of each crop soil samples were collected from 0-15 cm and 15-30 cm soil depths and were analyzed for physico-chemical properties. All the data were processed for statistical analysis using Duncan's Multiple Range Test (DMR Test, 1955).

Results and Discussion

Effect of compost and gypsum on soil properties
1-   Soil Electrical Conductivity (EC $_e$)

Soil EC_e is very important parameter that indicates an overall estimate of soluble salts. It is of prime importance in water relation of plants as well as nutrient uptake. If the quantity of soluble salts in soil increases over the critical limit that is specific for a plant (threshold), water and nutrients uptake suffers very badly; rather plants may die under environment of very high osmotic pressure (Munns et al., 2006). Original EC_e of the soil of experimental site was 8.52 dS m^{-1} that reduced to lesser than 3.0 dS m^{-1} (critical limit 4.0 dS m^{-1}) due to gypsum application @ 100 % GR or compost @ 20 t ha^{-1} alone or their combinations at reduced rates. The EC_e in control plots was 5.9 dS m^{-1} even after two years (Fig-1). The recorded reduction was 62% (T$_4$ gypsum 50 % GR+ compost 20 t ha^{-1}) to 51% (T$_5$ gypsum 50% GR + compost 10 t ha^{-1}) over control (Table-3). The application of amendments (gypsum or compost) and subsequent cropping may have improved physical properties that eased leaching of salts in to lower profile. Increase in EC_e of 15-30 cm depth during first year supported this view point also. Even the salt leached down from this layer during the last year. Sarwar (2005) also obtained a reduction in soil EC_e of saline sodic soil with the application of compost. Efficiency of gypsum in decreasing soil EC_e in the long run was recorded by Chaudhry et al. (1990) as well and reported that removal of salts was more in upper layer than that from 30-60 cm soil layer.

Soil pH

Soil pH is the sole property that indicates overall impact of many factors like parent material constitution, dominance of particular ion, natural climate, drainage, quality of irrigation water and main activities related to soil management. Initial pH of experimental soil was 9.3 that reduced significantly with application of gypsum or compost alone or with their different combinations (Figure 2). Maximum reduction of 13.33 % was due to application of gypsum @ 50 % GR + compost @ 20 t ha^{-1} whereas minimum reduction was recorded when gypsum was reduced to 25 % GR along with compost 10 t ha^{-1} (Table 3). The high pH of saline sodic soil is the result of sodium dominance (ESP >15 %) on the clay micelle. Application of Ca as gypsum (Ca SO$_4$.2H$_2$O) replaced the Na from clay complex that was pushed into soil solution and subsequently leached down into lower profile. Similarly organic acids released during decomposition of compost performed the same role. The other possibility in calcareous soil is dissolution of CaCO$_3$ with the reaction of organic acids released by plant roots (Abbas et al., 2016). The released Ca ultimately behaved like gypsum. Increase in dealkanized of soil depth from 12.5 to 32.5 cm was obtained during two years after application of chicken manure, water hyacinth, FYM and dry sludge (Rehman et al. 1996; Abd El-Rheem et al., 2016). Addition of organic matter in saline sodic soils would help to chelate Ca^{2+} and decrease soil pH leading to increase in solubility of CaCO$_3$ (Ghafoor et al., 2008; Zia-ur-Rehman et al., 2016). Wong et al (2009) determined that addition of organic materials increased soil microbial biomass while added gypsum decrease pH. Our results are supported by previous findings that application of gypsum with organic and inorganic amendments decreased the soil salinity and sodicity indicators (Nan et al., 2016; Qadir et al., 2017).

Soil sodium adsorption ratio (SAR)
Soil SAR is very important criteria to classify the soil as sodic or non-sodic. The degree of sodicity is also deciding yardstick about growth and success of crops in the sodic environment. The magnitude of this parameter at the initiation of the experiment was very high (96.0 m mol L^{-1})$^{1/2}$ and indicated that no crop could be successful unless it was reduced substantially. The devised strategies were found to be effective in this regard. Soil SAR was reduced to half of its original level or even lesser following the application of amendments (gypsum or compost) and subsequent leaching with water. Thus, the soil environment was converted into favorable one with

respect to rice and wheat crops. The growing of crops helped to continue positive effects of amendments due to root activity and crop residues decomposition. Ultimately, the soil SAR varied from 11.5 to18.0 (m mol L^{-1})$^{1/2}$ in different treatments as against the value of 50.9 (m mol L^{-1})$^{1/2}$ in control plots after two years (Figure 3). Thus, the treatments of the experiment were successful to bring this parameter in safe limits or nearer to it. The highest decrease (77.4%) in SAR was recorded with gypsum @ 50% GR + compost @ 20 t ha^{-1} (Table 3).

Soil SAR depends upon relative quantities of Na and Ca + Mg in the soil solution and clay complex. The increase in Ca^{2+} occurred due to direct application of gypsum or release during CaCO$_3$ dissolution through reaction with organic acids formed by decomposition of compost. This Ca^{2+} replaced Na$^+$ on exchange sites that was leached down during continuous irrigation. So, there was net increase in Ca and Mg content and very high decrease in the amount of Na from the soil solution that all resulted in significant reduction of SAR. Similar findings were also recorded by earlier researchers (Wright et al., 2008; El-Sanat et al., 2017). Furthermore, more positive effect on soil properties was noted when leaching began with gypsum and compost (Ashtar and ELEtreiby, 2006).

Bulk density (BD) and percent pore space
Bulk density of soil depends upon soil texture, structure, clay type, clay content, drainage and organic matter content of the soil. The effect of soil EC$_e$, pH and SAR is directly translated into increase or decrease of this soil characteristic. The higher value of BD indicates harder and less porous soil. Generally, BD of saline sodic soil with dominancy of Na$^+$ are higher than equivalent normal soil. Bulk density of the experimental soil was very high with the value of 1.72 Mg m^{-3} indicating its dispersed condition due to very high SAR and pH. Improvement of this soil would only be possible if its BD values decrease along with soil pH, EC$_e$ and SAR. It has been observed that a significant decrease in this soil property occurred in two years in all the treatments of the experiment (Figure 4). Resultantly, the porosity of the soil also increased (Table 4). A decrease of 7.69 to 12.8% in BD was recorded in different treatments of the experiment (Table 3). The most effective treatment was gypsum @ 50% GR + compost @ 20 t ha^{-1}. The decrease in soil pH and SAR decreased soil dispersion, increase soil porosity and resultant net reduction in BD was recorded. It was explained that amendments

improved soil physical condition by eliminating Na$^+$ dominance that caused undesirable changes in sodic soils through decrease in EC$_e$ and SAR ratio. An improvement in soil porosity through gypsum application was also obtained (Shainberg et al., 1989). The combined application of amendments in lesser quantity might be a good strategy in this regard (Hussain et al., 2001). Similarly decreased BD values with the use of calcium sources (Peters and Kelling, 2002) and compost (Wang et al., 2016) were previously reported.

Hydraulic conductivity (HC)
Hydraulic conductivity plays its role in water relation of soil and plants in general but is of prime importance in soil reclamation process. Because if water cannot permeable in to lower profile it will not take salt with it. The initial soil data presented very low value of HC. It was as low as 0.073 cm hr^{-1}. However, it increased manifold during two years with a maximum of 0.878 cm hr^{-1} (485% increase) when gypsum 50% GR + compost 20 t ha^{-1} were applied in combination (Figure 5). It might be due to reduction in SAR which resultantly decrease soil dispersion and encouraged coagulation of soil particles. The increase in pore spaces caused by aggregation increased the HC (Kauraw and Verma, 1982). Hence, a clear and highly significant increase in yield was recorded after application of amendments and subsequent cropping. Various research workers have reported an appreciable increase in HC due to application of inorganic or organic amendments, cultural practices and growing of crops (Qadir and Schubert, 2002; Carter et al., 2004; Evanylo et al., 2008).

Effect of gypsum and compost on paddy and wheat grain yield
Crops yields are ultimate result of soil factors, irrigation water quality, climatic conditions and crop varieties. At the first step, appropriate soil physical and chemical conditions are required for seed emergence and seedling establishment; subsequently the growth of the plant depends upon the conditions of the rhizosphere that are being faced by the plant. A good and healthy growth is translated into higher yields, when reclamation process is started in barren salt affected soils. The germination, establishment of seedlings, tillering, growth and yield all will depend upon magnitude of the reclamation occurred before sowing/transplantation of seedlings/seed or during the growth of crops. The quantum of reclamation depends on methods adopted for reclamation, type of amendment including its quality and quantity of irrigation water, cultural practices, soil fertility and salt tolerance of the plant. The most important factor, however, is the degree of rehabilitation prior to sowing of crops.

The data indicated that better yield with different compost combinations proved effective and efficient for decreasing soil pH, EC$_e$, SAR and BD while increasing porosity and HC. In the first step, plant height and tillering improved due to reduction of harmful elements through soil amelioration. The uptake of elements such as Na$^+$ decreased while the concentration of K increased to adjust Na and K ratio favorable for the plant growth. Eventually, the yield of rice and wheat crops increased significantly in the treatments that proved more effective for soil reclamation. The most effective treatment was the gypsum @ 50% GR+ compost @ 20 t ha^{-1} (Figure 6). Thus, combination of chemical and organic amendments was successful to increase paddy yield by 219% and wheat grain by 208% over control (Table 3). The other combinations of gypsum and compost proved inferior. Application of rice straw and FYM with 25% GR gave similar results as 100% GR in term of rice and wheat yield (Zaka et al., 2003). Mahdy, (2011) proved that sodium removal efficiency was the highest with soil+ NPK + compost + anthracite + coal powder + WTRs + FeSO$_4$.7H$_2$O and ultimately, biomass and yield of alfalfa was also more at the same treatment. Beneficial effect of compost on crop growth and yield have been reported by many researchers (Sarwar et al., 2017; Islam et al., 2017). However, combination of chemical amendments (gypsum) with compost is more beneficial to cut short the reclamation period and for achieving rapid rehabilitation (Ameen et al., 2017).

Table-3: Percent increase / decrease over control (soil parameters and yield of paddy and wheat) after harvesting of four crops

Treatments	Increase			Decrease			
	Paddy	Wheat grain	HC	BD	pH	EC_e	SAR
T$_2$- Gypsum 100% GR	200	162.4	423	7.69	7.78	60.0	76.4
T$_3$- Compost 20 t ha^{-1}	175.6	154.6	376	9.62	6.67	58.0	70.5
T$_4$- Gypsum 50% GR + Compost 20 t ha^{-1}	219.1	207.8	485	12.82	13.33	62.0	77.4
T$_5$- Gypsum 50% GR + Compost 10 t ha^{-1}	154.2	124.8	333	10.25	7.78	51.0	68.6
T$_6$- Gypsum 25% GR + Compost 20 t ha^{-1}	157.3	115.6	414	8.97	6.67	56.0	66.6
T$_7$- Gypsum 25% GR + Compost 10 t ha^{-1}	126.7	86.5	275	8.3	6.67	52.0	64.6

Table-4 Effect of compost and gypsum alone and their combinations on pore spaces (%)

Treatments	After amendments	After 1st Rice	After 1st Wheat	After 2nd Rice	After 2nd Wheat
T$_1$ Control	37.36D	39.62C	41.13D	40.76C	41.132D
T$_2$ Gypsum 100% GR	41.13BC	42.64 B	44.15C	44.15B	45.67C
T$_3$ Compost 20 t ha^{-1}	40.38C	42.64 B	45.28BC	44.52B	46.79BC
T$_4$ Gypsum 50% GR + Compost 20 t ha^{-1}	43.02A	43.78A	46.79A	47.17A	48.68A
T$_5$ Gypsum 50% GR + Compost 10 t ha^{-1}	41.51BC	43.40AB	44.15C	44.91B	47.17B
T$_6$ Gypsum 25% GR + Compost 20 t ha^{-1}	41.89AB	42.64B	46.42AB	44.91B	46.42BC
T$_7$ Gypsum 25% GR + Compost 10 t ha^{-1}	40.76BC	43.40AB	44.72C	44.53B	46.04BC

Original Pore spaces 35.09%

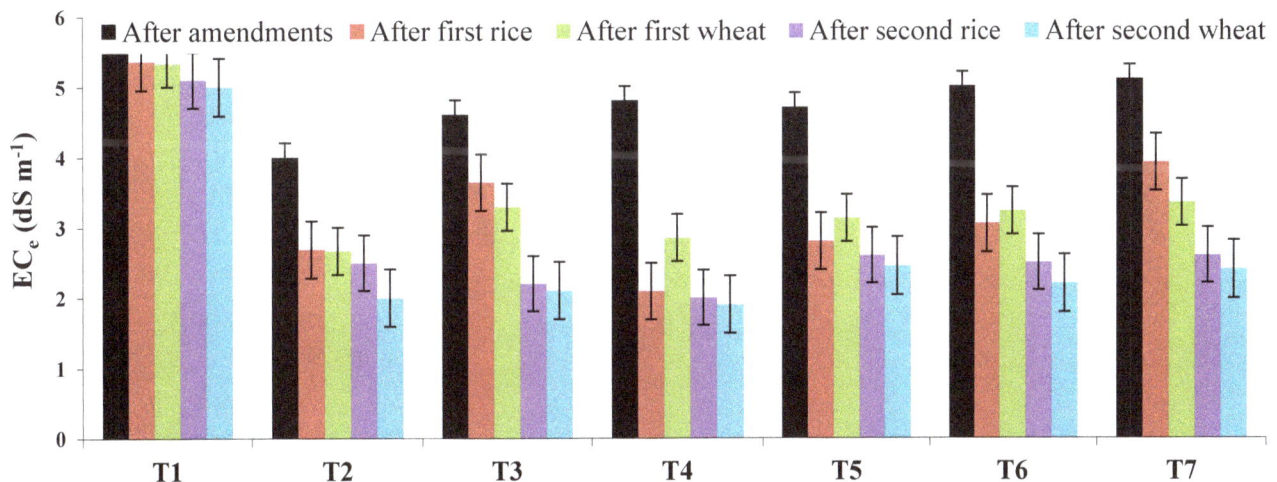

Fig. 1: EC_e changes with gypsum alone and in combination with compost

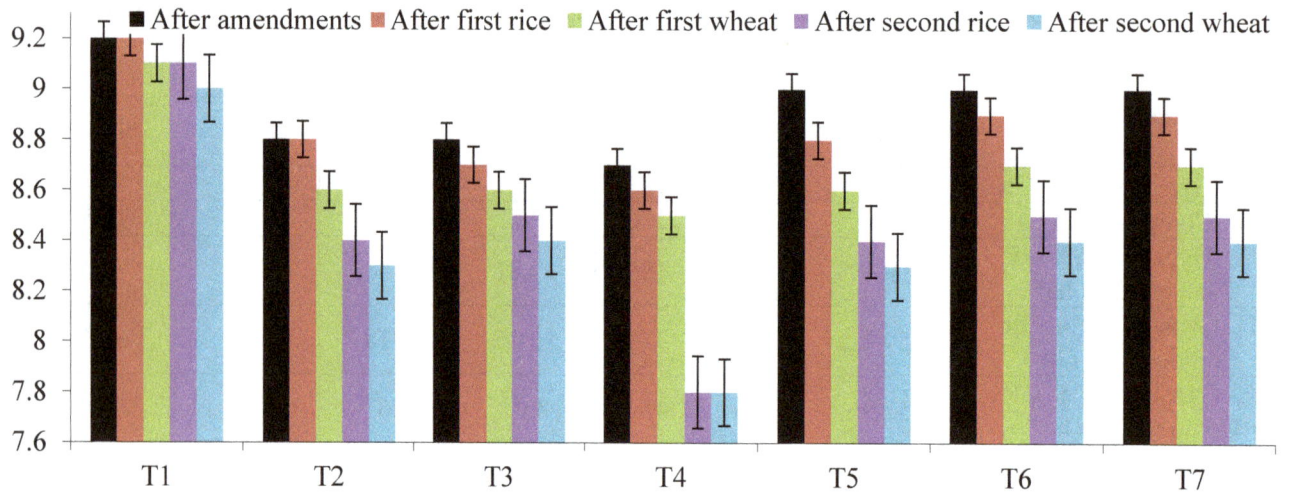

Fig. 2: pHs changes with gypsum alone and in combination with compost

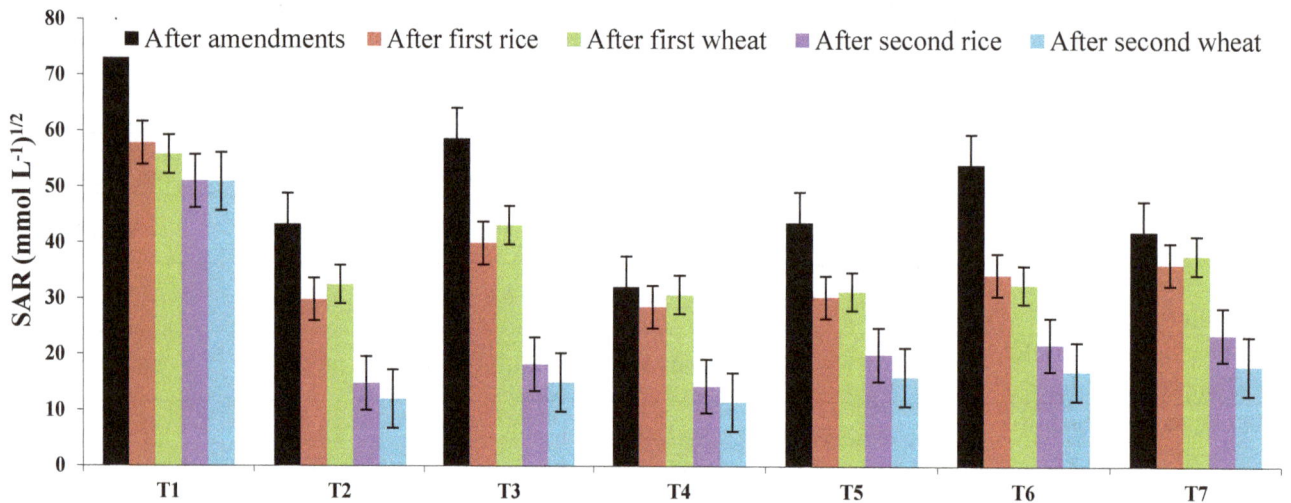

Fig. 3: SAR changes with gypsum alone and in combination with compost

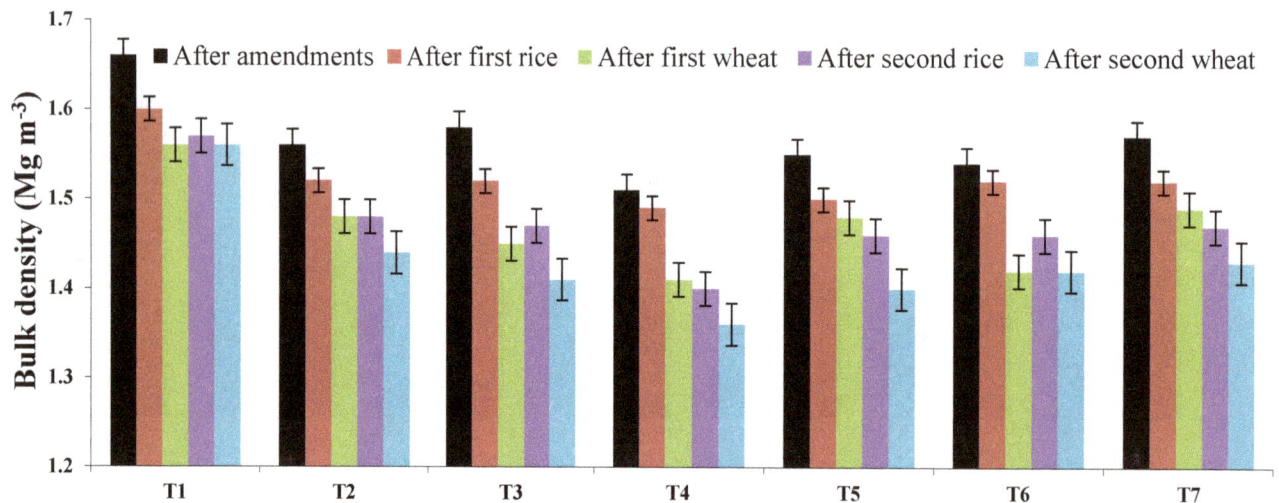

Fig. 4: Bulk densitychanges with gypsum alone and in combination with compost

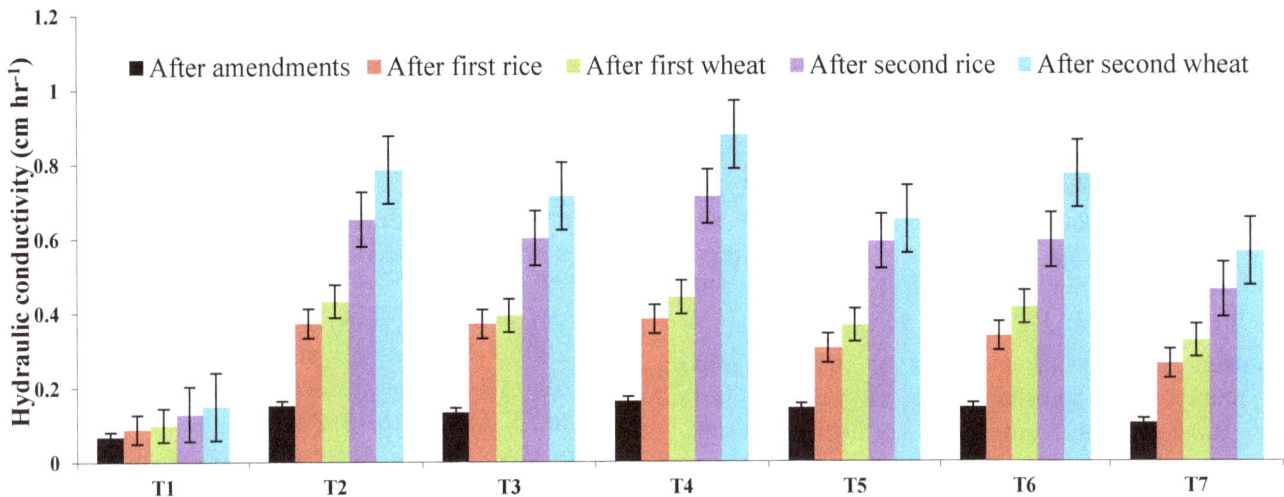

Fig. 5: Hydraulic conductivity changes with gypsum alone and in combination with compost

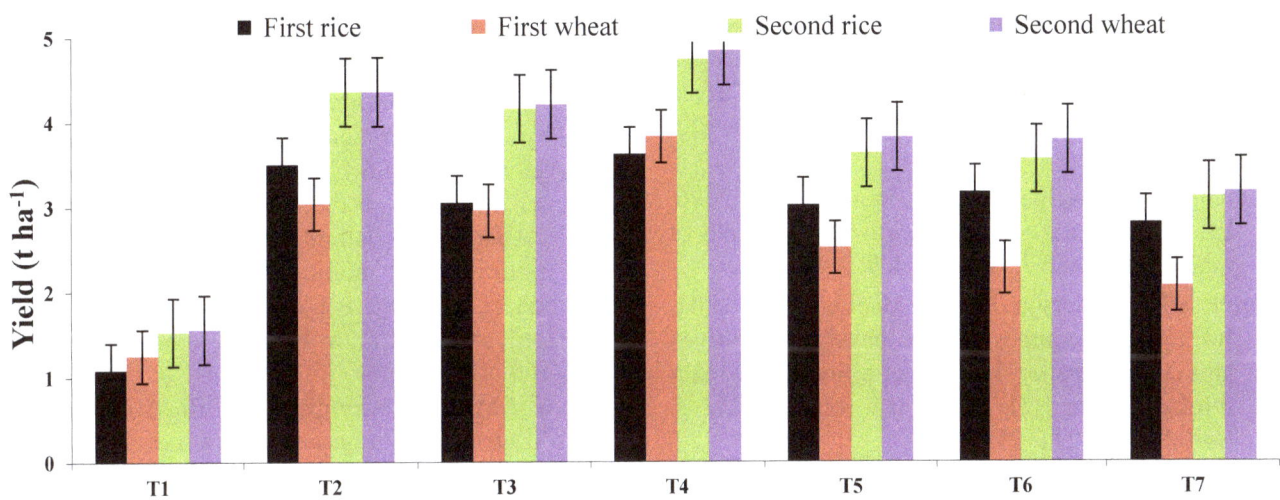

Fig. 6: Yield response to gypsum alone and in combination with compost

Conclusion

An effective reclamation procedure for saline sodic soils is removal of undesirable Na^+ by addition of some Ca^{2+} source paralleled with leaching of this Na^+ out of root zone. However, the combination of compost + gypsum proved to be the best soil amendment for reducing soil pH, salinity, SAR and improving bulk density, hydraulic conductivity. Resultantly, the paddy and grain yield of wheat increased with the improvement of soil health.

The message derived for the farmers from this study is that they can effectively rehabilitate their salt affected soils to obtain original potential if they apply gypsum at the full rate (100% GR) alone or they can decrease its quantity 25 or 50% by combining it with compost (20 or 10 t ha^{-1}).

References

Abbas G, Saqib M, Akhtar J, Murtaza G, Shahid M and Hussain A, 2016. Relationship between rhizosphere acidification and phytoremediation in two acacia species. J. Soil Sediments. 16 (4):1392–1399.

Abd El-Rheem KM, Zoghdan MGT and Hayam AAM, 2016. Effect of natural amendments as combined with different NPK fertilization combinations on nutrients content of wheat plant and sandy soil properties. Sci. Agric. 13 (3): 137-142.

Abdel- Fattah MK, 2012. Role of gypsum and compost in reclaiming saline-sodic soils. J. Agric. Vet. Sci. 1(3): 30-38.

Ahmad A, Fares A, Hue NV, Safeeq M, Radovich T, Abbas F and Ibrahim M, 2014. Root distribution of sweet corn (*Zea mays*) as affected by manure types, rates and frequency of applications. J. Animal Plant Sci. 24: 592-599.

Akhtar HK, Ashok KS, Mubeen SS, Najam WZ, Uma SS and Stephan MH, 2014. Response of Salt-Tolerant Rice Varieties to Biocompost Application in Sodic Soil of Eastern Uttar Pradesh. Amer. J. Plant Sci. 5: 7-13.

Ameen A, Ahmad J and Raza S, 2017. Effect of compost and gypsum on rice crop production in saline soil. Int. J. Adv. Sci. Res. 3(08): 99-100.

Ashtar AEL and Eletreiby F, 2006. Influence of leaching with gypsum and compost of rice straw on improvement of salt affected soil and rice growth. Alex. Sci. Exch. J. 27(2): 214-221.

Awaad MS, Rashad AA and Bayoumi MA, 2009. Effect of Farmyard Manure Combined with Some Phosphate Sources on the Productivity of Canola Plants Grown on a Sandy Soil. Res. J. Agri. Bio. Sci. 5(6): 1176-1181.

Boateng SA, Zickermann J and Kornaharens M, 2006. Effect of poultry manure on growth and yield of maize. West Afri J. App. Ecol. 9: 1-11.

Carter MR, Sanderson JB and MacLeod JA, 2004. Influence of compost on the physical properties and organic matter fractions of a fine sandy loam throughout the cycle of a potato rotation. Can. J. Soil. Sci. 84: 211–218.

Chaudhry MR, Ahmad B and Rafiq MS, 1990. Efficiency of chemical amendments in reclamation of saline sodic soil. Pak. J. Soil Sci. 5: 19-23.

Diacono M and Montemurro F, 2015. Effectiveness of Organic Wastes as Fertilizers and Amendments in Salt-Affected Soils. Agric. 5: 221-230.

El-Sanat GMA, Aiad MA and Amer MM, 2017. Impact of Some Soil Amendments and Different Tillage Depths on Saline Heavy Clay Soils Properties and Its Yield Water Productivity. Int. J. Plant Soil Sci. 14(2): 1-13.

Evanylo G, Sherony C, Spargo J, Starner D, Brosius M and Haering K, 2008. Soil and water environmental effects of fertilizer-, manure-, and compost-based fertility practices in an organic vegetable cropping system. Agric. Ecosys. Environ. 127: 50–58.

FAO. 2011. Production Year Book, Vol. 54, pp: 76–77.

Ghafoor A, Murtaza G, Ahmad B and Boers TM, 2008. Evaluation of amelioration treatments and economic aspects of using saline-sodic water for rice and wheat production on salt- affected soils under arid land conditions. Irrigat. Drain. 57: 424–434.

Grattan SR and Grieve CM, 1999. Salinity – mineral nutrient relations in horticultural crops. Sci. Hort. 78: 127-157.

Hussain N, Hassan G, Arshad Ullah M and Mujeeb F, 2001. Evaluation of amendments for the improvement of physical properties of sodic soil. Int. J. Agric. Biol. 3: 319-322.

Islam MA, Islam S, Akter A and Rahman MH, 2017. Effect of Organic and Inorganic Fertilizers on Soil Properties and the Growth, Yield and Quality of Tomato in Mymensingh, Bang. Agri. 7 (18): 2-7.

Kauraw DL and Verma GP, 1982. Improvement of soil structure of degraded soil with organic amendments. J. Indian Soc. Soil Sci. 30: 528-530.

Klute A, 1986. Methods of soil analysis (Part I). Physical and mineralogical methods (2nd Ed.) Agronomy 9. SSSA, Madison. WI, USA.

Lakhdar A, Hafsi C, Rabhi M, Debez A, Montemurro F, Abdelly C, Jedidi N and Ouerghi Z, 2009. Application of municipal solid waste compost reduces the negative effects of saline water in Hordeum maritimum L. Bioresour. Technol. 99: 7160–7167.

Mahdy AM, 2011. Comparative effects of different soil amendments on amelioration of saline sodic soil. Soil Water Res. 6(4): 205-216.

Mahmood IA, Ali A, Aslam M, Shahzad A, Sultan T and Hussain F, 2015. Phosphorus availability in different salt-affected soils as influenced by crop residue incorporation. Int. J. Agric. Biol. 15: 472–478.

Mikanova O, Simon T, Javurek M and Vach M, 2012. Relationships between winter wheat yields and soil carbon under various tillage systems. Plant Soil Environ. 12: 540–544.

Munns R, Richard AJ and Lauchli A, 2006. Approaches to increasing the salt tolerance of wheat and other cereals. J. Exp. Bot. 57(5): 1025-1043.

Nan J, Chen X, Wang X, Lashari MS, Wang Y, Guo Z and Du Z, 2016. Effects of applying flue gas desulfurization gypsum and humic acid on soil physicochemical properties and rapeseed yield of a saline-sodic cropland in the eastern coastal area of China. J. Soil Sedi. 16: 38-50.

NFDC, 2012. Fertilizer use Related Statistics. National Fertilizer Development Centre, Planning Division, Government of Pakistan, Islamabad, Pakistan.

Ould-Ahmed BA, Inoue M and Moritani S, 2010. Effect of saline water irrigation and manure application on the available water content, soil salinity, and growth of wheat. Agric. Water Manage. 97: 165–170.

Peters J and Kelling K, 2002. Should calcium be applied to Wisconsin Soils? Focus on Forage. 4 (3): 1-3.

Qadir G, Ahmad K, Qureshi MA, Saqib AI, Zaka MA, Sarfraz M, Warraich IA and Sana Ullah, 2017. Integrated use of inorganic and organic amendments for reclamation of salt affected soil. Int. J. Biol. 11(2): 1-10.

Qadir M and Schubert S, 2002. Degradation processes and nutrient constraints in sodic soils. Land Deg. Develop. 13: 275-294.

Rahman H, abdel A, Dahab MH and Mustafa MA, 1996. Impact of soil Amendments on intermittent evaporation, moisture distribution and salt redistribution in saline-sodic clay soil columns. Soil Sci. 161(11): 793 -802.

Sarwar G, 2005. Use of compost for crop production in Pakistan. Okologie und Umweltsicherung. Germany. 26/2005. Pp. 166-172.

Sarwar M, Ali A, Nouman W, Arshad MI and Patra JK, 2017. Compost and Synthetic Fertilizer Affect Vegetative Growth and Antioxidants Activities of *Moringa oleifera*. Int. J. Agric. Biol. 19(5): 1293–1300.

Shaaban M, Abid M and Abou S, 2013. Amelioration of salt affected soils in rice paddy system by application of organic and inorganic amendments. Plant Soil Environ. 59(5): 227–233.

Shainberg I, Summer ME, Miller WP, Farina MPW, Paran MA and Few MA, 1989. Use of gypsum on soils: a review. Adv. Soil Sci. 1: 1-111.

Tejada M, Garcia C, Gonzalez JL and Hernandez MT, 2006. Use of organic amendments as a strategy for saline soil remediation: Influence on the physical, chemical and biological properties of soil. Soil Biol. Biochem. 38: 1413-1421.

US Salinity Laboratory Staff, 1954. Diagnosis and Improvement of Saline and Alkali Soils. USDA Handbook 60, Washington, DC, USA.

Vance WH, Tisdell JM and MeKenzie BM, 1998. Residual effects of surface application of organic matter and calcium salts on the sub soil of a red brown earth. Aust. J. Exp. Agric. 38: 595-600.

Wang GJ, Xu ZW and Li Y, 2016. Effects of Biochar and Compost on Mung Bean Growth and Soil Properties in a Semi-arid Area of Northeast China. Int. J. Agric. Biol. 18 (5): 1056–1060.

Wong NL, Dalal RC and Greene RSB, 2009. Carbon dynamics of sodic and saline soils following gypsum and organic materials additions: A laboratory incubation. Appl. Soil Eco. 40: 29-40.

Wright AL, Provin TL, Hons FM, Zuberer DA and White RH, 2018. Compost Impacts on Sodicity and Salinity in a Sandy Loam Turf Grass Soil. Compost Sci. Util. 16(1): 30-35.

Wright AL, Provin TL, Hons FM, Zuberer DA and White RH, 2007. Soil micronutrient availability after compost addition to Saint Augustine grass. Compost Sci. Util. 15: 127-134.

Zaka MA, Mujeeb F, Sarwar G Hassan NM and Hassan G, 2003. Agromelioration of Saline Sodic soils. Online J. Biol. Sci. 3(3): 329-334.

Zia-ur-Rehman M, Murtaza G, Qayyum MF, Saifullah RM, Ali S, Akmal F and Khalid H, 2016. Degraded soils: origin, types and management. Springer International Publishing Switzerland. KR Hakeem (Eds.), Soil Science: agricultural and environmental prospective. http://dx.doi.org/10.1007/978-3-319-34451-5_2.

9

Comparative the impact of organic and conventional strawberry cultivation on growth and productivity using remote sensing techniques under Egypt climate conditions

Hassan A. Hassan[1], Sahar S.Taha[1], Mohamed A. Aboelghar[2], Noha A. Morsy[2]
[1]Department of Vegetables, Faculty of Agriculture, Cairo University, Cairo, Egypt
[2]National Authority for Remote Sensing and Space Sciences (NARSS), Cairo, Egypt

Abstract

Two years field experiment on strawberry plants (cv. Sweet Charlie) in Qalyubia Governorate, Egypt was carried out to study the effect of different growing conditions (organic and conventional) and the effect of some colors of plastic mulch such as clear, black, and silver on the quantitative and qualitative characteristics of the strawberry plantations using hyper spectral remotely sensed data. As the first step, spectral reflectance pattern for the different treatments (fertilization and colors of plastic mulch) was identified through in situ spectral measurements. It was found that silver plastic mulch recorded higher values with all observed vegetative and fruit traits, as compared with an organic strawberry growing systems without plastic mulch. Spectral reflectance parameters in form of vegetation indices (VIs) were examined as yield estimators and their correlation with leaf area index (LAI) was observed. Generated models with accuracy assessment were explained and the optimal vegetation index to estimate yield under each treatment was identified. Generally, it was found that fertilization has more effect on spectral characteristics than plastic mulch. Spectral vegetation indices (VIs) showed higher accuracy than LAI as yield estimators. (Spectral – yield) models showed the same trend with adequate correlation coefficient (r^2) exceeded (0.7) except the treatment of black plastic mulch conventional system that showed (r^2) less than (0.6) with two yield estimators. All generated models with an accuracy of each model are explained in the following sections.

Keywords: Strawberry, Spectral Reflectance, Vegetation Index, LAI, Yield

Introduction

In Egypt, Strawberry crop is the most important export vegetable crop. Egypt has increased its placement on global rank from number 9 in 2014 to number 5 in 2015 in terms of exports of frozen strawberries (OctoFrost Group 2016). Because of the importance of strawberry for local consumption and export, there is a high need for yield prediction system month/s before harvest. This system enables optimal management for strawberry production and exportation. Comparing conventional and organic production system and fruits quality depended on a large number of factors (Vallverdú-Queralt; Lamuela-Raventós, 2015). Palomaki et al. (2002) noticed that when they measured plant vegetative growth was less than in the conventional system .Organic growers employ cultural practices that support soil health for certification, to increase crop quality and yields, and to improve environmental sustainability. This increase is due to benefits of mulching such as an increase in soil temperature, reduced weed pressure, moisture conservation, reduction of certain insect pests, higher crop yields, and more efficient use of soil nutrients (Kyrikou and Briassoulis, 2007; Kasirajan and Ngouajio, 2012). The light color that is perceived by the plant can possibly influence the development of the plants including its physiological characteristics. Fatemi *et al.*, (2013) reported that the chlorophyll content of *Cucurbita pepo* was increased when grown with colored polyethylene mulch. Plastic mulch and fertilization are two vital factors for the growth and productivity of strawberry (Abo Sedera *et al.*, 2010a and b). Significant effects of organic and plastic mulches on vegetative growth, flowering traits and yield and its components of strawberry plants have been reported by several investigators (Hasanein *et al* 2011; Abou El-yazied and Mady, 2012; Haroon *et al* 2014). Remote sensing (RS) for crop monitoring is a vital requirement for agricultural development locally and globally. It has been used for the assessment of physiological conditions, biophysical and biochemical parameters of the plants and their effect on crop yield (Sims and Gamon, 2002). RS techniques include different tools that assess spectral characteristics (SC) of plants (Zhang et al., 2010). Changes in (SC) during the growing season are based on different parameters including plant pigments, chlorophyll and water content (Jorgensen et al., 2006; Maire et al., 2004). (SC) in form of vegetation indices (Vis) were used to identify spectral reflectance characteristics in different spectral regions specifically, red, near infrared and green (Gitelson et al., 2005). They were used to estimate chlorophyll content (Gitelson, 2004). Among different (VIs), normalized difference vegetation index (NDVI) is the most common ones (Cabrera-Bosquetet al., 2011). The objective of this study was to evaluate the response of strawberry plants for organic and conventional strawberry growing systems under different treatments of fertilization, some colors of plastic mulch such as clear, black, and silver and their interactions on the quantitative and qualitative characteristics of the strawberry cv. Sweet Charlie and their effect of these treatments on spectral reflectance characteristics. It is essential for the proposed method to be applicable under local agricultural conditions that might be different from a country to another and sometimes are different even within the same country.

Material and Methods

Two years field experiment on strawberry plants (cv. Sweet Charlie) in Qalyubia Governorate, Egypt was carried out to study the effect of different growing conditions (organic and conventional) and the effect of some colors of plastic mulch such as clear, black, and silver on the quantitative and qualitative characteristics of the strawberry plantations using hyper spectral remotely sensed data. The transplants were dipped in Rhizolex solution at a rate of 2.0 g /l for 20 minutes as recommended for pathogen disinfection before transplanting. A soil sample was collected from the experimental field at the beginning of the experiment, where some physical and chemical properties of the experimental field was sandy loam in texture with EC of 1.67 ds/m and pH 7.80, N was 22.7 mg/kg soil, P was 17.1 mg/kg soil and exchangeable K was 129.4 mg/kg soil. On the other hand, the chemical analysis of used compost is presented in Table (1).

Transplanting was done on 20 and 24th of September in 2014- 2015 and 2015- 2016, respectively. Sprinkler irrigation was used in the first month after transplanting. The drip irrigation was used until the end of the season. Treatments were arranged in a split-plot design with four replicates. The plot area was 14 m² included one bed (175cm width and 8 meters long) each bed included four rows and the planting distance was 25 cm between transplants.

Table 1. Average chemical analysis of compost during the two seasons of study.

Organic materials	Sources of compost (Delta Bio Tec.Co.)	pH	Ec dS /m	O.M %	N %	P %	K %	C/N	Humidity %	Weight of m³ (kg)
Botanical waste compost	AL wadi compost	6.6	1.6	58	1.4	0.65	0.79	18: 1	24	730

Treatments were as follows:

Growing conditions Systems:
Organic strawberry growing systems: 100 % recommended fertilizers. Compost 100 % N rate was used at (200 units N/ fed $^{-1}$ which were about 17.857 ton compost fed $^{-1}$ while potassium fertilizer rates (240 unit K_2O / fed $^{-1}$) which were about (17.857 ton compost fed $^{-1}$ plus 1199.5 kg from feldspar) while phosphorus rate was used at (90 units P_2O_5/ fed $^{-1}$.) which were about (17.857 ton compost fed $^{-1}$ plus 9.4 kg from rock phosphate). A source of P and K were added during soil preparation mixture with compost before agriculture. In each treatment, the content of compost, potassium, and phosphorus account calculated and completed to the required concentration by adding feldspar and rock phosphate.

I.

Conventional strawberry growing systems: 100 % recommended fertilizers (200 Kg N unit/fed., and80 Kg unit P2O5/fed., and 240 Kg unit K2O/fed.) Ditches were prepared in the sites of drip irrigation lines; calcium superphosphate added in the ditches then covered by soil. Ammonium sulfate (20.6 % N) was used as a source of nitrogen, calcium superphosphate (15.5 % P_2O_5) as a source of phosphorus and potassium sulfate (48 % K2O) was used as a source of potassium.

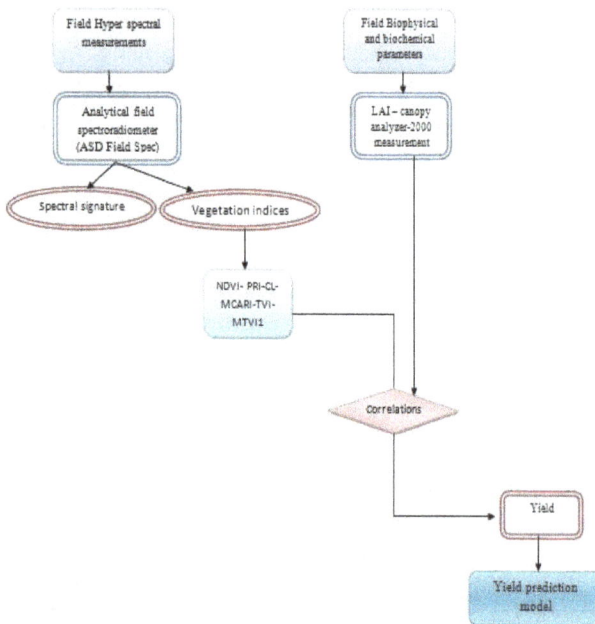

Figure1. Methodology Flowchart

Plastic culture strawberry production (using colored plastic mulches):
1. All plots were mulched with black plastic.
2. All plots were mulched with silver plastic.
3. All plots were mulched with clear plastic.
4. All plots were without Plastic Mulch.

The focus of this study is to identify spectroscopic characteristics of different samples under different fertilizer treatments and different colors of plastic mulches. The second step is to correlate spectroscopic parameters with biophysical and biochemical parameters and yield data to generate empirical models to retrieve these parameters through spectroscopic characteristics and to predict yield as well. Figure (1) is a flowchart that shows the whole methodology.

Study area
Experimental site was (933.8 square meters), a farm at El-Dair village, Qaluybia Governorate, Egypt, located between longitudes 30°22'10" and 30°22'12"E and latitudes 31°17'18" and 31°17'16" (figure 2).

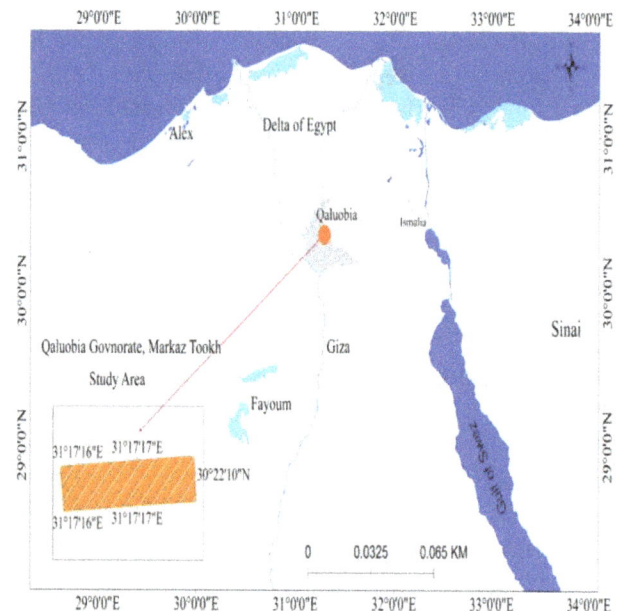

Figure2. Location of the study area

Data recorded: A random sample of five plants was randomly taken from each experimental plot on March 15 in the two seasons to determine the following data. Plant height (cm), Number of leaves/plant, Crown diameter (cm), Dry weight / plant (g) and Chemical compositions of plant foliage by Black et al.1981, Trough and Meyer 1939 and Brown and lilleland 1946 in both seasons, where early and total fruit yield (ton fed.ˉ1), was determined as the weight of all harvested fruits at the 80-90% color stage and Fruit quality by AOAC (2000).

Biophysical data recorded

Field spectroradiometer was the tool for spectral reflectance measurements of strawberry ''sweet Charli" plant at 10:00 – 14:00 local time under cloud free weather conditions. Measurements were performed in a wide spectral range (350 - 2500 nanometer (nm)) covering visible, near infrared (NIR) and shortwave infrared (SWIR). The final sampling intervals for the output data are: (1.37 nm) for the wavelength range (350 – 1000 nm), 2 nm for the range (1000 – 2500 nm). Plant canopy analyzer (LAI-2000)

was used for LAI measurements the same time with spectral measurements. The average LAI is calculated through measuring incoming radiation above the strawberry subplot and five below-canopy samples.

Spectral data analysis

The first step of the analysis was the identification of spectral reflectance pattern for strawberry plants cv. Sweet Charlie under different treatments. Second step, six (VIs) were calculated and used as estimators for crop yield. Finally, the optimal (VI) that could be used efficiently to predict yield with different treatments was identified after accuracy assessment process. Equations that were used to calculate the different vegetation indices and the references for each index are explained in table (2).

Table 2. Vegetation Indices (VI) calculated

No.	Index description	Equations	Reference
1	NDVI (Normalized difference vegetation index)	$R_{800}-R_{680}/R_{800}+R_{680}$	Rouse et al., 1974
2	PRI(Photochemical reflectance index)	$R_{531}-R_{570}/ R_{531+}R_{570}$	Gamon et al., 1992
3	CL (Chlorophyll Index)	$R_{750}/(R_{700} + R_{710})-1$	Gitelson et al., 2005
4	MCARI (Modified chlorophyll absorption ratio index)	$[(R_{700} - R_{670}) - 0.2*(R_{700} - R_{550})] (R_{700}/R_{670})$	Daughtry et al., 2000
5	TVI (Triangular vegetation index)	$0.5[120(R_{750}-R_{550})-200(R_{670}-R_{550})]$	Broge and Leblanc, 2000
6	MTVI1 (Modified Triangular Vegetation Index)	$1.2 \times [1.2 \times (R_{800} -R_{550}) - 2.5 \times (R_{670}-R_{550})]$	Haboudane et al., 2004

*R is relative reflectance at their respective wavelengths

Statistical analysis:

Data of the present study were statistically analyzed using (M Stat). The differences between the means of the treatments were considered significantly when they were more than least significant differences (LSD) at the confidence level of 5% level according to Snedecor and Cochran (1980). Spectral reflectance analysis and yield production modeling was carried out SPSS software.

Results and Discussion

Vegetative Growth Characters:
Effect of growing conditions Systems:

The effect of organic and conventional growing systems under different treatments of fertilization on some vegetative growth characteristics of strawberry plants is presented in tables (3 and 4) show that vegetative growth character of strawberry plants were

significantly affected by conventional strawberry growing systems under different treatments of fertilization. The highest values of the above-mentioned growth characters which in turn increased the number of leaves/ plant, crown diameter, leaf area and dry weight/plant were obtained with a conventional strawberry growing system under different treatments of fertilization. On the other hand, the lowest values were obtained with the organic strawberry growing systems under different treatments of fertilization. These findings were significant and true in both experimental seasons. These results are in agreement with Organic manure plays a direct role in plant growth as a source of all necessary macro and micronutrients in available forms during mineralization and improve physical and chemical properties of soils (Chaterjee et al., 2005). Rodas et al., 2013 Indicated that plant growth characters of the cultivar 'Aromas', such as plant height, a number of leaves/ plant, crown diameter, and dry weight/plant were influenced by combined doses of N and K applied through fertigation. On the other hand, those reported by Abo Sedera *et al.,* 2010, Abd El-Mawgoud *et al.,* 2010, Spinelli *et al.* 2010, Yadav *et al.,* 2010 and Hazarika *et al.,* 2015. On strawberry reported a positive response for using such organic compounds on plant growth characters as expressed by plant height, number of leaves/ plant, crown diameter and dry weight/plant of strawberry cultivars.

Effect of plastic culture strawberry production (using colored plastic mulches):
As shown in tables (3 and 4) show that the application treatments of strawberry plants with plastic culture strawberry production (using colored plastic mulches) significantly affected plant growth characters as expressed by plant height, number of leaves/ plant, crown diameter, and dry weight/plant compared with untreated plants (without Plastic mulch). The highest values of plant height, number of leaves/ plant, number of crown diameter and dry weight/plant were obtained with silver Plastic mulch compared all treatments in both seasons of study. These results, generally, are matched with those reported by Hasanein *et al.,* 2011; Abou El-yazied and Mady, 2012; Haroon *et al.,* 2014.

Effect of the interaction between growing conditions Systems and Plastic culture strawberry production (using colored plastic mulches):
Results in tables (3 and 4) show that vegetative growth parameters of strawberry plants were significantly responded to the interaction between growing conditions Systems and Plasticulture strawberry production (using colored plastic mulches) resulted in the highest values of plant growth characters as expressed by plant height, number of leaves/ plant, crown diameter, leaf area and dry weight / plant were obtained with a conventional strawberry growing systems under different treatments of fertilization combined with silver Plastic Mulch. On the other hand, the lowest values of the above-mentioned plant growth characters as expressed by plant height, number of leaves/ plant, crown diameter, leaf area and dry weight/plant were recorded with an organic strawberry growing systems under different treatments of fertilization combined with untreated plants (without Plastic mulch). The above-mentioned findings were true in both experimental seasons. Significant differences were detected among these

interaction treatments in both seasons. The obtained results seemed to complement with those reported by Abo Sedera *et al.,* 2010a and b; Abou El-yazied and Mady, 2012; Haroon *et al.,* 2014.

Chemical compositions of plant foliage:
Effect of growing conditions Systems:
It was clear from table 5 show that significant difference among the tested plant foliage in the percentage of total nitrogen, phosphorus and potassium increased by conventional strawberry growing systems under different treatments of fertilization. The highest values of the above-mentioned growth characters were obtained with a conventional strawberry growing system under different treatments of fertilization. On the other hand, the lowest values were obtained with an organic strawberry growing system under different treatments of fertilization. Preusch *et al.,* (2004) who reported that leaf- N, P, and K was greater in strawberry plants grown in a salty soil which was amended with composted and fresh poultry litter than synthetic fertilizer, but there were no differences in leaf- N in plants grown in clay and sandy soils. Brandt and Molgaard, 2001, Abo Sedera *et al.,* 2010, Abd El-Mawgoud *et al.,* 2010, Spinelli *et al.,* 2010, Yadav *et al.,* 2010 and Hazarika *et al.,* 2015. On strawberry reported a positive response for using such organic compounds on plant foliage in the percentage of total nitrogen, phosphorus, and potassium and leaf chlorophyll content of strawberry cultivar. On the other hand, Rodas et al. (2013) indicated that plant growth characters of the cultivar 'Aromas', such as plant height, a number of leaves/ plant, crown diameter, and dry weight/plant were influenced by combined doses of N and K applied through fertigation.

Plastic culture strawberry production (using colored plastic mulches):
It was obvious from table 5 show that the application plastic culture strawberry production (using colored plastic mulches) significantly affected chemical compositions of plant foliage as expressed by total nitrogen, phosphorus, and potassium content compared with untreated plants (without plastic mulch). The highest values in the percentage of total nitrogen, phosphorus and potassium were obtained with silver Plastic mulch compared all treatments in both seasons of study. These results were true and similar in the two seasons of the experiment. These

results seemed to be in general agreements with those reported by Hasanein et al., 2011; Abou El-yazied and Mady, 2012; Haroon et al., 2014.

Effect of the interaction between growing conditions Systems and Plastic culture strawberry production (using colored plastic mulches):

Interaction of interaction between growing conditions systems and plastic culture strawberry production (using colored plastic mulches) statistically affected N, P and K percentages in the plant foliage of strawberry. The highest N, P and K percentage in the plant foliage were obtained by the combined effect of conventional strawberry growing systems under different treatments of fertilization combined with silver plastic mulch in Table 5. On the other hand, the lowest N, P and K percentages were obtained by organic strawberry growing systems under different treatments of fertilization combined with untreated plants (without Plastic mulch). The obtained results are in general accordance with those reported by Abo Sedera et al., 2010a and b; Abou El-yazied and Mady, 2012; Haroon et al., 2014.

Total fruit yield
Effect of growing conditions Systems:
It was evident from data in tables (6 and 7) show that conventional strawberry growing systems under different treatments of fertilization had a significant effect on early fruit yield, exportable fruit yield, marketable fruit yield and total fruit yield. The highest values of early fruit yield, exportable fruit yield, marketable fruit yield and total fruit yield were obtained by application of conventional strawberry growing systems under different treatments of fertilization, while the lowest values in all studied fruit yield traits were obtained by application of organic strawberry growing systems under different treatments of fertilization. Meanwhile, the conventional strawberry growing systems under different treatments of fertilization decreased unmarketable fruit yield. It was clearly evident that all treatments that conventional strawberry growing systems under different treatments of fertilization recorded higher values in all studied fruit yield traits when compared with an organic strawberry growing systems under different treatments of fertilization. In this regard, Brandt and Molgaard, 2001 and Abo Sedera et al. (2010).

Plastic culture strawberry production (using colored plastic mulches):
Data presented in tables (6 and 7) show that the application of plastic culture strawberry production (using colored plastic mulches) significantly affected early fruit yield, exportable fruit yield, marketable fruit yield and total fruit yield.The highest values early fruit yield, exportable fruit yield, marketable fruit yield and total fruit yield compared with untreated plants (without plastic mulch) during both seasons of study. Whereas, silver plastic mulch recorded the highest early fruit yield, exportable fruit yield, marketable fruit yield and total fruit yield. Meanwhile, silver plastic mulch recorded decreased unmarketable fruit yield. These results were in agreement with those reported by Abu-Zahra and Tahboub (2008) and Abo Sedera et al., (2010).

Effect of the interaction between growing conditions Systems and Plasticulture strawberry production (using colored plastic mulches):
Interaction of interaction between growing conditions systems and plastic culture strawberry production (using colored plastic mulches) increased most of the total fruit yield and its components tables (6 and 7). the combined effect of conventional strawberry growing systems under different treatments of fertilization combined with silver plastic mulch gave the highest values of early fruit yield, exportable fruit yield, marketable fruit yield and total fruit yield compared with untreated plants (without plastic mulch). Early fruit yield, exportable fruit yield, marketable fruit yield and total fruit yield trait recorded the highest values with a conventional strawberry growing systems under different treatments of fertilization combined with colored plastic mulches comparing with untreated plants (without plastic mulch). Meanwhile, the combined effect of conventional strawberry growing systems under different treatments of fertilization and silver plastic mulch decreased unmarketable fruit yield. These findings appeared to be in general accordance with those reported by several investigators Abo Sedera et al., 2010a and b; Abou El-yazied and Mady, 2012; Haroon et al., 2014.

Fruit quality
Effect of growing conditions Systems:
The data tabulated in tables (8, 9, 10 and 11) show that conventional strawberry growing systems under different treatments of fertilization significantly

affected most of the physical quality, i.e., average length, diameter, weight and firmness of fruits and chemical constituents of fruit, i.e., TSS%, vitamin C, titratable acidity, anthocyanin and total sugars. The highest values in all measured fruit traits were obtained by application of conventional strawberry growing systems under different treatments of fertilization, while the lowest values in all measured fruit traits were obtained by application of organic strawberry growing systems under different treatments of fertilization. These results are in agreement with those reported by Brandt and Molgaard, 2001, Abu-Zahra and Tahboub, 2008 and Abo Sedera *et al.*, (2010).

Plastic culture strawberry production (using colored plastic mulches):
The results shown in tables (8, 9, 10 and 11) show plastic culture strawberry production (using colored plastic mulches) significantly effect on physical quality, i.e., average length, diameter, weight and firmness of fruits and chemical constituents of fruit, i.e., TSS%, vitamin C, titratable acidity, anthocyanin and total sugars. The highest values of average fruit length, diameter, weight, firmness, TSS%, vitamin C, titratable acidity, total sugars and anthocyanin compared with untreated plants (without plastic mulch) during both seasons of study. Whereas, silver plastic mulch recorded the highest values of average fruit length, diameter, weight, firmness, TSS%, vitamin C, titratable acidity, total sugars, and anthocyanin. These results were in agreement with those reported by Fatemi *et al.*, 2013 and Haroon *et a.,l* 2014.

Effect of the interaction between growing conditions Systems and Plasticulture strawberry production (using colored plastic mulches):
Interaction of interaction between growing conditions systems and plastic culture strawberry production (using colored plastic mulches) increased all measured fruit traits Tables (8, 9 and 10, 11) conventional strawberry growing systems under different treatments of fertilization and colored plastic mulches gave the highest values of physical quality, i.e., average length, diameter, weight and firmness of fruits and chemical constituents of fruit, i.e., TSS%, vitamin C, titratable acidity, anthocyanin and total sugars comparing with an organic strawberry growing systems under different treatments of fertilization plus untreated plants (without plastic mulch) during both seasons of study.

Average fruit length, diameter, weight, firmness, TSS%, vitamin C , titratable acidity, total sugars and anthocyanin traits recorded the highest values with a conventional strawberry growing systems under different treatments of fertilization and silver plastic mulch comparing with an organic strawberry growing systems under different treatments of fertilization combined with untreated plants (without plastic mulch) during both seasons of study. These results were obtained in the two seasons of the experiment. Similar results were obtained by Abo Sedera et al., 2010a and b; Abou El-yazied and Mady, 2012; Fatemi *et al.*, 2013 and Haroon *et al.*, 2014.

Spectral reflectance characteristics
Generally, (NIR) spectral region (700-1300 nm) is dependent on the internal leaf structure, the space amount in the mesophyll, cell shapes, number of cell layers, cell size and contents as reported by (Gausman, 1974; Gausman,1977; Slaton et al., 2001). Shortwave infrared (SWIR-1 and SWIR-2) regions (1300 - 2500 nm) are characterized by the light absorption by the leaf water. Shortwave infrared (SWIR-1 and SWIR-2) regions (1300 -2500 nm) are characterized by the light absorption through leaf water content. Reflectance's increase when leaf liquid water content decreases. Visible spectral region (400-700 nm) is the absorption region of leaf pigments.
Analysis of the spectral reflectance characteristics for cultivar Sweet Charlie in the organic system under three colored plastic mulch with control (without mulch) is shown in figure (3).
Plants under black plastic mulch showed the highest spectral reflectance in (NIR) and (SWIR) spectral regions. Plants under silver plastic mulch showed the highest reflectance in visible and moderate in (NIR) and (SWIR). Plants with clear plastic mulch showed the lowest reflectance in (NIR) and moderate reflectance in (SWIR) and visible. Plants without plastic mulch showed the lowest spectral reflectance in visible and (SWIR) spectral regions and moderate reflectance in (NIR).
Analysis of the spectral reflectance characteristics for cultivar Sweet Charlie in the conventional system under three colored plastic mulch with control (without mulch) is shown in figure (4).
Plants under silver plastic mulch showed the highest spectral reflectance in visible and (NIR) and (SWIR) spectral regions. Plants without plastic mulch showed the lowest reflectance in (NIR) and moderate spectral reflectance in visible and (SWIR). Plants under clear

and black plastic mulch showed moderate spectral reflectance in all regions.

Generated models
Generated models NDVI and yield under different treatments.
Were explained in the table (12). As shown from these results that the highest correlation coefficient (0.99) was found with the treatment of yield prediction of the treatment (organic with black mulch), however, other treatment showed also sufficient correlation coefficient.

Generated models PRI and yield under different treatments
Were explained in table (13) showed high correlation coefficient except with the treatment of (clear plastic mulch conventional) that showed low correlation coefficient (0.24).The highest correlation coefficient was observed with the treatment (without plastic mulch organic).

Generated models CI and yield under different treatments
Were explained in table (14) as shown that the relatively low correlation coefficient was found with the treatments of black plastic mulch conventional and silver plastic mulch conventional while other treatment showed acceptable correlation coefficients.

Generated models MCARI and yield under different treatments
Correlation coefficients of the generated (MCARI – yield) models were high with all treatments (higher than 0.95) with all treatments as shown in the table (15).

Generated models TVI and yield under different treatments
Among generated (TVI – yield) models, treatment of (black plastic mulch conventional) showed the lowest correlation coefficient (0.582) while the highest correlation coefficients were found with the treatments of (silver plastic mulch organic and black plastic mulch organic) as shown in the table (16).

Generated models MTVI1 and yield under different treatments
The Same trend was found also with (MTVI1 – yield) models. Treatments of (black plastic mulch conventional)and (silver plastic mulch conventional) showed relatively low correlation coefficient (0.627) and (0.7924) while the highest correlation coefficients were found with the treatments of (silver plastic mulch organic and black plastic mulch organic) as shown in the table (17).

Generated models LAI and yield under different treatments
Generated models to predict yield through measured LAI are shown in the table (18). The lowest correlation coefficient was found with the treatment (black plastic mulch conventional). Treatment (silver plastic mulch conventional) showed relatively low correlation coefficient while the high correlation coefficient was found with the treatment (clear plastic mulch organic).

Generated models LAI and NDVI under different treatments
Models to retrieve LAI from NDVI are shown in the table (19). A sufficient model to retrieve LAI from NDVI was found with the treatment (clear plastic mulch organic) while the treatments (black plastic mulch conventional) and (silver plastic mulch conventional) were not sufficient to retrieve LAI from NDVI as they showed low correlation coefficients (0.500) and (0.630).

As shown from all generated models that organic treatments were more correlated with spectral characteristics than conventional treatments. The optimal vegetation index to observe plants under each treatment is explained in the table (20). It is clear that fertilizer treatment may have more effect on spectral reflectance characteristics than plastic mulch as the same color of plastic mulch showed different sensitivity to spectral characteristics according to different fertilization treatments. Field observation of the different vegetative parameters was carried out with field spectral measurements to understand the correlation between different biological and biophysical factors that may affect crop yield separately or binary.

Table 3. Effect of organic and conventional strawberry growing systems under different treatments of fertilization, some colors of plastic mulch and their interactions on vegetative growth of strawberry plants in 2014/2015season.

Treatments		Plant Length (cm)	Number of leaves /plant	Crown diameter (cm)	Dry weight (g)
Strawberry growing systems (A)	Colors of plastic mulch (B)				
Organic cultivation		23.96	18.71	2.87	13.67
Conventional cultivation		28.36	24.11	3.22	15.81
LSD 0.05 value		1.94	0.88	0.88	1.18
	Without	23.18	20.13	2.72	13.38
	Clear	25.93	20.90	3.01	14.45
	Black	26.77	21.42	3.07	15.07
	Sliver	28.75	23.18	3.38	16.07
LSD 0.05 value		1.48	2.07	0.31	1.45
Organic cultivation	Without	20.30	17.20	2.50	12.33
	Clear	23.43	18.00	2.89	13.97
	Black	25.37	18.90	2.89	13.40
	Sliver	26.73	20.73	3.22	15.00
Conventional cultivation	Without	26.07	23.07	2.94	14.43
	Clear	28.43	23.80	3.12	14.93
	Black	28.17	23.93	3.25	16.73
	Sliver	30.77	25.63	3.56	17.13
LSD 0.05 value		2.10	2.92	0.44	2.05

Table 4: Effect of organic and conventional strawberry growing systems under different treatments of fertilization, some colors of plastic mulch and their interaction on vegetative growth of strawberry plants in 2015/2016 season.

Treatments		Plant Length (cm)	Number of leaves /plant	Crown diameter (cm)	Dry weight (g)
Strawberry growing systems (A)	Colors of plastic mulch (B)				
Organic cultivation		24.92	19.45	3.28	14.64
Conventional cultivation		29.66	26.67	4.28	14.96
LSD 0.05 value		1.16	2.75	0.39	1.95
	Without	25.77	20.77	3.29	14.25
	Clear	26.32	21.68	3.52	14.52
	Black	27.50	24.43	4.04	14.97
	Sliver	29.57	25.37	4.27	15.46
LSD 0.05 value		1.58	2.86	0.41	3.04
Organic cultivation	Without	23.17	17.27	2.85	14.83
	Clear	22.20	17.50	3.17	13.00
	Black	25.63	21.23	3.42	15.20
	Sliver	28.67	21.80	3.69	15.53
Conventional cultivation	Without	28.37	24.27	3.73	13.67
	Clear	30.43	25.87	3.87	16.03
	Black	29.37	27.63	4.67	14.73
	Sliver	30.47	28.93	4.85	15.40
LSD 0.05 value		2.23	4.04	0.57	4.29

Table 5. Effect of organic and conventional strawberry growing systems under different treatments of fertilization, some colors of plastic mulch and their interaction on Chemical compositions of plant foliage of strawberry plants in Two seasons.

Treatments		Season 2014/2015			Season 2015/2016		
Strawberry growing systems (A)	Colors of plastic mulch (B)	Nitrogen %	Phosphorus %	Potassium %	Nitrogen %	Phosphorus %	Potassium %
Organic cultivation		2.67	0.52	1.52	2.92	0.61	1.64
Conventional cultivation		3.01	0.66	1.71	3.36	0.73	1.78
LSD 0.05 value		0.08	0.04	0.11	0.05	0.04	0.04
	Without	2.65	0.49	1.49	2.83	0.54	1.58
	Clear	2.74	0.55	1.56	3.05	0.63	1.65
	Black	2.88	0.62	1.66	3.17	0.71	1.75
	Sliver	3.09	0.71	1.76	3.51	0.81	1.87
LSD 0.05 value		0.10	0.04	0.04	0.10	0.04	0.04
Organic cultivation	Without	2.5	0.41	1.41	2.72	0.44	1.52
	Clear	2.59	0.47	1.46	2.83	0.57	1.56
	Black	2.69	0.55	1.56	2.92	0.66	1.68
	Sliver	2.89	0.64	1.68	3.19	0.76	1.81
Conventional cultivation	Without	2.8	0.58	1.58	2.94	0.64	1.63
	Clear	2.88	0.63	1.66	3.26	0.69	1.74
	Black	3.06	0.68	1.76	3.41	0.76	1.82
	Sliver	3.30	0.76	1.84	3.83	0.85	1.92
LSD 0.05 value		0.14	0.06	0.06	0.14	0.06	0.06

Table 8. Effect of organic and conventional strawberry growing systems under different treatments of fertilization, some colors of plastic mulch and their interaction on Physical Fruit quality of strawberry plants in 2014/2015season.

Treatments		Average fruit weight (gm)	Average Fruit diameter (cm)	Average Fruit length (cm)	Fruit firmness (g/cm²)
Strawberry growing systems (A)	Colors of plastic mulch (B)				
Organic cultivation		18.58	3.29	4.88	11.08
Conventional cultivation		21.14	3.95	4.92	10.81
LSD 0.05 value		5.09	1.18	0.44	0.25
	Without	17.08	3.04	4.05	9.94
	Clear	20.64	3.21	4.65	10.59
	Black	20.188	4.08	5.05	11.16
	Sliver	21.53	4.17	5.86	12.09
LSD 0.05 value		2.57	0.89	0.89	0.26
Organic cultivation	Without	16.22	2.62	4.35	10.18
	Clear	20.27	3.08	4.58	10.66
	Black	18.04	3.63	5.12	11.42
	Sliver	19.81	3.87	5.64	12.07
Conventional cultivation	Without	17.96	3.45	3.75	9.71
	Clear	21.02	3.32	4.73	10.51
	Black	22.32	4.54	4.98	10.90
	Sliver	23.26	4.48	6.08	12.11
LSD 0.05 value		3.63	1.26	1.26	0.37

Table 9. Effect of organic and conventional strawberry growing systems under different treatments of fertilization, some colors of plastic mulch and their interaction on Physical fruit quality of strawberry plants in 2015/2016 season.

Treatments		Average fruit weight (gm)	Average Fruit diameter (cm)	Average Fruit length (cm)	Fruit firmness g/cm2
Strawberry growing systems (A)	Colors of plastic mulch (B)				
Organic cultivation		20.34	3.93	5.17	11.44
Conventional cultivation		20.78	4.38	5.45	10.94
LSD 0.05 value		2.27	1.47	0.01	0.16
	Without	18.59	3.23	4.22	9.91
	Clear	20.01	4.32	4.87	10.79
	Black	20.24	4.17	5.82	11.6
	Sliver	23.42	4.89	6.33	12.47
LSD 0.05 value		3.30	0.74	0.56	0.16
	Without	18.14	3.22	3.82	10.37
Organic cultivation	Clear	20.46	3.90	4.91	11.21
	Black	20.22	3.84	5.76	11.68
	Sliver	22.55	4.74	6.21	12.51
	Without	19.04	3.24	4.63	9.46
Conventional cultivation	Clear	19.54	4.74	4.83	10.36
	Black	20.27	4.52	5.88	11.51
	Sliver	24.29	5.04	6.46	12.43
LSD 0.05 value		4.67	1.04	0.79	0.23

Table 10. Effect of organic and conventional strawberry growing systems under different treatments of fertilization, some colors of plastic mulch and their interaction on Chemical fruit quality of strawberry plants in 2014/2015 season.

Treatments		TSS %	Titratable acidity %	Total sugar (mg/ g F.W)	Anthocyanin (mg/100g F.W)	Vitamin C (mg/100g F.W.)
Strawberry growing systems (A)	Colors of plastic mulch (B)					
Organic cultivation		9.84	0.73	7.45	85.55	46.13
Conventional cultivation		10.1	0.81	7.65	86.47	46.52
LSD 0.05 value		0.34	0.04	0.05	0.29	0.04
	Without	9.4	0.69	7.19	84.61	45.71
	Clear	9.76	0.74	7.44	85.04	46.10
	Black	10.18	0.80	7.65	85.94	46.46
	Sliver	10.55	0.86	7.91	88.45	47.04
LSD 0.05 value		0.25	0.04	0.21	0.19	0.33
	Without	9.32	0.64	7.12	84.51	45.63
Organic cultivation	Clear	9.72	0.70	7.28	84.63	45.87
	Black	9.99	0.76	7.55	85.42	46.23
	Sliver	10.34	0.83	7.84	87.62	46.79
	Without	9.49	0.74	7.27	84.71	45.78
Conventional cultivation	Clear	9.79	0.78	7.61	85.45	46.32
	Black	10.37	0.84	7.75	86.46	46.68
	Sliver	10.77	0.89	7.97	89.29	47.30
LSD 0.05 value		0.36	0.06	0.30	0.27	0.47

Table 11. Effect of organic and conventional strawberry growing systems under different treatments of fertilization, some colors of plastic mulch and their interaction on Chemical fruit quality of strawberry plants in 2015/2016 season.

Treatments		TSS %	Titratable acidity %	Total sugar (mg/ g F.W)	Anthocyanin (mg/100g F.W)	Vitamin C (mg/100g F.W.)
Strawberry growing systems (A)	Colors of plastic mulch (B)					
Organic cultivation		10.1	0.75	7.84	87.46	46.62
Conventional cultivation		10.5	0.82	8.13	88.87	46.88
LSD 0.05 value		0.08	0.04	0.16	0.09	0.09
	Without	9.51	0.70	7.59	86.51	45.97
	Clear	10.03	0.77	7.76	87.70	46.54
	Black	10.51	0.81	8.10	88.45	46.88
	Sliver	11.15	0.87	8.50	90.01	47.62
LSD 0.05 value		0.28	0.04	0.16	0.45	0.25
Organic cultivation	Without	9.38	0.69	7.55	86.31	45.62
	Clear	9.64	0.74	7.70	86.93d	46.44
	Black	10.27	0.77	7.88	87.39	46.75
	Sliver	11.09	0.83	8.22	89.23	47.67
Conventional cultivation	Without	9.63	0.71	7.63	86.71	46.31
	Clear	10.43	0.81	7.81	88.47	46.63
	Black	10.72	0.84	8.31	89.52	47.00
	Sliver	11.20	0.91	8.77	90.78	47.57
LSD 0.05 value		0.39	0.06	0.23	0.64	0.36

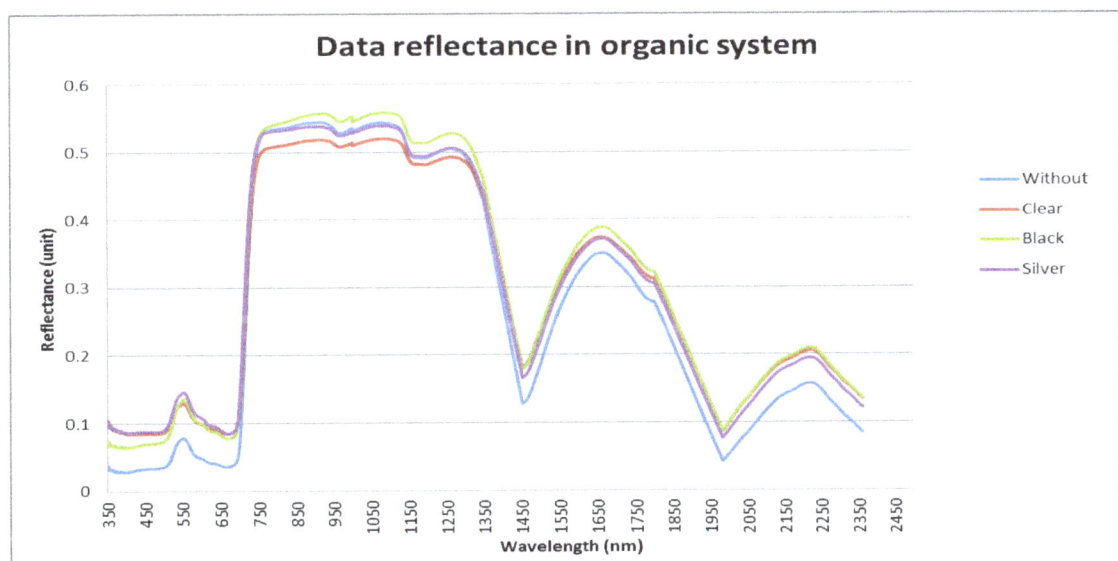

Figure 3. Spectral reflectance of organic system under different plastic mulch

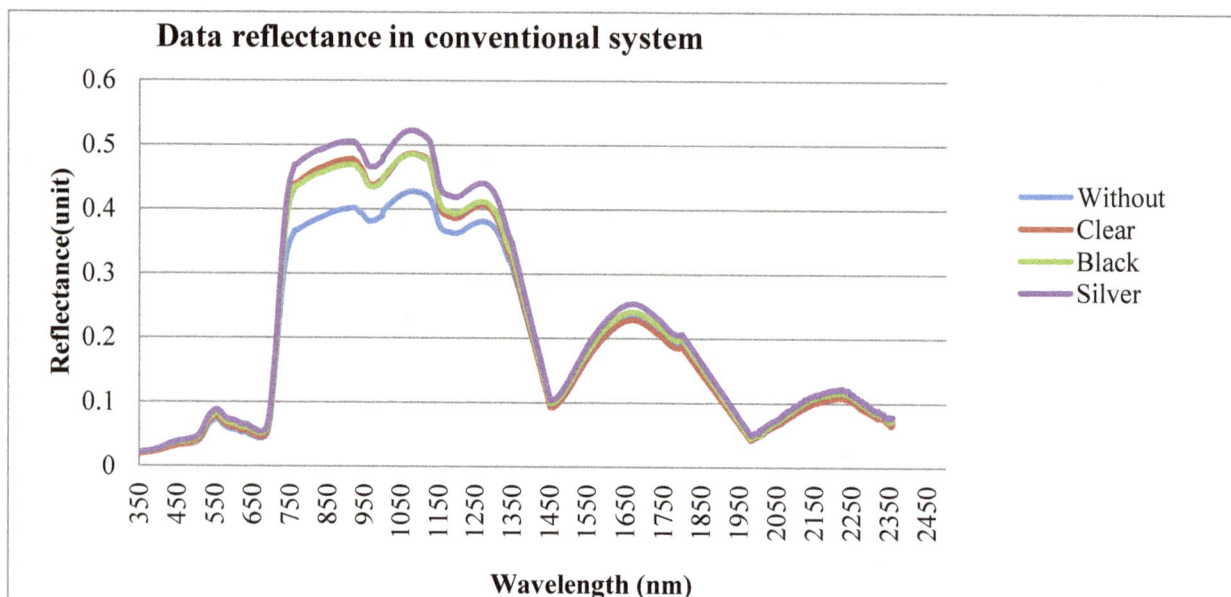

Figure 4. Spectral reflectance of conventional system under different plastic mulch

Table 12. Generated models between vegetation index NDVI and yield under different treatments

Treatments	Prediction equation	R^2
Without plastic mulch organic system	$y = 2157.2_{NDVI} - 1847.5$	0.917
Clear plastic mulch organic system	$y = 57.802_{NDVI} - 18.363$	0.962
Black plastic mulch organic system	$y = 120.23_{NDVI} - 66.699$	0.995
Silver plastic mulch organic system	$y = 78.051_{NDVI} - 32.667$	0.729
Without plastic mulch conventional system	$y = 44.47_{NDVI} - 13.626$	0.809
Clear plastic mulch conventional system	$y = 54.873_{NDVI} - 22.089$	0.951
Black plastic mulch conventional system	$y = 153.16_{NDVI} - 98.399$	0.974
Silver plastic mulch conventional system	$y = 297.18_{NDVI} - 211.95$	0.935

Table13. Generated models between vegetation index PRI and yield under different treatments

Treatments	Prediction equation	R^2
Without plastic mulch organic system	$y = 1003.6_{PRI} + 5.8193$	0.996
Clear plastic mulch organic system	$y = 2974.7_{PRI} - 37.704$	0.971
Black plastic mulch organic system	$y = 874.47_{PRI} + 12.8$	0.982
Silver plastic mulch organic system	$y = 214.73_{PRI} + 21.183$	0.754
Without plastic mulch conventional system	$y = 266.34_{PRI} + 9.6987$	0.985
Clear plastic mulch conventional system	$y = 179.72_{PRI} + 15.247$	0.240
Black plastic mulch conventional system	$y = 3.9519_{PRI} + 22.123$	0.942
Silver plastic mulch conventional system	$y = 558.75_{PRI} - 0.6093$	0.687

Table 14. Generated models between vegetation index CI and yield under different treatments

Treatments	Prediction equation	R^2
Without plastic mulch organic system	$y = 200.5_{CI} - 195.59$	0.996
Clear plastic mulch organic system	$y = 24_{CI} + 12.488$	0.956
Black plastic mulch organic system	$y = 53.578_{CI} + 2.8012$	0.996
Silver plastic mulch organic system	$y = 43.477_{CI} + 8.9684$	0.966
Without plastic mulch conventional system	$y = 14.641_{CI} + 14.564$	0.984
Clear plastic mulch conventional system	$y = 8.3732_{CI} + 16.577$	0.990
Black plastic mulch conventional system	$y = 12.127_{CI} + 15.611$	0.719
Silver plastic mulch conventional system	$y = 28.245_{CI} + 6.9157$	0.771

Table15. Generated models between vegetation index MCARI and yield under different treatments

Treatments	Prediction equation	R^2
Without plastic mulch organic system	$y = 1085.3_{MCARI} - 70.513$	0.960
Clear plastic mulch organic system	$y = 385.05_{MCARI} - 2.0054$	0.961
Black plastic mulch organic system	$y = 379.94_{MCARI} - 16.059$	0.995
Silver plastic mulch organic system	$y = 96.512_{MCARI} + 14.093$	0.982
Without plastic mulch conventional system	$y = 62.305_{MCARI} + 16.06$	0.973
Clear plastic mulch conventional system	$y = 50.743_{MCARI} + 18.008$	0.964
Black plastic mulch conventional system	$y = 79.554_{MCARI} + 16.351$	0.988
Silver plastic mulch conventional system	$y = 202.27_{MCARI} + 5.4801$	0.994

Table 16. Generated models between vegetation index TVI and yield under different treatments

Treatments	Prediction equation	R^2
Without plastic mulch organic system	$y = 36.473_{TVI} - 1078.1$	0.985
Clear plastic mulch organic system	$y = 2.0657_{TVI} - 31.403$	0.953
Black plastic mulch organic system	$y = 3.1156_{TVI} - 67.09$	0.996
Silver plastic mulch organic system	$y = 2.0883_{TVI} - 35.609$	0.998
Without plastic mulch conventional system	$y = 0.344_{TVI} + 14.692$	0.832
Clear plastic mulch conventional system	$y = 0.571_{TVI} + 8.4664$	0.861
Black plastic mulch conventional system	$y = 0.4196_{TVI} + 13.29$	0.582
Silver plastic mulch conventional system	$y = 0.9272_{TVI} + 1.4279$	0.791

Table 17. Generated models between vegetation index MTVI1 and yield under different treatments

Treatments	Prediction equation	R^2
Without plastic mulch organic system	$y = 1638.6_{MTVI1} - 1264.4$	0.972
Clear plastic mulch organic system	$y = 81.168_{MTVI1} - 32.663$	0.955
Black plastic mulch organic system	$y = 120.35_{MTVI1} - 68.55$	0.995
Silver plastic mulch organic system	$y = 76.446_{MTVI1} - 33.014$	0.999
Without plastic mulch conventional system	$y = 12.717_{MTVI1} + 14.743$	0.848
Clear plastic mulch conventional system	$y = 23.034_{MTVI1} + 7.2616$	0.835
Black plastic mulch conventional system	$y = 17.039_{MTVI1} + 12.427$	0.628
Silver plastic mulch conventional system	$y = 35.323_{MTVI1} + 0.825$	0.792

Table 18. Generated models between vegetation index LAI and yield under different treatments

Treatments	Prediction equation	R^2
Without plastic mulch organic system	$y = 2.4404_{LAI} + 12.19$	0.818
Clear plastic mulch organic system	$y = 4.1172_{LAI} + 6.0093$	0.926
Black plastic mulch organic system	$y = 10.036_{LAI} - 24.893$	0.899
Silver plastic mulch organic system	$y = 15.971_{LAI} - 53.904$	0.893
Without plastic mulch conventional system	$y = 1.793_{LAI} + 15.548$	0.869
Clear plastic mulch conventional system	$y = 2.3336_{LAI} + 18.975$	0.807
Black plastic mulch conventional system	$y = 9.4204_{LAI} + 2.4941$	0.578
Silver plastic mulch conventional system	$y = 32.383_{LAI} - 28.459$	0.738

Table 19. Generated models between vegetation index LAI and NDVI under different treatments

Treatments	Prediction equation	R^2
Without plastic mulch organic system	$y = 0.0011_x + 0.8622$	0.815
Clear plastic mulch organic system	$y = 0.0676_x + 0.4363$	0.868
Black plastic mulch organic system	$y = 0.0818_x + 0.3558$	0.867
Silver plastic mulch organic system	$y = 0.1575_x - 0.0439$	0.726
Without plastic mulch conventional system	$y = 0.0337_x + 0.6778$	0.752
Clear plastic mulch conventional system	$y = 0.0417_x + 0.7495$	0.818
Black plastic mulch conventional system	$y = 0.0536_x + 0.676$	0.500
Silver plastic mulch conventional system	$y = 0.0974_x + 0.6364$	0.630

Table 20. Optimal vegetation index for different treatments

Treatment	Optimal VI
Without plastic mulch organic system	PRI and CI
Clear plastic mulch organic system	PRI
Black plastic mulch organic system	CI and TVI
Silver plastic mulch organic system	MTVI
Without plastic mulch conventional system	PRI
Clear plastic mulch conventional system	CI
Black plastic mulch conventional system	MCARI
Silver plastic mulch conventional system	MCARI

Conclusions

This study observed the effect of different growing conditions of strawberry cv. Sweet Charlie (organic and conventional under different treatments of fertilization and some colors of plastic mulch such as clear, black, and silver on the quantitative and qualitative characteristics of plantations through remote sensing tools. Conventional strawberry growing systems under different treatments of fertilization and silver plastic mulch recorded higher values with all observed plants and fruit traits comparing with an organic strawberry growing systems under different treatments of fertilization combined with untreated plants (without plastic mulch). Field remotely sensed hyperspectral measurements were carried out to identify spectral reflectance signature for different samples. Different (VIs) were generated from spectral measurements and were examined with measured leaf area index (LAI) as estimators for yield through statistical-empirical models. Three replicates for the measurements (spectral reflectance measurements and LAI) of each treatment during the two growing seasons (2014 – 2015) and (2015 – 2016) represented the dataset for the statistical analysis. It was found that fertilization has more effect on spectral characteristics than plastic mulch. All (LAI – yield) and (spectral- yield) models were tested to spectrally identify the optimal model for each treatment. Adequate accuracy was observed with most of the generated yield prediction models as the correlation coefficient between observed and modeled yield reached more than (0.7) with most of the generated models except (LAI – yield) and (TVI - yield) models that showed accuracy less than (0.7) with the treatment (black plastic mulch conventional system). According to the correlation coefficient between modeled and observed yield, the optimal model to predict yield for each treatment was identified. The generated model could be the base of using remotely sensed data for yield prediction of strawberry yield under local conditions and local agricultural treatments in Egypt.

References

Abo Sedera FA, Bader LAA and Rezk SM, 2010a. Effect of NPK mineral fertilizer levels and foliar application with humic and amino acids on yield and quality of strawberry. Egypt. J. Appl. Sci. 25(4):154-169.

Abo Sedera FA, Bader LAA and Rezk SM, 2010b. The response of strawberry plants to the foliar application with different sources of calcium and potassium. Egypt. J. Appl. Sci. 25(4):170-185.

Abou El-Yazied A and Mady MA, 2012. Plastic color and potassium foliar application affect growth and productivity of strawberry (Fragaria ×annanassa Dush). J. App. Sci. Res. 8(2):1227-1239.

Abu-Zahra TR and Tahboub AA, 2008. Strawberry (Fragaria x annanassa Duch) growth, flowering and yielding as affected by different organic matter sources. Int. J. Bot. 4(4):481-485.

AOAC, 2000. Official Methods of Analysis.14[th]ed. Association of Official Analytical Chemists. (Ed. S. Williams) Washington D.C., USA, pp. 152-164.

Black CA, Evans DD, Ensmingern LE, White GL and Clark FE, 1981. Methods of Soil Part2, pp. 1-100.Agron. Inc. Madison. Wisc., USA.

Broge NH and Leblanc E, 2001. Comparing prediction power and stability of broadband and hyperspectral vegetation indices for estimation of green leaf area index and canopy chlorophyll density. Rem. Sens. Environ. 76:156-172.

Brown JG and Lilleland O, 1946. Rapid determination of potassium and sodium in plant materials and soil extracts by flame photometry. In Proceedings of the American Society for Horticultural Science. 48(DEC):341-346.

Cabrera-Bosquet L, Molero G, Stellacci AM, Bort J, Nogués S and Araus JL, 2011. NDVI as a potential tool for predicting biomass, plant nitrogen content and growth in wheat genotypes subjected to different water and nitrogen conditions. Cereal Res. Commun. 39(1):147–159

Chaterjee B, Ghanti P, Thapa U and Tripathy P, 2005. Effect of organic nutrition in sport broccoli (Brassica aleraceae var. italicaplenck), Vegetab. Sci. 33(1): 51-54.

Daughtry CST, Walthall CL, Kim MS, DeColstoun EB and McMurtrey JE, 2000. Estimating corn leaf chlorophyll concentration from leaf and canopy reflectance. Rem. Sens. Environ. 74(2):229-239.

FAO, 2014. Food and Agriculture Organization of the United Nations. (http://www.fao.org/faostat/en/#data/QC).

Fatemi H, Aroiuee H, Azizi M and Nemati H, 2013. Influenced of quality of light reflected of colored mulch on Cucurbita pepo var rada under field condition. Int. J. Agric. 3(2):374.

Gamon JA, Peñuelas J and Field CB, 1992. A narrow-waveband spectral index that tracks diurnal changes in photosynthetic efficiency. Rem. Sens. Environ. 41: 35–44.

Gausman HW, 1974. Leaf reflectance of near-infrared. Photogrammetric Engin. 40(2): 183–191

Gausman HW, 1977. Reflectance of leaf components. Rem. Sens. Environ. 6(1):1-9.

Gitelson AA, 2004. Wide dynamic range vegetation index for remote quantification of biophysical characteristics of vegetation. J. Plant Physiol. 161:165–173.

Gitelson AA, Vina A, Ciganda V, Rundquist DC and Arkebauer TJ, 2005. Remote estimation of canopy chlorophyll content in crops. Geophysical Res. Letters. 32:L08403. doi:10.1029/2005GL022688.

Haboudane D, Miller JR, Pattey E, Zarco-Tejada PJ and Strachan IB, 2004. Hyperspectral vegetation indices and novel algorithms for predicting green LAI of crop canopies: Modeling and validation in the context of precision agriculture. Rem. Sens. Environ. 90(3):337-352.

Haroon M, Saeed M, Ahmad I, Khan R, Bibi S and Ullah Khattak S, 2014. Weed density and strawberry yield as affected by herbicides and mulching techniques. Pak. J. Weed Sci. Res. 20(1): 67-75.

Hasanein NM, Manal MH, El-Mola G and Mona AM, 2011. Studies on influence of different colors of shade on growth and yield of strawberries and different color of plastic mulch under low tunnels conditions. Res. J. Agric. Biol. Sci.7(6):483-490.

Hazarika TK, Ralte Z, Nautiyal BP and Shukla AC, 2015. Influence of bio-fertilizers and bio-regulators on growth, yield and quality of strawberry (Fragaria× ananassa). The Indian J. Agric. Sci. 85(9):1201–1205.

Kasirajan S and Ngouajio M, 2012. Polyethylene and biodegradable mulches for agricultural applications: a review. Agronom. Sustainable Develop. 32(2):501-529.

Kyrikou I and Briassoulis D, 2007. Biodegradation of agricultural plastic films: a critical review. Journal of Polymers and the Environment. 15(2):125-150.

Maire G, Francois Cand Dufrene E, 2004. Towards universal broad leaf chlorophyll indices using PROSPECT simulated database and hyperspectral reflectance measurements. Rem. Sens. Environ. 89(1):1–28.

OctoFrost Group, 2016. A Glimpse into The Fresh And Frozen Strawberry Markets.www.octofrost.com.

Palomaki V, Mansikka-Aho AM and Etelamaki M, 2002. Organic fertilization and technique of strawberry grown in greenhouse. Acta Hortic. 567: 597-599.

Preusch PL, Takeda, F and Tworkoski TG, 2004. N, P and K uptake by strawberry plants grown with composted poultry litter. Sci. Hort. 102(1):91-103.

Rodas CL, da Silva IP, Coelho VAT, Ferreira DMG, de Souza RJ and de Carvalho JG, 2013. Chemical properties and rates of external color of strawberry fruits grown using nitrogen and potassium fertigation. IDESIA (Chile). 31:53-58.

Rouse JW, Haas RH, Schell JA, Deering DW and Harlan JC, 1974. Monitoring the Vernal Advancements and Retrogradation of Natural Vegetation. NASA/GSFC, Final Report, Greenbelt, MD, USA. 1-137.

Sims DA and Gamon JA, 2002. Relationships between leaf pigment content and spectral reflectance across a wide range of species, leaf structures and developmental stages. Rem. Sens. Environ. 81:337–354.

Slaton MR, Hunt ER and Smith WK, 2001. Estimating near-infrared leaf reflectance from leaf structural characteristics. Am. J. Bot. 88(2):278-284.

Snedecor GW and Cochran WG, 1991. Statistical Methods. 8th Ed., Iowa State Univ. Press, Iowa, USA.

Spinelli F, Fiori G, Noferini M, Sprocatti M and Costa G, 2010. A novel type of seaweed extract as a natural alternative to the use of iron chelates in strawberry production. Scientia Horticulturae. 125:263-269.

Trough E and Mayer AH, 1939. Improvement in the deiness calorimetric method for phosphorus and areseni. Indian Eng. Chem. Annual, 1:136-139.

Vallverdú-Queralt A and Lamuela-Raventós RM, 2016. Foodomics: A new tool to differentiate between organic and conventional foods. Electrophoresis. 37:1784-1794.

Yadav J, Verma JP and Tiwari KN, 2010. Effect of plant growth promoting Rhizobacteria on seed germination and plant growth Chickpea (Cicer arietinum L.) under in vitro conditions. Biol. Forum - An Int. J. 2(2):15-18.

Zhang XH, Tian QJ and Shen RP, 2010. Analysis of directional characteristics of winter wheat canopy spectra. Spectrosc. Spectr. Anal. 30(6):1600–1605.

10

Identification of Resistance Sources to Mungbean Yellow Mosaic Virus among Mungbean Germplasm

Saeed Ahmad[1], Muhammad Sajjad[1], Rabia Nawaz[2], Muhammad Arshad Hussain[1] and Muhammad Naveed Aslam[3]

[1] Regional Agricultural Research Institute, Bahawalpur, Pakistan
[2] Government Sadiq Women University, Bahawalpur, Pakistan
[3] University College of Agriculture and Environmental Sciences, The Islamia University of Bahawalpur, Pakistan

Corresponding author email:
naveed.aslam@iub.edu.pk

Abstract

Mungbean yellow mosaic virus (MYMV) is an important production constraint in mungbean cultivation in Pakistan. The yield further decreases if susceptible varieties are cultivated. By using cultivars resistant to MYMV, the losses can be reduced. As the host status of mungbean genotypes grown in Pakistan is not known, therefore, in the present study 23 varieties/lines of mungbean collected from various sources were tested for their relative resistance or susceptibility to MYMV under field conditions. The results revealed that none of the entries was found highly resistant. Six entries viz. BRM-325, BRM-345, BRM-363, BRM-364, BRM-366 and NM-2011 were found to be resistant and ten genotypes/lines namely BRM-311, BRM-312, BRM-321, BRM-331, BRM-335, BRM-365, BRM-378, BRM-382, BRM-343 and BRM-353 appeared as moderately resistant. On the contrary, five genotypes Chakwal-06, BRM-334, BRM-348, BRM-354 and BRM-356 were rated as moderately susceptible to the disease. Likewise, two entries each (BRM-349 and BRM-350) and (Mash bean and Pigeon pea) showed susceptible and highly susceptible responses to the virus respectively.

Keywords: *Vigna radiata*, Yellow mosaic virus, Tolerance, Resistance

Introduction

Mungbean (*Vigna radiata* L.) belonging to family Fabaceae, is grown extensively in major tropical and subtropical countries of the world. Its grains contain significantly high amount of carbohydrates (51%), protein (26%), moisture (10%), and vitamins A, B and C (3%) (Asaduzzaman et al., 2008). It also increases the soil fertility though symbiotic interaction with *Rhizobium* species which fix the atmospheric nitrogen (Karamany, 2006). Mungbean has been extensively cultivated on thousands of hectares in Asia including Pakistan, India, Burma, Bangladesh, Thailand, and Philippines (Hafeez et al., 1988). It has been cultivated over an area of 13.6 thousand hectares in Pakistan with a total annual production of 89.3 thousand tons (GoP,

2012). The average per hectare yield of the crop in Pakistan is less than other mungbean producing countries of the world. Different biotic and abiotic factors have been found responsible for this low productivity. Among biotic factors diseases caused by fungi (Iqbal and Mukhtar, 2014; Iqbal et al., 2014), bacteria (Shahbaz et al., 2015), nematodes (Hussain et al., 2014; 2016; Kayani et al., 2017; Mukhtar et al., 2014, 2017a, b; Tariq-Khan et al., 2017) and viruses (Ashfaq et al., 2014a, b) affect the sustainable production of this crop.

Of several viral diseases, the yellow mosaic disease caused by Mungbean yellow mosaic virus (MYMV) is of prime significance. This is a serious and widespread disease in many Asian countries including Pakistan (Biswas et al., 2008; John et al., 2008). The virus belongs to the family *Geminiviridae* and the genus *Begomovirus*. The virus was first identified in 1955 and is transmitted by whitefly (*Bemisia tabaci*). High incidence of the disease has been reported throughout the world (Nene, 1972; Bansal et al., 1984; Pathak and Jhamaria, 2004; Salam et al., 2011). MYMV causes serious yield losses in Pakistan and other countries of the world (Bashir et al., 2005). In the beginning, small yellow specks appear along the veins and then spread over the leaf. In severe conditions, the whole leaf becomes chlorotic which later on turns into necrotic region and ultimately results in the formation of shrunk and shriveled seeds (Qazi et al., 2007; Ilyas et al., 2010; Mohan et al., 2014). In Pakistan, its strains were reported for the first time in 1971 which cannot be transmitted though mechanical inoculation; however, some strains of MYMV can be mechanically transmitted in Thailand (Ahmad and Harwood, 1973; Shad et al., 2005).

Although management of disease by controlling its vector chemically was considered to be an effective way but due to development of resistance in insect vectors against chemicals, the latter are losing attraction and the incidence of the disease is still increasing. Furthermore, the use of chemicals is being discouraged as their continuous use creates hazardous impacts on environment and human health (Lopez et al., 2008).

In the recent years, research on mungbean enhancement aims to develop high yielding varieties with resistance to this disease. By using virus resistant or tolerant varieties, the incidence of MYMV can be significantly reduced. Though, some researchers have reported resistance among mungbean lines in other parts of the world (Pathak and Jhamaria, 2004; Salam et al., 2009; Iqbal et al., 2011), there is no information regarding the resistance among available mungbean germplasm in the country. Therefore, in the present studies evaluations were made to identify resistant genotypes to the disease.

Materials and Methods

The screening of mungbean germplasm was carried out in mungbean field at Regional Agricultural Research Institute (RARI), Bahawalpur during mungbean growing season i.e. April to July, 2016. Twenty three mungbean varieties/advance lines were collected from Pulses Section of RARI to identify disease incidence under natural condition (Table 1). The experiment was conducted in Randomized

Complete Block Design with 3 replications. During the first week of April, each test entry was sown in 4 rows of 3meter maintaining row to row distance of 30cm and plant to plant distance of 10cm. The distance between the blocks was 60cm. Two susceptible checks (Mash bean and Pigeon Pea) were sown in the trial to provide the maximum inoculum pressure. One susceptible check (Mash bean) was sown at the one side and the other susceptible check (Pigeon Pea) was sown at the other side of each test entry.

Recommended cultural practices were adopted throughout the growing season. No insecticide was applied to increase the white fly population for the spread of the disease by natural means. Fertilizers were applied at the time of final land preparation as per recommended doses. After germination, the crop was regularly monitored for the presence of whitefly and the development of yellow mosaic disease. Assessment of selected germplasm was carried out on the basis of percent disease infection at reproductive stage and was scored by using recommended 0-5 disease rating scale (Bashir et al., 2005).

Whitefly population was recorded with the help of wooden split cage (65 × 35 × 25 cm) (Chhabra and Kooncer, 1998). The box was kept in such a way that it covered 3 to 4 plants in each plot. Population was assessed at the mid period of the experiment. The disease rating scale used to find out the disease severity is shown in Table 2.

Results and Discussion

Evaluation of twenty-three mungbean varieties/lines under field condition of RARI, Bahawalpur against mungbean yellow mosaic virus (MYMV) carried out on the basis of scale (Table 2) showed wide variations in reaction to MYMV disease under field conditions. The severity of tested mungbean varieties/lines increased with the increase in the age of plants. The results revealed that none of the entries was found to be highly resistant. Six entries viz. BRM-325, BRM-345, BRM-363, BRM-364, BRM-366, NM-2011 were found to be resistant and ten genotypes/lines namely BRM-311, BRM-312, BRM-321, BRM-331, BRM-335, BRM-365, BRM-378, BRM-382, BRM-343, BRM-353 appeared as moderately resistant. On the contrary, five genotypes Chakwal-06, BRM-334, RM-348, BRM-354, and BRM-356 were rated as moderately susceptible to the disease. Likewise, two entries each (BRM-349 and BRM-350) and (Mash bean and Pigeon pea) exhibited susceptible and highly susceptible responses to the virus respectively. The population of whitefly was also recorded which subsequently increased along the crop season. The population ranged from 2 to 3.5 per leaf on six resistant mungbean lines/varieties whereas population ranging from 4.1 to 5.7 adults per leaf was observed in majority of the tested lines. In general no relationship was found between whitefly population and MYMV infection.

Earlier many researchers have screened mungbean germplasm against the virus and reported similar findings. Iqbal et al. (2011) found only four lines and Salam et al. (2009) three genotypes resistant to the disease. On the other hand, Habib et al. (2007) and Shad et al. (2006) screened mungbean germplasm and reported no resistance sources. Datta et al. (2012) also reported a resistant genotype IPM-02-03. Similarly, Asthana (1998) and Paul et al. (2013) also found PDM-139 (Samrat) as a resistant variety to yellow mosaic and recommended for use in disease resistance breeding programs. Mohan et al. (2014) screened 120 germplasm lines of mungbean under field conditions at two locations during kharif, 2013. Results revealed that most of the genotypes studied were categorized as moderately susceptible to highly susceptible in both the locations. None of the test entries appeared to be immune. It was observed that the genotype showed differential response against MYMV at these locations i.e. the genotype found to be resistant at one location was found to be susceptible at another location. The differential response of MYMV at different sites might be due to the presence of various strains of the pathogen. In addition to this, other factors like population pressure of the vector, genetic characteristics of the germplasm and environmental conditions at that location are also responsible for differential response of germplasm.

Table 1: List of germplam evaluated for resistance to MYMV at RARI

Sr. No.	Germplasm	Sr. No	Germplasm
1	Chakwal-06	13	BRM-350
2	BRM-311	14	BRM-353
3	BRM-312	15	BRM-354
4	BRM-321	16	BRM-356
5	BRM-325	17	BRM-363
6	BRM-331	18	BRM-364
7	BRM-334	19	BRM-365
8	BRM-335	20	BRM-366
9	BRM-343	21	BRM-378
10	BRM-345	22	BRM-382
11	BRM-348	23	NM-2011
12	BRM-349		

Table 2: Disease rating scale for categorization of mungbean germplasm to MYMV (Bashir et al., 2005)

Disease severity scale	Percent infection	Reaction
0	0%	Highly Resistant
1	1-10%	Resistant
2	11-20%	Moderately Resistant
3	21-30%	Moderately Susceptible
4	31-50%	Susceptible
5	More than 50%	Highly Susceptible

Table 3. Response of mungbean germplasm to Mungbean yellow mosaic virus

Disease severity Scale	%age infection	Reaction	No. of entries	Names of Varieties/lines
0	0%	Highly resistant	0	Nil
1	1-10%	Resistant	6	BRM-325, BRM-345, BRM-363, BRM-364, BRM-366, NM-2011
2	11-20%	Moderately resistant	10	BRM-311, BRM-312, BRM-321, BRM-331, BRM-335, BRM-365, BRM-378, M-382, BRM-343, BRM-353
3	21-30%	Moderately susceptible	5	Chakwal-06, BRM-334, BRM-348, BRM-354, BRM-356
4	31-50%	Susceptible	2	BRM-349, BRM-350
5	>50%	Highly Susceptible	2	Mash bean, Pigeon pea

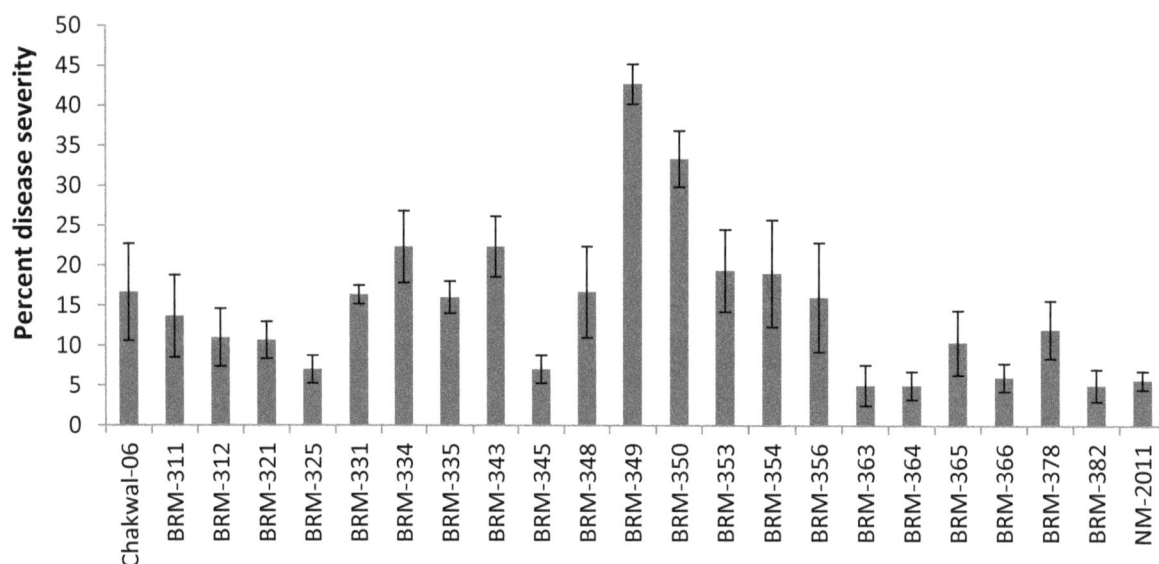

Figure 1. Mean percent disease severity on different mungbean germplasm

References

Ahmad M and Harwood RF, 1973. Studies on a whitefly-transmitted yellow mosaic of urd bean (*Phaseolus mungo*). Plant Dis. Rep. 57(9): 800-802.

Asaduzzaman M, Karim F, Ullah MJ and Hasanuzzaman M, 2008. Response of mungbean (*Vigna radiata* L.) to nitrogen and irrigation management. Am-Euras. J. Sci. Res. 3(1): 40-43.

Ashfaq M, Iqbal S, Mukhtar T and Shah H, 2014a. Screening for resistance to cucumber mosaic cucumovirus in chilli pepper. J. Anim. Plant Sci. 24 (3): 791-795.

Ashfaq M, Khan MA, Mukhtar T and Sahi ST, 2014b. Role of mineral metabolism and some physiological factors in resistance against urdbean leaf crinkle virus in black gram genotypes. Int. J. Agric. Biol. 16 (1): 189-194.

Asthana AN, 1998. Pulse crops research in India. Indian J. Agril. Sci. 68 (8): 448–52.

Bansal RD, Khatri HL, Sharma OP and Singh IP,

1984. Epidemiological studies on viral diseases of mung and mash in Punjab. J. Res. Punjab Agri. Univ. 21(1): 54-58.

Bashir M, Zahoor A and Ghafoor A, 2005. Sources of genetic resistance in mungbean and blackgram against Urdbean leaf crinkle virus (ULCV). Pak. J. Bot. 37(1): 47-51.

Biswas KK, Malath VG and Varma A, 2008. Diagnosis of symptomless Yellow mosaic begomovirus infection in pigeonpea by using cloned Mungbean yellow mosaic India virus as probe. J. Plant Biochem. Biotechnol. 17(1): 9-14.

Chhabra KS and Kooner BS, 1998. Insect pest management in mungbean and black gram: Status andstrategies. In: Upadhyay RK, Mukherji KG and RajakRL (eds): IPM System in Agriculture, Vol. 4, Pulses, Aditya Books Pvt. Ltd., New Delhi. pp. 233–310.

Datta S, Gangwar S, Kumar S, Gupta S, Rai1 R, Kaashyap M, Singh P, Chaturvedi SK, Singh BB, Nadarajan N, 2012. Genetic diversity in selected Indian Mungbean [*Vignaradiata* (L.)Wilczek] Cultivars Using RAPD Markers. American J. Plant Sci. 3: 1085–91.

GOP, 2012. Agricultural statistics of Pakistan. Ministry of National Food Security and Research, Economies Wing. Government of Pakistan. Islamabad.

Habib S, Shad N, Javaid A, Iqbal U, 2007. Screening of mungbean germplasm for resistance/tolerance against yellow mosaic disease. Mycopath. 5(2): 89–94.

Hafeez FY, Aslam Z and Malik KA, 1988. Effect of salinity and inoculation on growth, nitrogen fixation and nutrient uptake of *Vigna radiata* (L.) Wilczek. Plant Soil. 106(1): 3-8.

Hussain MA, Mukhtar T and Kayani MZ, 2014. Characterization of susceptibility and resistance responses to root-knot nematode (*Meloidogyne incognita*) infection in okra germplasm. Pak. J. Agri. Sci. 51 (2): 319-324.

Hussain MA, Mukhtar T and Kayani MZ, 2016. Reproduction of *Meloidogyne incognita* on resistant and susceptible okra cultivars. Pak. J. Agri. Sci. 53(2): 371-375.

Ilyas M, Qazi J, Mansoor S and Briddon RW, 2010. Genetic diversity and phylogeography of begomoviruses infecting legumes in Pakistan. J. Gen. Virol. 91(8): 2091-2101.

Iqbal U and Mukhtar T, 2014. Morphological and pathogenic variability among *Macrophomina phaseolina* isolates associated with mungbean (*Vigna radiata* L.) Wilczek from Pakistan. Sci. World J. 2014, http://dx.doi.org/10.1155/2014/950175.

Iqbal U, Iqbal SM, Afzal R, Jamal A, Farooq MA, Zahid A, 2011. Screening of mungbean germplasm against Mungbean Yellow Mosaic Virus (MYMV) under field conditions. Pak. J. Phytopathol. 23(1): 48–51.

Iqbal U, Mukhtar T and Iqbal SM, 2014. *In vitro* and *in vivo* evaluation of antifungal activities of some antagonistic plants against charcoal rot causing fungus, *Macrophomina phaseolina*. Pak. J. Agri. Sci. 51 (3): 689-694.

John P, Sivalingam PN, Haq QMI, Kumar N, Mishra A, Briddon RW and Malathi VG, 2008. Cowpea golden mosaic disease in Gujarat is caused by a Mungbean yellow mosaic India virus isolate with a DNA B variant. Arch. Virol. 153(7): 1359-1365.

Karamany EL, 2006. Double purpose (forage and seed) of mung bean production 1-effect of plant density and forage cutting date on forage and seed yields of mung bean (*Vigna radiata* (L.) Wilczck). Res. J. Agric. Biol. Sci. 2: 162-165.

Kayani MZ, Mukhtar T and Hussain MA, 2017. Effects of southern root knot nematode population densities and plant age on growth and yield parameters of cucumber. Crop Prot. 92: 207-212.

López E, Schuhmacher M and Domingo JL, 2008. Human health risks of petroleum-contaminated groundwater. Environ. Sci. Pollut. R. 15(3): 278-288.

Mohan S, Sheeba A, Murugan E and Ibrahim SM, 2014. Screening of mungbean germplasm for resistance to Mungbean yellow mosaic virus under natural condition. Indian J. Sci. Technol. 7(7): 891-896.

Mukhtar T, Arooj M, Ashfaq M and Gulzar A, 2017a. Resistance evaluation and host status of selected green gram genotypes against *Meloidogyne incognita*. Crop Prot. 92: 198–202.

Mukhtar T, Hussain MA and Kayani MZ, 2017b. Yield responses of twelve okra cultivars to southern root-knot nematode (*Meloidogyne incognita*). Bragantia. DOI: http://dx.doi.org/10.1590/1678-4499.005.

Mukhtar T, Hussain MA, Kayani MZ and Aslam MN, 2014. Evaluation of resistance to root-knot nematode (*Meloidogyne incognita*) in okra cultivars. Crop Prot. 56: 25-30.

Nene YL, 1972. A survey of viral diseases of pulse crops in Uttar Pradesh. Research Bulletin No. 4, Uttar Pradesh Agri. Univ., Pantnagar, p.191.

Pathak AK and Jhamaria SL, 2004. Evaluation of mungbean (*Vigna radiate* L.) varieties to yellow mosaic virus.J. Mycol. Plant Pathol. 34(1): 64-65.

Paul PC, Biswas MK, Mandal D, Pal P, 2013. Studies on host resistance of Mungbean against Mungbean Yellow Mosaic Virus in the agro-ecological condition of lateritic zone of West Bengal.The Bioscan. 8 (2): 583–87.

Qazi J, Ilyas M, Mansoor S and Briddon RW, 2007. Legume yellow mosaic viruses: genetically isolated begomoviruses. Mol. Plant Pathol. 8(4): 343-348.

Salam SA, Patil MS and Byadgi AS, 2011. Status of mungbean yellow mosaic virus disease incidence on green gram. Karnataka J. Agric. Sci. 24 (2): 247-248.

Salam SA, Patil MS, Salimath PM, 2009. Evaluation of mungbean cultures against MYMV in Karnataka under natural conditions. Legume Res. 32(4): 286–289.

Shad N, Mughal SM and Bashir M, 2005. Transmission of mungbean yellow mosaic Begomovirus (MYMV). Pak. J. Phytopathol. 17(2): 141-143.

Shad N, Mughal SM, Farooq K, Bashir M, 2006. Evaluation of Mungbean germplasm for resistance against Mungbean Yellow Mosaic Begomovirus. Pak. J. Bot. 38(2): 449–57.

Shahbaz MU, Mukhtar T, Haque MI and Begum N, 2015. Biochemical and serological characterization of *Ralstonia solanacearum* associated with chilli seeds from Pakistan. Int. J. Agric. Biol. 17 (1): 31-40.

Tariq-Khan M, Munir A, Mukhtar T, Hallmann J and Heuer H, 2016. Distribution of root-knot nematode species and their virulence on vegetables in northern temperate agro-ecosystems of the Pakistani-administered territories of Azad Jammu and Kashmir. J. Plant Dis. Prot. DOI: 10.1007/s41348-016-0045-9.

11

Prevalence and incidence of Tikka disease (*Cercospora* spp.) of groundnut in Pothwar region of Punjab

Muhammad Aslam, Khola Rafique*

Pest Warning and Quality Control of Pesticides, Department of Agriculture, Govt. of Punjab, Pakistan

Abstract

To monitor Tikka or *Cercospora* leaf spot disease of groundnut, surveys were carried out in the groundnut areas of four major districts of Pothwar region viz. Chakwal, Attock, Jhelum and Rawalpindi during main crop season in 2017. To assess prevalence and incidence of Tikka disease, 997 farmer's fields were visited. Clear disease symptoms were noted in the fields during the season. The disease prevalence ranged from 7.85 to 45% where highest prevalence was recorded in district Jhelum (45%) and lowest in Attock (7.85%). The disease incidence also varied and ranged from 9.35 to 22.48%. The highest mean disease incidence (22.48%) was observed in district Chakwal, whereas the lowest (10.57%) was observed in district Rawalpindi. The results of the study indicated that Tikka disease is significantly distributed in all the major groundnut producing districts of Pothwar region therefore, timely and possible management strategies are of vital importance to control this potential threat.

Corresponding author email: khola_47@yahoo.com

Keywords: *Cercospora*, Tikka disease, Leaf spot, Incidence, Prevalence

Introduction

Groundnut (*Arachis hypogaea* L.) is a major cash crop of Pakistan being mostly planted in the Pothwar region during summer season (Ali et al., 2002). The major groundnut producing areas are Chakwal, Attock, Jhelum, Rawalpindi, Karak, Swabi and Sanghar (Khan, 2005). It is cultivated on an area of about 81.5 thousand ha, with a production of 91.4 thousand tons, 85% of which is contributed by the Pothohar region of Punjab, 12% by Khyber Pakhtunkhwa and 3% by Sindh (Asad et al., 2017). It is a traditional oilseed crop of the country, containing 43-55% oil contents (Shad et al., 2009). It is widely used in pharmaceuticals medicines, livestock, fuels (Ndiame et al., 2004), confectionery, snacks etc. (Atasie et al., 2009). However, the productivity levels are much lower to meet the requirements of edible oil production (Shahid et al., 2010), mainly due to a number of production constraints including increased incidence of insect pests and diseases, weeds infestation, its cultivation under rainfed conditions, prevalence of drought stress due to variation in rainfall, low input-use (improved seed material), inefficient fertilizer use, shortage of irrigation water, and insufficient research efforts (Ashfaq et al., 2003; Ashraf, 2004).

Groundnut is susceptible to a number of fungal diseases among which a known leaf spot disease of groundnut, commonly called Tikka disease is a widespread foliar disease that causes severe losses in crop (Ijaz et al., 2008). The disease infects crop directly as well as indirectly and results in huge losses due to leaf defoliation, disruption of photosynthesis and fewer pods that are inferior in quality (Waliyar et al., 2000). Losses are even more when crop is unsprayed (Anonymous, 2000). Infection starts as a result of irregular rains during flowering stage to pod formation. The temperature requirement for disease development ranges from 31 to 35 °C (maximum) and 18 to 23 °C (minimum) (Pande et al., 2000). It is generally of two types (early and late) caused by two different species of the genus *Cercospora* namely *Cercospora arachidicola* S. Hori (early leaf spot) and *Cercosporidium personatum* (Berk and Curt) Deighton (late leaf spot) (Mehrotra and Aggarwal, 2003). Disease symptoms generally appear at 30-36 days after sowing and increase in intensity up to crop harvest time. In the early stages, the disease is characterized by the appearance of light brown spots surrounded by yellow halo on the leaves on one to two months old plants due to the attack of *C. arachidicola*. In the later stages, black spots usually without yellow halo appear on the stem caused by *C. personatum* (Subrahmanyam et al., 1995; Van and Cilliers, 2000). The spots weaken the plant and lead to defoliation which adversely affects fruit size as well as the quality. Eventually, fewer leaves result in less photosynthesis and reduced yield. Both leaf spots can be controlled by growing disease resistant varieties and through chemicals which is not widely practiced due to high costs of fungicides and unawareness of farmers to chemicals. Tikka disease is a major restriction to increased groundnut productions and can cause higher yield losses up to 50% worldwide (Tshilenge et al., 2012). In India, yield losses of 15-59% have been estimated due to this disease. In Florida, a 10% reduction in groundnut yield has been reported due to tikka disease epidemics (Alderman and Nutter, 1994). According to Walls and Wynne (1985), loss in yield of up to 70% has been indicated worldwide. Varied incidence levels of disease have been reported in different countries. In Pakistan, 87.16% incidence of disease was reported in district Attock while 2.12% in district Bhakar (Ijaz et al., 2008). In Nigeria, highest incidence of 43.34% was observed during 2011 and 45.36% during 2012 (Richard et al., 2017).

Keeping in view the economic importance of groundnut in Pothwar region of Punjab and significant losses caused by Tikka disease, the present study was conducted to assess the prevalence and incidence of Tikka disease of ground nut.

Material and Methods

Study area for disease survey

A survey of Tikka disease of groundnut was conducted during crop season (May to October) of 2017 in four major groundnut growing districts of Pothwar region viz. Rawalpindi (33°37'N; 73°4'E),

Jhelum (31°20'N; 72°10'E), Chakwal (32°56'N; 72°53'E) and Attock (33°52'N; 72°20'E). The altitude of the area is 508 meters above sea level. During the survey, 997 farmer's fields in 11 tehsils were assessed during the crop season (Figure 1). In district Chakwal, 295 fields were surveyed, in district Attock 525 fields, in district Jhelum 133 fields while 44 fields were surveyed during the year.

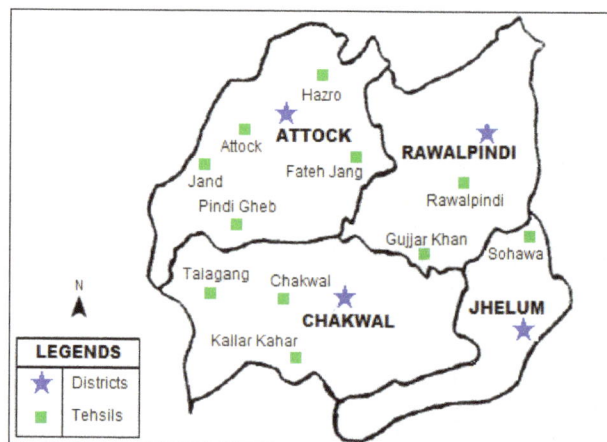

Fig. – 1: Districts and tehsils of Pothwar region surveyed for groundnut Tikka disease during 2017 crop season.

Disease assessment

Each groundnut field was surveyed randomly and disease prevalence and incidence were documented. Disease assessment was performed using a quadrate (1 m2) and 4 spots were randomly selected from each field. Plants infected with Tikka disease were observed showing particular symptoms and an attempt was made to identify the symptoms of both early leaf spot and late leaf spot. Fields were visited regularly for the identification and characterization of the disease symptoms caused by two species viz. *C. arachidicola* and *C. personatum*.

Disease prevalence was observed to assess the distribution of Tikka disease in surveyed areas of Pothwar region and calculated by using the following formula:

$$\text{Disease Prevalence \%} = \frac{\text{No. of infected fields}}{\text{Total No. of Fields}} \times 100$$

For disease incidence, the number of total groundnut plants and infected plants in 1 m2 were counted. These observations were used to calculate the average Tikka disease incidence in each field visited using the following formula (Kanade et al., 2015):

Disease Incidence % = <u>No. of diseased plants </u>x 100
                                 Total No. of plants observed

Statistical analysis

Data collected during the study was analyzed through ANOVA using software program SPSS. The data was analyzed statistically and significance of results was expressed at 5% level.

Results and Discussion

Disease symptoms in field

For disease assessment, the disease symptoms were carefully observed in the surveyed fields from the start of appearance of symptoms. In the beginning around 30 to 35 days after sowing, small and chlorotic spots were noticed on the plant leaves. At early stages i.e. associated with *C. arachidicola*, the spots turned into black lesions surrounded by chlorotic or yellow halo on upper leaf surfaces while light brown spots on lower surface. During late leaf spot stage caused by *C. personatum*, small, circular, rough and darker lesions were identified on lower leaflet surfaces, petioles and stems. The symptoms observations were in accordance with the description by Van and Cilliers (2000) and Subrahmanyam et al. (1995).

Disease prevalence and incidence

The study showed widespread prevalence of Tikka disease as shown in Figure 2. Results of the survey revealed variations among the farmer's fields assessed for disease prevalence and incidence. It was noticed that peak period for Tikka infestation was generally confined between the months of August and October during the pod formation of groundnut. The prevalence Tikka disease was apparent in all the fields under groundnut crop plantation in the four districts surveyed in 2017. The mean disease prevalence was 20% in the region where highest prevalence was witnessed in district Jhelum with 45% prevalence percentage followed by Rawalpindi (11.93%) and Chakwal (14.8%) while minimum prevalence was noted in district Attock (7.85%). Tikka was prevalent with varied percentages in various tehsils of surveyed districts as shown in Table 1. In district Chakwal, ground nut was found in three tehsils where maximum prevalence was

recorded in tehsil Chakwal (31.4%) followed by Kallar Kahar (7.74%) and Talagang (5.26%). In district Attock, groundnut plantation was observed in five tehsils and disease prevalence varied among them i.e. Pindigheb (15.09%), Hazro (14.70%), Jand (7%), Fateh Jang (1.96%) and Attock (0.52%). In district Jhelum, ground nut was found in tehsil Sohawa where highest prevalence of 45% was recorded. Groundnut found in two tehsils of district Rawalpindi viz. Rawalpindi and Gujjar Khan showed 27.78% and 8% prevalence, respectively. It is shown that both leaf spot diseases caused by two species viz. *C. arachidicola* and *C. personanta* are highly prevalent in the Pothwar region and can pose a serious threat to groundnut production. Similarly, Ambang et al. (2011) reported prevalence of both leaf spot disease every year on groundnut in the southeastern of the United States. In India, earlier it has been reported that early leaf spot (*C. arachidicola*) is generally more prevalent in northern groundnut growing states with varied levels of incidence and severity (Joshi, 2005). On the contrary, Macedo- Nobile et al. (2008) indicated that late leaf spot (*C. personanta*), is the most prevalent and destructive disease on groundnut in Brazil.

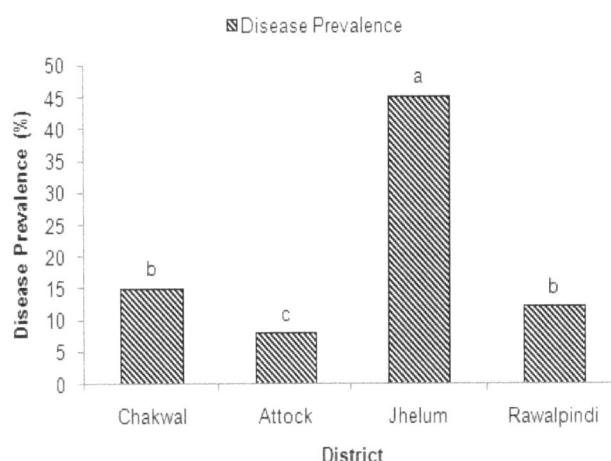

Fig. - 2: Percentage of disease prevalence in major groundnut growing districts of Pothwar region of Punjab during 2017. Different letters indicated on bars represent significant differences in Tikka prevalence values (P=0.05).

The disease incidence varied among the districts and mean disease incidence in the region was 16% during the crop season (Figure 3). Tikka disease appeared in low to moderate intensity at district Chakwal during the pod formation stage and the mean disease incidence observed was 22.48%. In rest of the districts, low disease intensities were recorded where

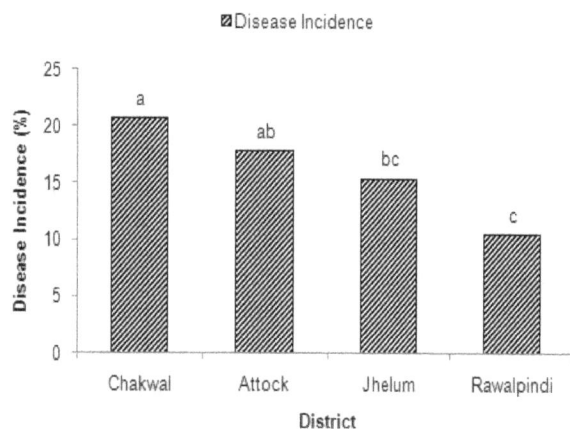

maximum incidence was observed in Attock with 17.9% incidence percentage followed by Jhelum (15.45%). **Fig. - 3: Percentage of disease incidence in major groundnut growing districts of Pothwar region during 2017. Different letters indicated on bars represent significant differences in Tikka incidence values (P=0.05).**

Table - 1: Percentage of Tikka disease prevalence and incidence in various tehsils of surveyed districts

District	Tehsil	Disease Prevalence (%)	Disease Incidence (%)
Chakwal	Chakwal	31.4 [b]	19.28 [b]
	Talagang	5.26 [de]	25.16 [a]
	Kallar Kahar	7.74 [d]	23 [a]
Attock	Attock	0.52 [e]	11.67 [c]
	Pindigheb	15.09 [c]	20.76 [b]
	Hazro	14.70 [c]	20.8 [b]
	Jand	7 [d]	14.17 [bc]
	Fateh Jang	1.96 [e]	7 [d]
Jhelum	Sohawa	45 [a]	15.45 [bc]
Rawalpindi	Rawalpindi	27.78 [b]	6.5 [d]
	Gujjar Khan	8 [d]	12.2 [c]
LSD value at α=0.05		4.52	3.62

At α=0.05 level of significance means sharing same letters are non-significant

District Rawalpindi had least disease incidence (10.57%) among all the districts surveyed. Significant difference in disease incidence was observed among tehsils of surveyed district as depicted in Table 1. In district Chakwal, maximum incidence was recorded in tehsil Talagang (25.16%) followed by Kallar Kahar (23%) and Chakwal (19.28%). In district Attock, the disease incidence was 20.8%, 20.76%, 14.17%, 11.67% and 7% in Hazro, Pindigheb, Jand, Attock and Fateh Jang, respectively. In district Jhelum, 15.45% incidence was observed in tehsil Sohawa. Two tehsils of district Rawalpindi viz. Rawalpindi and Gujjar Khan showed 6.5% and 12.2%, incidence, respectively. The study showed that groundnut crop in the region is prone to disease incidence throughout the season. Earlier, Ijaz et al. (2008) reported highest (87.16%) Tikka incidence in Attock while lowest (2.12%) incidence in Bhakar. Richard et al. (2017) recorded highest leaf spot incidence during 2011 (43.34%) and 2012 (45.36%) in Nigeria.

Conclusion

It can be concluded from the study that Tikka or leaf spot is a potential threat to groundnut production. The study reports widespread distribution of leaf spot disease in the major groundnut producing districts of Pothwar region of Punjab with varied incidence percentages. Therefore, detailed studies are necessary to utilize appropriate practices to minimize the damage and future losses caused by this disease.

References

Alderman SC and Nutter FJ, 1994. Effect of temperature and relative humidity on development of *Cercosporidium personatum* on peanut in Georgia. Plant Dis. 78: 690‑694.

Ali S, Schwanke GD, People MB, Scott JF and Herridge DF, 2002. Nitrogen, yield and economic benefits of summer legumes for wheat production in rainfed Northern Pakistan. Pak. J. Agron. 1: 15-19.

Ambang Z, Ndongo B, Essono G, Ngoh JP, Kosma P, Chewachong GM and Asanga A, 2011. Control of leaf spot disease caused by *Cercospora* sp on groundnut (*Arachis hypogaea*) using methanolic extracts of yellow oleander (*Thevetia peruviana*) seeds. Aust. J. Crop Sci. 5: 227‑232.

Anonymous, 2000. Groundnut production technology training course (March, 2000). Natural Resources

College, Lilongwe, Malawi.

Asad S, Munir A, Malik SN and Nawaz N, 2017. Evaluation of ground nut material against Tikka leaf spot disease under natural field conditions at NARC. Pak. J. Phytopathol. 29(1): 23-27.

Ashfaq A, Ashraf M and Kokhar MIA, 2003. Integrated land and water resources management in Pothwar: Issues and options. Sci. Intern. 15(1): 87-92.

Ashraf M, 2004. Impact evaluation of water resources development in the command area of small dams. Pakistan Council of Research in Water Resources. Res. Rep. 5: 35-66.

Atasie VN, Akinhami TF and Ojiodu CC, 2009. Proximate analysis and physico-chemical properties of groundnut (*Arachis hypogaea* L.). Pak. J. Nutr. 8: 194-197.

Ijaz M, Rauf CA, Haque IU, Hussan FU and Mahmood A, 2008. Distribution and severity of Cercospora leaf spot of peanut in rain fed region of Punjab. Pak. J. Phytopath. 20: 165-172.

Joshi NH, 2005. Characterization of groundnut (*Arachis hypogaea* L.) germplasm in relation to major foliar pests and diseases. Ph.D. dissertation, Saurashtra University, India.

Kanade SG, Shaikh AA, Jadhav JD and Chavan CD, 2015. Influence of weather parameters on tikka (*Cercospora* spp.) and rust (*Puccinia arachidis*) of groundnut (*Arachis hypogaea* L.). Asian J. Envir. Sci. 10(1): 39-49.

Khan A, 2005. Physiological response of groundnut to growth regulators under drought stress. Ph.D. dissertation, Quaid-e-Azam University, Islamabad, Pakistan.

Macedo‑Nobile PM, Lopes CR, Cavallari CB, Quecinia V, Coutinhod LL, Hoshino AA and Gimenesc MA, 2008. Peanut genes identified during initial phase of *Cercosporidium personatum* infection. Plant Sci. 174: 78‑87.

Mehrotra RS and Aggarwal A, 2003. Leaf spots, leaf blights and anthracnoses, pp. 505-548. In R.S. Mehrotra and A. Aggarwal (eds.), Plant Pathology, ed. 2. Tata McGraw-Hill Publishing Company Limited, New Delhi, India.

Ndiame D, Beghin J and Sewadeh M, 2004. Groundnut policies, global trade dynamics and the impact of trade liberalization. World Bank Policy Research.

Pande S, Rao JN and Kumar E, 2000. Survey of groundnut diseases in India. Survey Report. html//www.icrisat.org/gt3/r3.

Richard BI, Ukwela MU and Avav T, 2017. Major diseases of groundnut (*Arachis hypogaea* L.) in Benue State of Nigeria. Asian J. Adv. Agric. Res. 4(3): 1-12.

Shad MA, Perveez H, Nawaz H, Khan H and Aman Ullah M, 2009. Evaluation of biochemical and phytochemical composition of some groundnut varieties grown in arid zone of Pakistan. Pak. J. Bot. 41(6): 2739-2749.

Shahid LA, Saeed MA and Amjad N, 2010. Present status and future prospects of mechanized production of oilseed crops in Pakistan-A review. Pak. J. Agric. Res. 23(1-2): 83-93.

Subrahmanyam P, McDonald D, Raddy LJ, Nigam SN, Gibbons RW, Rammanatha RV, Singh AK, Pande S, Reddy PM and Subba RPV, 1995. Screening methods and sources of resistance to rust and late leaf spot of groundnut. Inf. Bull. 47: 24.

Tshilenge LL, Nkongolo KKC, Kalonji MA and Kizungu RV, 2012. Epidemiology of the groundnut (*Arachis hypogaea* L.) leaf spot disease: Genetic Analysis and developmental cycles. Am. J. Plant Sci. 3: 582-588.

Van, WPS and Cilliers, AJ, 2000. Groundnut diseases and pests. ARC Grain Crops Institute, Potchefstroom, SA. pp. 80-88.

Waliyar F, Adomou M and Traore A, 2000. Rational use of fungicide applications to maximize peanut yield under foliar disease pressure in West Africa. Plant Dis. 84: 120-1211.

Walls SB and Wynne JC, 1985. Combining ability for resistance to *Cercosporidium personatum* for five late leaf spot‑resistant peanut Germplasm lines. Oleagineux. 40: 389‑394.

12

Bridelia cathartica Bertol. f. (Phyllanthaceae): a review of its pharmacological properties and medicinal potential

Alfred Maroyi*

Medicinal Plants and Economic Development (MPED) Research Centre, Department of Botany, University of Fort Hare, Private Bag X1314, Alice 5700, South Africa

Abstract

Bridelia cathartica is an important medicinal plant throughout its distributional range in sub-Saharan Africa. The aim of the current study was to evaluate the botany, ethnomedicinal uses, chemical and biological properties of *B. cathartica*. Information on the medicinal, phytochemistry and biological properties of *B. cathartica* was undertaken using electronic databases such as Medline, Pubmed, SciFinder, SCOPUS, Google Scholar, Science Direct, EThOS, ProQuest, OATD and Open-thesis. Pre-electronic literature was sourced from the University library. Literature search revealed that *B. cathartica* is mainly used as a charm and to cast spells, as herbal medicine used by women during child bearing and pregnancy, remedy for fever and malaria, gastro-intestinal, headache, haemorrhoids, menstrual problems, pain, sores and wounds, reproductive, respiratory disorders and sexually transmitted infections. Pharmacological studies of *B. cathartica* extracts revealed that the species has antibacterial, antifungal, antimalarial and antioxidant properties. Based on its wide use as herbal medicine in tropical Africa, *B. cathartica* should be subjected to detailed phytochemical and pharmacological evaluations aimed at elucidating its chemical, pharmacological and toxicological properties.

Corresponding author email:
amaroyi@ufh.ac.za

Keywords: *Bridelia cathartica*, Ethnopharmacology, Phyllanthaceae, Traditional medicine, Tropical Africa

Introduction

The medicinal and pharmacological properties of *Bridelia cathartica* Bertol. f. were recognized not only by traditional healers but also by taxonomists who gave the species the specific name "*cathartica*" which means cathartic or purgative, in reference to the purgative properties of the species (Schimdt et al., 2002; Palmer and Pitman, 1972). *Bridelia cathartica* is a popular medicinal plant throughout its distributional range in sub-Saharan Africa with its bark, leaves and stems marketed as traditional medicines in informal herbal medicine markets and

other informal markets in Gauteng and KwaZulu Natal provinces in South Africa (Williams et al., 2001; Cunningham, 1993). *Bridelia cathartica* is a large shrub to small tree belonging to the family Phyllanthaceae, previously included in the Euphorbiaceae family. The name of the genus, "*Bridelia*" was derived from the name of a Swiss-German muscologist, Samuel Elisée Bridel-Brideri (1761-1828) (Palmer and Pitman, 1972; Schimdt et al., 2002; Maroyi, 2017). The Phyllanthaceae family was a sub-family of Euphorbiaceae until Hoffmann et al. (2006) separated the two families using molecular results that utilized DNA sequence data of nuclear

PHYC and plastid atpB, matK, ndhF and rbcL as well as morphological characteristics. Globally, the family has 59 genera and 2 000 species (Hoffmann et al., 2006). *Bridelia* is a genus of about 60 to 70 species that have been recorded in tropical and subtropical Africa and Asia (Smith, 1987; Ngueyem et al., 2009; Maroyi, 2017). Research by Ngueyem et al. (2009) revealed that *Bridelia* species are characterized by antihelmintic, antiamebic, antioxidant, antianemic, antiplasmodial, antibacterial, antinociceptive, anticonvulsant, antiarrhythmic, antidiarrhoeal, anti-hypertensive, muscle relaxant, anti-inflammatory, analgesic, stimulant, antimalarial, anti-cholinergic and antiviral properties and therefore, widely used as herbal medicines against abdominal pain, cardio-vascular, gynecological and sexual diseases. It is within this context that the current study was undertaken aimed at reviewing the botany, ethnomedicinal uses, chemical and pharmacological properties of *B. cathartica* so as to provide baseline data required for evaluating the therapeutic potential of the species.

Material and Methods

Information relevant to the botany, ethnomedicinal uses, chemical and pharmacological properties of *B. cathartica* was carried out from September 2017 to March 2018. Online electronic databases including Google Scholar, SciFinder, ScienceDirect, Medline, Pubmed, SCOPUS, EThOS, ProQuest, OATD and Open-thesis were used to search for relevant literature. Pre-electronic literature of conference papers, scientific articles, books, book chapters, dissertations and theses were carried out at the University of Fort Hare library. The key words used in the electronic search criteria were "*Bridelia cathartica*", synonyms of the plant species "*B. melanthesoides* (Baill.) Klotzsch, *B. fischeri* Pax, *B. lingelsheimii* Gehrm., *B. niedenzui* Gehrm., *B. schlechteri* Hutch., *Pentameria melanthesoides* Baill.", English common names "blue sweetberry and knobbly bridelia". The following keywords were used in combination with the species name, synonyms and English common names to search for relevant information: "biological properties", "ethnobotany", "ethnomedicinal uses", "ethnopharmacological properties", "medicinal uses", "pharmacological properties" and "phytochemistry". Total number of publications included 38 articles published between 1941 and 2017, and these were in agreement with the literature search criteria (Figure 1).

The sources of data included research articles results published in international journals (23), books (eight), other scientific publications (three), dissertations (two), book chapter and conference proceeding (one each).

Botanical profile and description of *Bridelia cathartica*

Bridelia cathartica is an evergreen, small multi-stemmed tree with a flat, spreading crown growing to a maximum height of 9 meters (Smith, 1987). The trunk is grey, greyish-brown or black in colour, rough, reticulate, fissured or stringy. Leaves are alternate, obovate to elliptic-oblanceolate in shape with rounded, obtuse to sub-acute at the apex, grey-green to blue-green on the underside and darker green and glossy on the upper side. The small greenish to yellowish flowers are male or female, formed in terminal axillary clusters carried on short stalks. The fruits are small, round berries, changing colour from red to black as they mature (Van Wyk and Van Wyk, 1997; Palmer and Pitman, 1972). *Bridelia cathartica* is native to Swaziland, Somalia, Malawi, the Democratic Republic of Congo (DRC), Mozambique, Botswana, Ethiopia, Zambia, South Sudan, Sudan, South Africa, Namibia, Tanzania, Zimbabwe and Kenya. *Bridelia cathartica* grows in woodland, bushland, along stream banks, in riverine fringe thicket and rocky places, persisting in secondary associations at an altitude ranging from 0 to 2000 m above sea level (Smith, 1987).

Bridelia cathartica is divided into two subspecies, *B. cathartica* subsp. *cathartica* and *B. cathartica* subsp. *melanthesoides* (Klotzsch) J. Léon. The two subspecies are separated mainly on a small difference in the veining, in the subspecies *cathartica*, the lateral veins are slender, clearly marked, extend to the edge of the leaf margin while in the subspecies *melanthesoides*, the lateral veins are branched, forming loops before they reach the edge of the leaf (Coates Palgrave, 2002; Palmer and Pitman, 1972). The majority of ethnobotanical literature does not separate *B. cathartica* into specific subspecies, but rather *B. cathartica* and this is the plant name that is going to be used throughout this study.

Medicinal uses of *Bridelia cathartica*
Several parts of the plant including bark, fruits, leaves and root bark are used to prepare herbal concoctions used to treat 44 human diseases in tropical Africa (Table 1). The major diseases and ailments include the

species being used as a charm and to cast out evil spells, used as herbal medicine during child bearing and pregnancy, remedy for fever and malaria, gastro-intestinal, headache, haemorrhoids, menstrual problems, pain, sores and wounds, reproductive, respiratory disorders and sexually transmitted infections (Figure 2). There is a cross-cultural agreement of medicinal usage of *B. cathartica* as herbal medicine against these diseases recorded in at least three countries and literature records (Figure 2). *Bridelia cathartica* is also used in multi-therapeutic applications taken by pregnant women, as remedy for headache, menstrual problems, reproductive and respiratory disorders (Table 1). *Bridelia cathartica* is used to manage and treat some ailments and diseases listed by the World Health Organization (2014) as the leading causes of disease burden in tropical and sub-tropical Africa, and these include (in descending order of importance) respiratory infections, human immunodeficiency virus / acquired immune deficiency syndrome (HIV/AIDS), diarrhoeal diseases, malaria, birth asphyxia and trauma, and preterm birth complications. There is therefore, need to validate the ethnomedicinal applications of *B. cathartica* through phytochemical and pharmacological evaluations of both the crude extracts and compounds associated with the species. World Health Organization has recognized the role of traditional medicines in resource-poor regions like the sub-Saharan Africa where usage of herbal medicines has been scientifically validated (WHO, 2013; Hughes et al., 2015). Therefore, the significance of herbal medicines in the face of increasing global practice of using both the orthodox and traditional medicines cannot be ignored (WHO, 2013; Hughes et al., 2015).

Phytochemical and nutritional composition of *Bridelia cathartica*

Several phytochemical compounds and minerals have been identified from leaves, stems, roots, root and stem bark of *B. cathartica*. Van Valen (1978) identified triglochinin (Figure 2, Table 2), a cyanogenic glycoside from the seeds of *B. cathartica*. Chhabra et al. (1984) identified tannins, flavonoids, anthracene glycosides, fatty acids, steroids, emodins, triterpenoids, volatile oils and anthocyanins from leaves of *B. cathartica*. Similarly, Madureira et al. (2012) identified alkaloids, flavonoids, phenolics and terpenes from roots of *B. cathartica*. Cumbane and Munyemane (2017) identified flavonoids, phenolics, condensed and hydrolysable tannins from leaves and

stems of *B. cathartica* (Table 2). Azimova and Glushenkova (2012) identified five fatty acids (Figure 1) from the seeds of *B. cathartica*. The major fatty acids in the seeds were linolenic acid (44.0%), oleic acid (23.0%), linoleic acid (15.0%), palmitic acid (10.0%) and stearic acid (8.0%) (Azimova and Glushenkova, 2012). The macronutrients identified from leaves, root and stem bark of *B. cathartica* included phosphorus (P), calcium (Ca), potassium (K) and magnesium (Mg), while micronutrients included copper (Cu), iron (Fe), manganese (Mn) and zinc (Zn) (Ouma, 1994; Ouma et al., 1997). Some of these phytochemical compounds and minerals may not confirm the medicinal applications of *B. cathartica* but will provide ethnopharmacological evidence of the therapeutic potential of the species. For example, Ouma et al. (1997) argued that the iron content exhibited by the leaves, root and stem bark of *B. cathartica* ranging from 3.5 to 35.7 mg/100g was adequate to justify the use of the species against anaemia in Kenya (Ouma, 1994; Ouma et al., 1997).

Pharmacological activities of *Bridelia cathartica*

Some of the pharmacological activities of *B. cathartica* listed in literature include antibacterial (Madureira, 2012; York, 2012; Cumbane et al., 2017), antifungal (Sawhney, 1978; York, 2012), antimalarial (Jurg, 1991; Ramalhete, 2008), antioxidant (Cumbane and Munyemana, 2017) and cytotoxicity (Moshi et al., 2004) activities. These pharmacological activities of various parts of the species are summarized below.

Antibacterial activities

Madureira (2012) assessed antibacterial properties of dichloromethane, ethyl acetate, methanol and n-hexane root extracts of *B. cathartica* against *Staphylococcus aureus*, *Escherichia coli*, *Enterococcus faecalis*, *Pseudomonas aeruginosa*, *Mycobacterium smegmatis* and *Klebsiella pneumoniae* using broth microdilution method with gentamicin, rifampicin and vancomycin as positive controls. All extracts showed activities with minimum inhibitory concentration (MIC) values stretching from 7.5 µg/ml to 250 µg/ml (Madureira et al., 2012). Similarly, York (2012) assessed antibacterial properties of aqueous and dichloromethane-methanol (1:1) root extracts of *B. cathartica* against *Klebsiella pneumoniae*, *Moraxella catarrhalis*, *Mycobacterium smegmatis* and *Staphylococcus aureus* using microdilution assay with ciprofloxacin as positive control. York et al. (2012) also evaluated the sum of the fractional inhibitory concentration ($\sum$FIC) which was assessed for *B. cathartica* when used mixed with *Lippia javanica*. The extract showed activities with MIC values stretching from 0.5 mg/ml to 16.0 mg/ml. The combination of aqueous and organic extracts and essential oil extract of *Lippia javanica*, and organic extract of *B. cathartica* against *Klebsiella pneumoniae*, *Moraxella catarrhalis*, *Mycobacterium smegmatis* and *Staphylococcus aureus* resulted in additive interactions with $\sum$FIC values stretching from 0.53 to 0.88 (York, 2012). Antibacterial evaluations of *B. cathartica* combined with *Lippia javanica* support this common practice of mixing these remedies for chills, cough, headache and runny nose in South Africa (York, 2011; York, 2012). Cumbane and Munyemana (2017) assessed antibacterial properties of ethyl acetate and hydroethanol leaf and root extracts of *B. cathartica* against *Pseudomonas aeruginosa*, *Escherichia coli*, *Staphylococcus aureus*, *Enterococcus faecalis* and *Streptococcus pneumoniae* using the disc diffusion method with dimethylsulfoxide (DMSO) as negative control, ciprofloxacin and gentamycin as positive controls. The extracts demonstrated activities with MIC values stretching from 250 μg/mL to >1000 μg/mL (Cumbane and Munyemana, 2017). These antibacterial properties exhibited by different extracts of *B. cathartica* corroborate the traditional application of the species as traditional medicine against bacterial infections causing diarrhoea in Tanzania and Zambia (Chhabra et al., 1990; Chinsembu, 2016), gonorrhoea in DRC, Tanzania and Zambia (Mbayo et al., 2016; Chhabra et al., 1990; Chinsembu, 2016), oral infections in Zambia (Chinsembu, 2016), sexually transmitted infections in South Africa (De Wet, 2012), stomachache and stomach ailments in Kenya and Malawi (Kokwaro, 1993; Morris,1996), syphilis and syphilitic sores in DRC and Malawi (Morris, 1996; Mbayo et al., 2016) and wounds in Malawi (Morris, 1996).

Antifungal activities

Sawhney (1978) assessed antifungal properties of methanol root bark extract of *B. cathartica* against *Trichophyton mentagrophytes* and *Candida albicans*. The extract exhibited activities against the tested pathogens (Sawhney, 1978). Similarly, York (2012) assessed antifungal properties of aqueous and dichloromethane/methanol (1:1) root extracts of *B. cathartica* against *Cryptococcus neoformans* using microdilution assay with amphotericin B as positive control. York et al. (2012) also evaluated the summation of the fractional inhibitory concentration ($\sum$FIC) which was assessed for *B. cathartica* used when mixed with *Lippia javanica*. The dichloromethane/methanol (1:1) extract demonstrated the best activity with MIC value of 0.67 mg/ml, while aqueous extract exhibited activities with MIC value of 8.0 mg/ml (York et al., 2012). The combination of aqueous extracts of *B. cathartica* and *Lippia javanica* as well as essential oils of *Lippia javanica* and *B. cathartica* resulted in additive interactions with $\sum$FIC values of 0.73 and 0.92, respectively. The antifungal evaluations of *B. cathartica* combined with *Lippia javanica* support the traditional practice of mixing these remedies for microbial infections in South Africa (York et al., 2011; York, 2012) and also monotherapeutic applications against fungal and microbial infections such as oral infections in Zambia (Chinsembu, 2016) and wounds in Malawi (Morris, 1996).

Antimalarial activities

Jurg et al. (1991) evaluated the antimalarial activities of ethanol, petroleum ether and aqueous leaf, root and stem extracts of *B. cathartica* against *Plasmodium falciparum*. The aqueous and ethanol root extracts caused 50% inhibition (ID_{50}) of parasite growth at an incubation concentration of 0.05 μg/mL (Jurg et al., 1991). Ramalhete (2008) assessed antimalarial properties of n-hexane, ethyl acetate, dichloromethane and methanol extracts of root extracts of *B. cathartica* against *Plasmodium falciparum*. The extracts showed weak moderate to no significant activity with half maximal inhibitory concentration (IC_{50}) values stretching from 44.0 ± 1.30 mg/mL to > 100 mg/mL (Ramalhete, 2008). These results corroborate the use of *B. cathartica* as traditional medicine for fever in Tanzania (Chhabra, 1990) and malaria in Mozambique (Jurg, 1991; Bandeira et al., 2001) and Zambia (Chinsembu, 2016) and lack of significant *in vitro* antimalarial activity could be explained by the fact that the species may act as antipyretic or may enhance the immune system rather than having direct antiparasitic properties (Phillipson and Wright, 1991).

Table 1: Medicinal applications of *Bridelia cathartica* in tropical Africa

Medical problems	Plant parts used	Countries	References
Charms and casting of spells			
Bewitchment, love charm and spiritual ailments	Leaves and roots	Zimbabwe, South Africa and Kenya	Gerstner, 1941; Gelfand et al., 1985; Pakia and Cooke, 2003
Child bearing and pregnancy problems			
Cleanse the blood when pregnant, ease labour, prevent pre-mature birth and pre-natal care	Leaves and roots mixed with those of *Crotalaria monteiroi* Baker f., *Grewia occidentalis* L., *Opuntia stricta* (Haw.) Haw., *Garcinia livingstonei* T. Anderson, *Rhoicissus digitata* (L. f.) Wild & R. B. Drumm., *Ochna natalitia* (Meisn.) Walp., *Rhus nebulosa* Schönland and *Commiphora neglecta* Verd.,	Kenya, Malawi and South Africa	Morris, 1996; Pakia and Cooke, 2003; De Wet and Ngubane, 2014
Convulsions and epilepsy			
Convulsions and epilepsy	Leaves and roots	Kenya and Zimbabwe	Gelfand et al., 1985; Kokwaro, 1993; Pakia and Cooke, 2003
Fever and malaria			
Fever and malaria	Fruits, leaves and roots	Mozambique, Tanzania and Zambia	Bandeira et al., 2001; Chhabra et al., 1990; Chinsembu, 2016; Jurg et al., 1991
Gastro-intestinal problems			
Amoebic dysentery, anorexia, constipation, diarrhoea and stomach ailments	Fruits, leaves, roots and root bark	Kenya, Malawi, Tanzania and Zambia	Chhabra et al., 1990; Kokwaro, 1993; Morris, 1996; Chinsembu, 2016
Headache			
Headache	Roots and leaves mixed with those of *Lippia javanica* (Burm. f.) Spreng.	Zimbabwe, Malawi and Mozambique	Gelfand et al., 1985; York, 2012; Bandeira et al., 2001; York et al., 2011; Morris, 1996
Haemorrhoids			
Haemorrhoids and protruding rectum	Roots and root bark	Tanzania, Kenya and Malawi	Hedberg et al., 1983; Morris, 1996; Kokwaro, 1993
Menstrual problems			
Amenorrhoea, dysmenorrhoea, menorrhagia and oligomenorrhoea	Bark, leaves and roots mixed with those of *Acalypha brachiata* Krauss, *Erythrina humeana* Spreng., *Garcinia livingstonei*, *Hyphaene coriacea* Gaertn., *Ochna natalitia, Commiphora neglecta, Peltophorum africanum* Sond., *Crotalaria monteiroi, Rhoicissus digitata, Grewia occidentalis, Rhus nebulosi* and *Tabernaemontana elegans* Stapf.	South Africa, Zimbabwe and DRC	Gelfand et al., 1985; De Wet and Ngubane, 2014; Mbayo et al., 2016
Pain, sores and wounds			
Abdominal pain, sores and wounds	Bark, leaves and roots	DRC, Malawi, Tanzania and South Africa	Chhabra et al., 1990; Mbayo et al., 2016; Morris, 1996; De Wet et al., 2012
Purgative			
Purgative and to prevent vomiting	Roots and root bark	Tanzania and Zambia	Watt and Breyer-Brandwijk, 1962; Chhabra et al., 1984, 1990
Reproductive problems			
Aphrodisiac and infertility	Root bark and roots mixed with those of *Acalypha brachiata, Erythrinia humeana, Hyphaene coriacea, Ozoroa engleri, Peltophorum africanum, Rhoicissus digitata, Rhus nebulosa* and *Tabernaemontana elegans*	Zimbabwe, Tanzania and South Africa	Palmer and Pitman, 1972; De Wet and Ngubane, 2014; Gelfand et al., 1985; Chhabra et al., 1990; Maroyi, 2011
Respiratory problems			
Asthma, chills, cough, respiratory infections and runny nose	Roots, root bark and leaves mixed with those of *Lippia javanica*	Malawi, Tanzania and South Africa	Chhabra et al., 1990; York, 2012; Morris, 1996; York et al., 2011
Sexually transmitted infections			
Gonorrhoea and syphilis	Bark, fruits, leaves and roots	Zambia, DRC, Tanzania, Malawi and South Africa	Chhabra et al., 1990; Mbayo et al., 2016; Morris, 1996; De Wet et al., 2012; Chinsembu, 2016
Miscellaneous			
Anaemia, bilharzia, cardiac pains, depression, heartburn, hernia, kwashiorkor and oral infections	Fruits, leaves, roots and root bark	Kenya, Malawi, Tanzania and Zambia	Chhabra et al., 1990; Ouma, 1994; Morris, 1996; Ouma et al., 1997; Chinsembu, 2016

Antioxidant activities

Cumbane and Munyemana (2017) evaluated antioxidant properties of ethyl acetate and hydroethanol leaf and root extracts of B. cathartica using the phosphomolybdenum method and the 1,1-diphenyl-2-picrylhydrazyl (DPPH) radical scavenging assay. The phosphomolybdenum method showed total antioxidant activities ranging from 42.4% to 57.9% against 100% exhibited by quercetin, the standard. The extracts in DPPH assay showed half maximal effective concentration (EC$_{50}$) values stretching from 3.63 μg/mL to 14.60 μg/mL, against 1.50 μg/mL exhibited by quercetin, the standard (Cumbane and Munyemana, 2017). These antioxidant properties of B. cathartica are probably due to the flavonoids and phenolics that have been identified from the leaves and stems of the species (Cumbane and Munyemane, 2017).

Cytotoxicity activities

Moshi et al. (2004) evaluated the cytotoxicity activities of aqueous ethanol stem bark extract of B. cathartica using the brine shrimp lethality test. The concentrations killing 50% of the shrimps (LC$_{50}$) was 58.5 μg/ml for the extract (Moshi et al., 2004). These findings imply that extracts of the species may have deleterious health implications and detailed toxicological evaluations are required to determine toxicity and/or any side effects associated with consumption of plant extracts and other products derived from the species.

Table 2: Phytochemical and nutritional composition of *Bridelia cathartica*

Caloric and nutritional composition	Values	Plant parts	Reference
Ash (g/100g)	9.3 - 13.0	Leaves, root and stem bark	Ouma, 1994
Ca (mg/100g)	1566.8 - 2367.0	Leaves, root and stem bark	Ouma, 1994
Condensed tannins (mg cyanidin 3-glucoside equivalents/g dry extract)	34.9 - 65.3	Leaves and stem	Cumbane and Munyemane, 2017
Cu (mg/100g)	0.6 - 0.8	Leaves, root and stem bark	Ouma, 1994
Fe (mg/100g)	3.5 - 35.7	Leaves, root and stem bark	Ouma, 1994
Flavonoids (mg rutin equivalents/g dry extract)	0.5 - 25.5	Leaves and stem	Cumbane and Munyemane, 2017
Hydrolysable tannins (mg tannic acid equivalents/g dry extract)	1.4 - 2.7	Leaves and stem	Cumbane and Munyemane, 2017
K (mg/100g)	600.8 -1305.5	Leaves, root and stem bark	Ouma, 1994
Mg (mg/100g)	131.2 - 555.3	Leaves, root and stem bark	Ouma, 1994
Mn (mg/100g)	3.1 - 10.9	Leaves, root and stem bark	Ouma, 1994
Moisture (g/100g)	11.8 - 12.2	Leaves, root and stem bark	Ouma, 1994
Na (mg/100g)	90.6 - 142.4	Leaves, root and stem bark	Ouma, 1994
P (mg/100g)	50.0 - 181.4	Leaves, root and stem bark	Ouma, 1994
Phenolic (mg gallic acid equivalent/g dry weight)	427.5 - 437.0	Leaves and stem	Cumbane and Munyemane, 2017
Zn (mg/100g)	1.3 - 4.4	Leaves, root and stem bark	Ouma, 1994
Cyanogenic glycoside			
Triglochinin	-	seed	Van Valen, 1978
Fatty acids			
Linoleic acid (%)	15.0	Seed	Azimova and Glushenkova, 2012
Linolenic acid (%)	44.0	Seed	Azimova and Glushenkova, 2012
Oleic acid (%)	23.0	Seed	Azimova and Glushenkova, 2012
Palmitic acid (%)	10.0	Seed	Azimova and Glushenkova, 2012
Stearic acid (%)	8.0	Seed	Azimova and Glushenkova, 2012

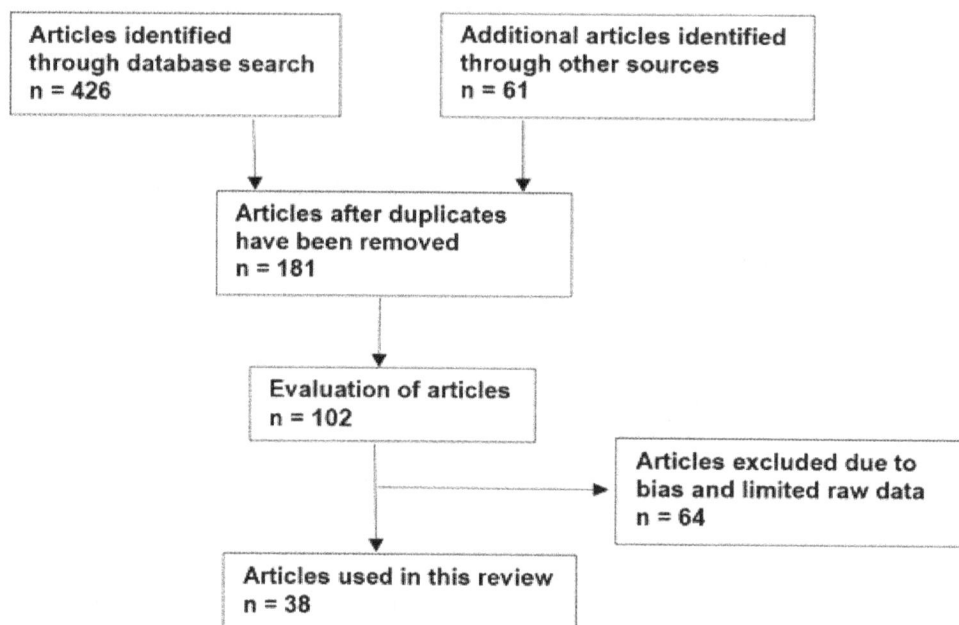

Figure 1: Flow diagram showing literature search and selection processes

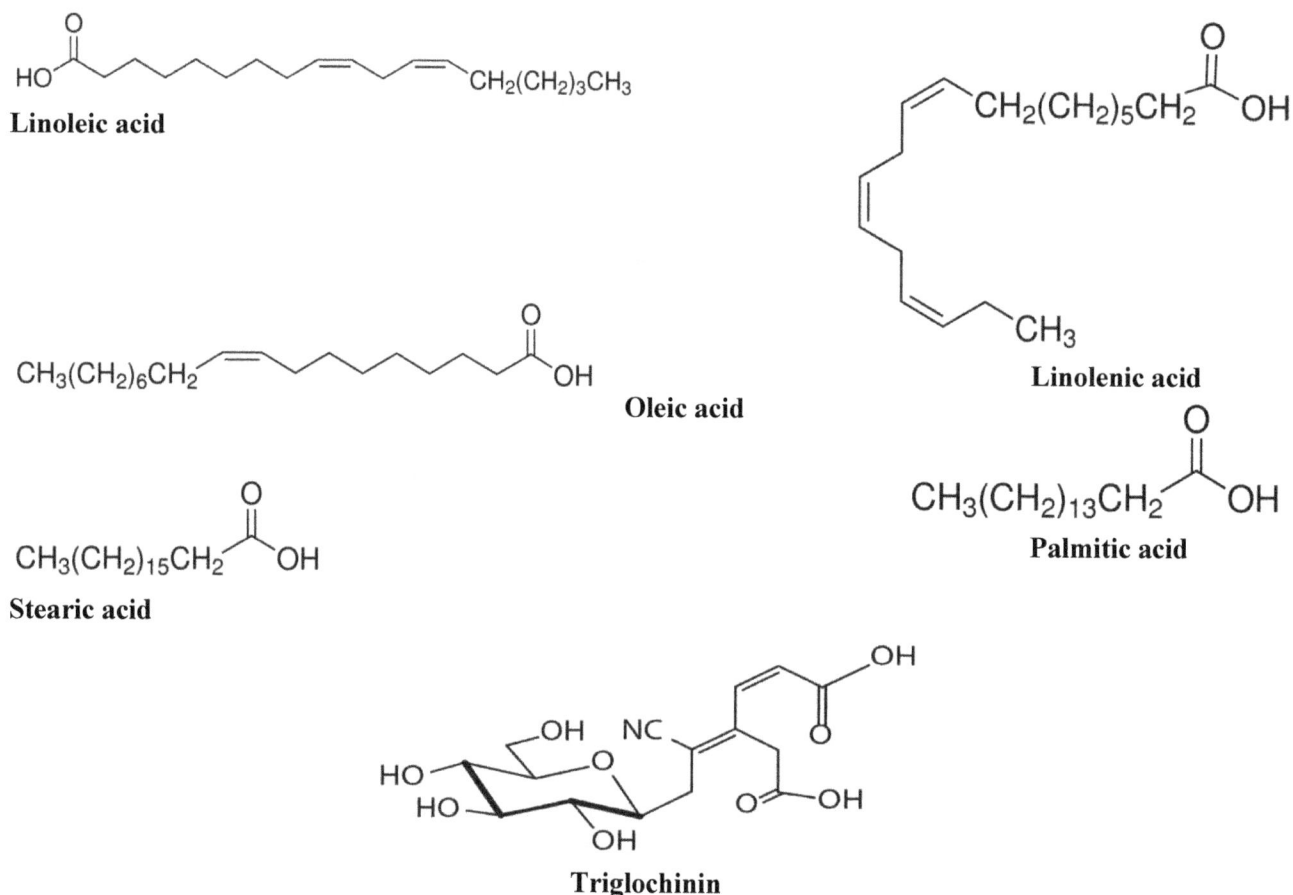

Figure 2: Chemical structures of fatty acids and triglochinin

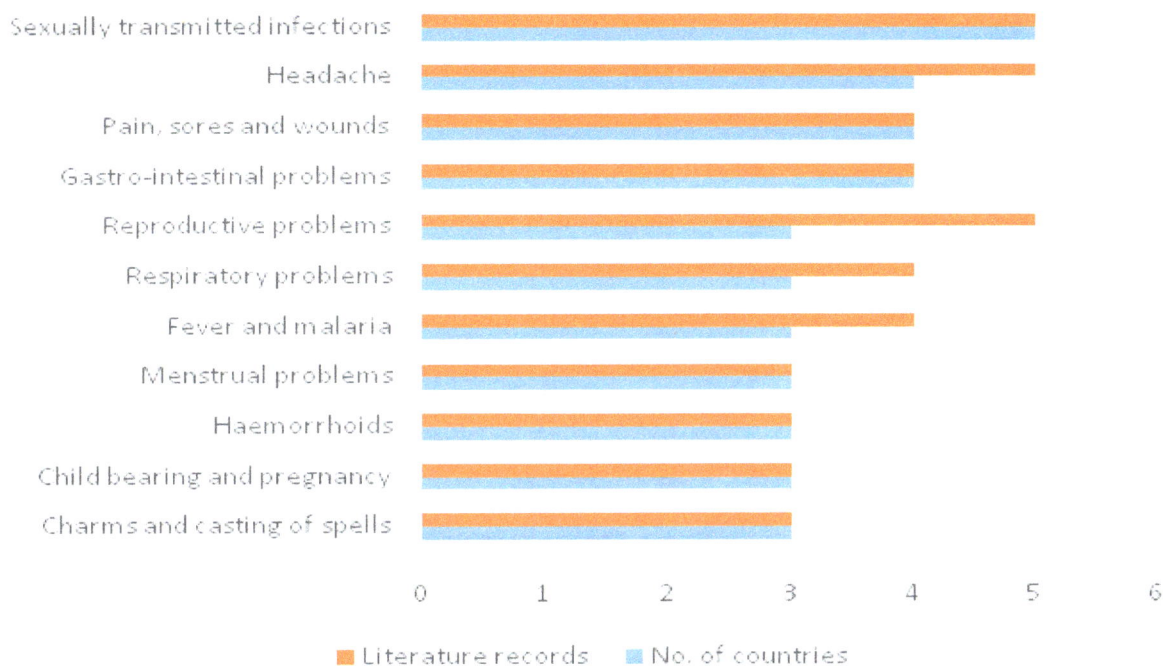

Figure 3: Major medicinal uses of *B. cathartica* in tropical Africa based on literature records

Conclusions

Bridelia cathartica is an important medicinal plant throughout its distributional range. Few studies carried out so far provided supporting evidence for most of the documented ethnomedicinal uses of the species focusing on child bearing and pregnancy, fever and malaria, gastro-intestinal, headache, haemorrhoids, menstrual, pain, sores and wounds, reproductive, respiratory disorders and sexually transmitted infections. Therefore, it seems premature to draw firm conclusions about the alleged therapeutic effects of *B. cathartica*. More detailed research is needed aimed at assessing several plant parts of the species used as traditional medicines, evaluating their chemical compounds, biological and toxicological properties. Detailed clinical trials are also required aimed at evaluating the efficacy of crude extracts of *B. cathartica* or compounds isolated from the species. Detailed pharmacological, molecular and cellular mechanisms of action are needed for *B. cathartica* when used alone or mixed with other plant species to confirm synergetic, additive, efficacy and safety of such combinations and usage. Therefore, evaluation of the chemical, biological and toxicological properties of *B. cathartica* are important as this baseline data is required for future research on the species.

Acknowledgment

The author would like to express his gratitude to the National Research Foundation (NRF), South Africa and Govan Mbeki Research and Development Centre (GMRDC), University of Fort Hare for financial support to conduct this study.

Conflict of interest

The author declares that there is no conflict of interest regarding the publication of this paper

References

Azimova SS and Glushenkova AI, 2012. *Bridelia cathartica* Bertol. In lipids, lipophilic components and essential oils from plant sources. pp. 359. Azimova SS and Glushenkova AI (eds.). Springer, London, UK.

Bandeira SO, Gaspar F and Pagula FP, 2001. African ethnobotany and healthcare: Emphasis on Mozambique. Pharmaceut. Biol. 39: 70-73.

Chhabra SC, Mahunnah RLA and Mshiu EN, 1990. Plants used in traditional medicine in eastern Tanzania. III. Angiosperms (Euphorbiaceae to

Menispermaceae). J. Ethnopharmacol. 28: 255-283.

Chhabra SC, Uiso FC and Mshiu EN, 1984. Phytochemical screening of Tanzanian medicinal plants. I. J. Ethnopharmacol. 11(2): 157 - 179.

Chinsembu KC, 2016. Ethnobotanical study of plants used in the management of HIV/AIDS-related diseases in Livingstone, southern province, Zambia. Evidence-Based Complem. Altern. Med. Volume 2016, Article ID 4238625.

Coates Palgrave M, 2002. Keith Coates Palgrave trees of southern Africa. Struik Publishers, Cape Town, South Africa.

Cumbane P and Munyemana F, 2017. Antioxidant and antibacterial activity of leaves and stem extracts of Bridelia cathartica Bertol. Sky J. Microbiol. Res. 5(2): 18-26.

Cunningham AB, 1993. African medicinal plants: setting priorities at the interface between conservation and primary healthcare. People and Plants Working Paper, UNESCO-United Nations Educational Scientific and Cultural Organisation, Paris, France.

De Wet H and Ngubane SC, 2014. Traditional herbal remedies used by women in a rural community in northern Maputaland (South Africa) for the treatment of gynaecology and obstetric complaints. S. Afr. J. Bot. 94: 129-139.

De Wet H, Nzama VN and Van Vuuren SF, 2012. Medicinal plants used for the treatment of sexually transmitted infections by lay people in northern Maputaland, KwaZulu-Natal Province, South Africa. S. Afr. J. Bot. 78: 12-20.

Gelfand M, Mavi S, Drummond RB and Ndemera B, 1985. The traditional medical practitioners in Zimbabwe: his principles of practice and pharmacopoeia. Mambo Press, Gweru, Zimbabwe.

Gerstner J, 1941. A preliminary checklist of Zulu names of plants with short notes. Bantu Stud. 15(3): 277-301.

Hedberg I, Hedberg O, Madatic PJ, Mshigeni KE, Mshiu EN and Samuelsson G, 1983. Inventory of plants used in traditional medicine in Tanzania. ii. Plants of the families Dilleniaceae - Opiliaceae. J. Ethnopharmacol. 9: 105-128.

Hoffmann P, Kathriarachchi H and Wurdack KJ, 2006. A phylogenetic classification of Phyllanthaceae (Malpighiales: Euphorbiaceae sensu lato). Kew Bull. 61: 37-53.

Hughes GD, Aboyade OM, Beauclair R, Mbamalu ON and Puoane TR, 2015. Characterizing herbal medicine use for noncommunicable diseases in urban South Africa. Evidence-Based Complem. Altern. Med. Volume 2015, Article ID 736074.

Jurg A, Tomás T and Pividal J, 1991. Antimalarial activity of some plant remedies in use in Marracuene, southern Mozambique. J. Ethnopharmacol. 33: 79-83.

Kokwaro JO, 1993. Medicinal plants of East Africa. Kenya Literature Bureau, Nairobi, Kenya.

Madureira AM, Ramalhete C, Mulhovo S, Duarte A and Ferreira M-JU, 2012. Antibacterial activity of some African medicinal plants used traditionally against infectious diseases. Pharmaceut. Biol. 50(4): 481-489.

Maroyi A, 2017. Ethnopharmacology and therapeutic value of Bridelia micrantha (Hochst.) Baill. in tropical Africa: a comprehensive review. Molecules 22: 1493.

Maroyi A, 2011. Ethnobotanical study of medicinal plants used by people in Nhema communal area, Zimbabwe. J. Ethnopharmacol. 136: 347-354.

Mbayo KM, Kalonda ME, Tshisand TP, Kisimba KE, Mulamba M, Richard MK, Sangwa KG, Mbayo KG, Maseho MF, Bakari S, Mpiana TP, Kahumba BJ and Lumbu SJ-B, 2016. Contribution to ethnobotanical knowledge of some Euphorbiaceae used in traditional medicine in Lubumbashi and its surroundings (DRC). J. Adv. Bot. Zool. 4(2): 1-16.

Moshi MJ, Cosam JC, Mbwambo ZH, Kapingu M and Nkunya MHH, 2004. Testing beyond ethnomedical claims: brine shrimp lethality of some Tanzanian plants. Pharmaceut. Biol. 42(7): 547-551.

Morris B, 1996. Chewa medical botany: a study of herbalism in southern Malawi. International African Institute, Lit Verlag, Hamburg, Germany.

Ngueyem TA, Brusotti G, Caccialanza G and Finzi PV, 2009. The genus Bridelia: a phytochemical and ethnopharmacological review. J. Ethnopharmacol. 12: 339-349.

Ouma OJ, Chhabra SC and Nyagah G, 1997. Determination of iron content in different parts of herbs used traditionally for anemia treatment in East Africa. J. Ethnopharmacol. 58: 97-102.

Ouma OJ, 1994. Determination of the mineral composition of some medicinal plants used to treat anaemia in traditional African societies. M.S. thesis, Kenyatta University, Nairobi.

Pakia M and Cooke JA, 2003. The ethnobotany of the

Midzichenda tribes of the coastal forest areas in Kenya: 2. Medicinal plant uses. S. Afr. J. Bot. 69(3): 382-395.

Palmer E and Pitman P, 1972. Trees of southern Africa covering all known indigenous species in the Republic of South Africa, South West Africa, Botswana, Lesotho and Swaziland. A.A. Balkema, Cape Town, South Africa.

Phillipson JD and Wright CW, 1991. Medicinal plants against protozoal diseases. Trans. R. Soc. Trop. Med. Hyg. 85(1): 155-165.

Ramalhete C, Lopes D, Mulhovo S, Rosario VE and Ferreira M-JU, 2008. Antimalarial activity of some plants traditionally used in Mozambique. In: Workshop Plantas Medicinais e Fitoterapêuticas nos Trópicos, Lisbon; 2008. Available from: 2.iict.pt/archive/doc/ C_Ramalhete_wrkshp_plts_medic.pdf.

Sawhney AN, Khan MR, Ndaalio G, Nkunya MHH and Wevers H, 1978. Studies on the rationale of African traditional medicine. Part III. Preliminary screening of medicinal plants for antifungal activity. Pak. J. Sci. Ind. Res. 21: 193-196.

Schmidt E, Lotter M and McCleland W, 2002. Trees and shrubs of Mpumalanga and Kruger National Park. Jacana Media, Johannesburg, South Africa.

Smith AR, 1987. Euphorbaceae. In Flora of tropical East Africa. pp. 1-407. Polhill RM (ed.). AA Balkema, Rotterdam, The Netherlands.

Van Valen F, 1978. Contribution to the knowledge of cyanogenesis in angiosperms. 10. Communication. Cyanogenesis in Euphorbiaceae.

Pl. Med. 34: 408-413.

Van Wyk B and Van Wyk P, 1997. Field guide to trees of southern Africa. Struik Publishers, Cape Town, South Africa.

Watt JM and Breyer-Brandwijk MG, 1962. The medicinal and poisonous plants of southern and eastern Africa. Edinburgh: E and S Livingstone Ltd, London, UK.

Williams VL, Balkwill K and Witkowski ETF, 2001. A lexicon of plants traded in the Witwatersrand umuthi shops, South Africa. Bothalia 31: 71-98.

World Health Organization (WHO), 2014. The health of the people: what works - the African regional health report 2014. World Health Organization, Geneva, Switzerland.

World Health Organization (WHO), 2013. Global action plan for the prevention and control of noncommunicable diseases 2013-2020. World Health Organization, Geneva, Switzerland.

York T, 2012. An ethnopharmacological study of plants used for treating respiratory infections in rural Maputaland. M.S. thesis, University of Zululand, KwaDlangezwa.

York T, van Vuuren SF and de Wet H, 2012. An antimicrobial evaluation of plants used for the treatment of respiratory infections in rural Maputaland, KwaZulu-Natal, South Africa. J. Ethnopharmacol. 144: 118-127.

York T, de Wet and van Vuuren SF, 2011. Plants used for treating respiratory infections in rural Maputaland, KwaZulu-Natal, South Africa. J. Ethnopharmacol. 135: 696-710.

13

Modeling the potassium requirements of potato crop for yield and quality optimization

Farheen Nazli[1]*, Bushra[2], Muhammad Mazhar Iqbal[3], Fatima Bibi[4], Zafar-ul-Hye[5], Muhammad Ramzan Kashif[1] and Maqshoof Ahmad[6]

[1]Pesticide Quality Control Laboratory Bahawalpur-63100, Pakistan
[2]Institute of Soil and Environmental Sciences, University of Agriculture Faisalabad, Pakistan
[3]Soil and Water Testing Laboratory, Chiniot, Pakistan
[4]Mango Research Station, Multan, Pakistan
[5]Department of Soil Science, College of Agriculture, Bahauddin Zakariya University, Multan, Pakistan
[6]Department of Soil and Environmental Sciences, the Islamia University of Bahawalpur, Bahawalpur-63100, Pakistan

Abstract

The intensity, quantity as well as capacity factors are important to predict the amount of nutrient in soil required for maximum plant growth. Sorption isotherm considers these three factors so believed to be one of the most important techniques in soil which control the fate and mobility of nutrients. The field experiment was conducted to find out site-specific and crop-specific potassium requirement for potato crop. The potassium adsorption isotherm was constructed and Freundlich model was used to theoretically work out different soil solution K levels (0, 5, 10, 15, 20, 25, and 30 mg L^{-1}). The K fertilizer doses were calculated against these specific soil solution levels. Field experiment was conducted with seven model based K fertilizer treatments (0, 49, 94, 139, 183, 228, and 273 kg K ha^{-1}) and three replications in Randomized Complete Block Design (RCBD) using potato as test crop. The results showed that growth parameters like plant height, leaf area and chlorophyll significantly contributed to potato tuber yield. Different yield response models were tested and it was observed that linear plus plateau and quadratic plus plateau predicted equally well the optimum fertilizer K rate both for yield and quality attributes of potato. For maximum potato tuber yield i.e.34.41 Mg ha^{-1} the economic optimum K was 100 kg ha^{-1}. Optimum fertilizer K rates (at 95 % relative yield) for potato tuber yield, dry matter percentage, protein, starch contents and vitamin C contents were 100, 103, 180, 230 and 200 kg K ha^{-1}, respectively. So, it is suggested that adsorption isotherm technique should be used to calculate site specific and crop specific fertilizer requirements of crops and 100 kg ha^{-1} is recommended as optimum potassium fertilizer for potato crop. Moreover, the K fertilizer application would improve crop quality that would support the quality based marketing system in Pakistan.

Corresponding author email:
farheenmaqshoof@gmail.com

Keywords: Potato, Quadratic plus plateau, Adsorption isotherm, Yield, Quality

Introduction

Soil testing methods are essentially used as tool for monitoring soil fertility. However, they fail frequently, if used to predict optimal fertilizer requirements for crops. In Pakistan, the fertilizer recommendations are usually generalized and are given as a range (FAO, 2016) which are predicted from simple experiments in the fields and are extrapolated on all soils which is one of main reasons that crop responses to potassium fertilizers are very irregular and erratic (Mengel et al., 1998). Maximum profitability and environmental sustainability is only possible through fertilizer recommendations for maximizing the yield and quality (Thompson et al., 2017). These recommendations are developed through extensive field trials by studying the effect of different rates of fertilizers on crop yield and quality. To solve this problem, attempts should be made to predict fertilizer requirements using nutrient adsorption models. These models are based on different soil nutrient concentrations such as intensity, quantity and capacity factors. These all are important to predict the amount of soil nutrient requirements to adjust model based soil solution level required for maximum plant growth (Louison et al., 2015). The nutrient adsorption isotherms can be used for the estimation of fertilizers needed to adjust the soil solution nutrients to the level optimum for maximum yield (Kenyanya et al., 2014). The use of these models can only be possible through identification of critical solution level to get maximum plant growth which is specific for that particular soil and for that particular crop (Samadi, 2003, 2006). The adsorption characteristics are linked with type and quality of minerals in soil as well as other soil chemical properties, and vary greatly from soil to soil, in particular, the amount and type of clay minerals, cation exchange capacity and organic matter contents. So, the knowledge of the adsorption characteristics of soils can be an accurate and precise source of estimation for fertilizer requirement of potato crop.

In Pakistan, the use of potassium fertilizer for potato crop is not common (Hannan et al., 2011). The farmers either use generalized fertilizer for potato crop of even they do not use K fertilizer. So, the crop meets the potassium requirements from native soil K supply in tube well irrigated soils, and potassium from irrigation water in canal commanded areas. As potato is heavy K feeder and in most of the soils under potato crop the K is deficient so both sources cannot meet the K requirements of crop. The ratio of different nutrients in soil is affected by low or even no use of fertilizers for specific nutrients like P and K that restricts the crop growth and keeps it far below the genetic potential of the crop. Moreover, the potassium has pivotal role in the induction of resistance against different biotic and abiotic stresses

including pests and disease resistance. The inadequate K supply to crop plants results in weakened plants which become more susceptible to various kinds of biotic and abiotic stresses such as disease stress, pests attack, heavy metal stress, frost, salinity, drought, nutritional stress, etc. These stresses affect the plant phenology and lower the ultimate yield and profitability.

Potato crop labeled as heavy feeder of K has high nutrient requirements and uptake of over 300 kg K ha^{-1} is common under optimum K supply (Perrenoud, 1993). As potassium deficiency has been reported by Directorate of Soil Fertility, Pakistan in most of the soils under potato crop in Pakistan so there exists a great opportunity to increase yield and quality of potato crop by improving nutrient management. Therefore, it appears rational that soil test results and fertilizer recommendations must be site-specific, model based (Wang et al., 2016) and calibrated scientifically. Scanty information is available on determination of optimum potassium fertilizer rates for vegetables in Pakistan so it is imperative to optimize the potassium fertilizer rate for maximizing potato yield by comparing and evaluating different yield response models in potato cropping sequence. In view of this background, field experiment was conducted to see the effect of potassium fertilizer rate using adsorption isotherm technique for potato crop by improving nutrient management to work out the site specific fertilizer recommendations for potato crop in Pakistan.

Material and Methods

A field study was conducted in potato growing tract of Pakistan (Farmer's field in Kassowal) to compare the effect of different potassium levels on the growth, yield and quality of potato by using sorption isotherm technique. The composite soil sample was collected and analyzed for the physical and chemical characteristics of the experimental field following the standard protocols (Ryan et al., 2001). The same was used to develop adsorption isotherm for the calculation of K fertilizer rates.

The potassium adsorption isotherm was constructed by using 2.50 g soil samples. These samples were equilibrated for 24 hours under shaking conditions at 25 ± 1 °C with different K levels viz. 0, 25, 50, 75, 100, 125, 150, 175, 200, 225 and 250 µg mL^{-1} in 25 mL CaCl$_2$ (0.01 M) solution. After achieving the steady state condition, the amount of K adsorbed was determined. Freundlich model was used to theoretically work out different soil solution K levels (0, 5, 10, 15, 20, 25, and 30 mg L^{-1}). The K fertilizer doses were calculated against these specific soil solution levels. Field experiment was conducted with seven model based K fertilizer treatments (0, 49, 94, 139, 183, 228, and 273 kg K ha^{-1}) and three replications in Randomized Complete Block Design (RCBD) using potato as test crop.

Field studies

The field experiment was conducted in Kassowal for the evaluation of different K fertilizer rates on the growth, yield and quality of potato tuber to optimize the fertilizer doze under field conditions. For this purpose, three replications were made and adsorption model based K fertilizer rates (0, 49, 94, 139, 183, 228, and 273 kg K ha^{-1}) were applied to potato crop as sulphate of potash using randomized complete design (RCBD). Potato seed tubers (cultivar Cardinal) were treated with fungicide Topsin-M and planted on ridges 75 cm apart at 20 cm spacing between plants. Plot size was 5m x 3m and there were four ridges per plot. All plots received basic application of 250 kg P$_2$O$_5$ (DAP) and 300 kg N ha^{-1}(Urea). Nitrogen was applied in three splits i.e. 1/3 at the time of sowing, second and third dose after 45 and 60 days of planting, respectively. An overall check (K=0) was also kept. The crop was irrigated with good quality irrigation water (canal water) with first irrigation after three days of planting. The subsequent irrigations were applied as and when required by the crop without any stress. The other plant protection measures and agronomic practices were carried out according to crop needs. Growth parameters like plant height, leaf area and chlorophyll contents were determined after 60 days of plant emergence. The crop was harvested at maturity on the development of potato tubers and data were recorded. Potato tubers were then graded manually into large (>75 g), medium (75-25 g) and small (< 25 g) sizes. Marketable tuber yield was determined and only marketable tuber data were reported in this manuscript. Internal and external K requirements of potato were worked out.

Physiological parameters

The physiological parameters of potato tubers were determined. Leaf area meter MK2 (Delta-T Devices Ltd, Cambridge, UK) was used for the measurement of leaf area. For the determination of chlorophyll contents, the chlorophyll meter (SPAD-502, Minolta Camera Co., Ltd, Japan) was used. The dry matter contents were determined by drying a known weight (W$_1$) of the sample in an oven at 105 °C to a constant weight (W$_2$), (AOAC, 1995).

Plant quality parameters

Total protein concentration was determined by following the method of Chapman and Parker (1961). For this purpose, one gram of well prepared (dried and ground sample) plant sample was digested in Kjeldahl flask by following the standard protocol. The digested material was distillated on micro Kjeldahl distillation apparatus and titrated against 0.1 N sulphuric acids. Data were used to calculate total protein concentration. The starch contents were determined by using the recommended method of Blankensh et al. (1993). Reducing sugars in the extract were estimated as described by Hortwitz (1960) while the vitamin C contents were estimated by following the method of Ruck (1961).

Statistical analysis

Data were subjected to statistical analyses using simple and multiple regression equations. Quadratic, square root, linear plus plateau, exponential and Quadratic plus plateau yield response models were tested using Graph pad ver.4.1. The treatment means were compared through analysis of variance techniques at 5% level of probability (Steel el al., 1997).

Results

Field studies were conducted to see the effect of potassium fertilizer rate using adsorption isotherm technique for potato crop. The basic analysis of the soil indicated that it was normal (EC$_e$ = 0.75 dS m^{-1}), alkaline in reaction (pH$_s$ = 7.81), low in organic matter (0.72 %), deficient in available nutrients (N = 0.04 %, P = 5.80 ppm and K = 71.00 ppm), calcareous in nature (CaCO$_3$ = 8.71 %) and loam in texture (Table 1).

Freundlich adsorption isotherm of the selected soil

Results showed that Freundlich isotherm of equilibrium K concentration against adsorbed K gave a highly significant linear relationship (Fig. 1). The potassium sorption data were fitted in the Freundlich equation that gave good results with highest value of coefficient of determination (0.96**). The potassium adsorption isotherm was constructed and Freundlich model was used to theoretically work out different soil solution K levels (0, 5, 10, 15, 20, 25, and 30 mg L^{-1}). The K fertilizer doses were calculated against these specific soil solution levels. For the adjustment of same soil solution K levels in the field experiment, the equivalent K fertilizer rates were calculated which varied from 0 to 237 kg ha^{-1} (Table 2). All potassium fertilizer rates were applied as basal dose of sulphate of potash during potato planting / at the time of sowing.

Effect of K fertilizer on growth, physiological and yield parameters

The results of the present study showed that plant height increased significantly with the application of K fertilizer (Table 3). It was improved from 40.67 cm with no K (T$_0$) to 56.33 cm with 139 kg K ha^{-1} (T$_3$) against soil solution level of 15 mg K L^{-1} which is 22.47 % increase as compared to control. Further increase in plant height with increasing K rate in T$_4$, T$_5$ and T$_6$ was statistically non-significant. The results of our study showed that potassium fertilization improved leaf area of potato crop up to 3438 cm^2 plant^{-1} with K application @ 228 kg ha^{-1} (T$_5$) as indicated in Table 3 and it was 39.44 % higher when compared with control. Adequate supply of K improved chlorophyll contents of potato plant significantly. It increased from 35.00 % with native K to a maximum level of 43.67 % with K supply @ 273 kg ha^{-1} (T$_6$) equivalent soil solution level for this treatment was 30 mg K L^{-1} (Table 3).

Data (Table 4) showed that tuber yield was increased with increasing rate of potassium fertilizer and the maximum potato tuber yield (34.05 Mg ha^{-1}) was observed in the treatment T$_3$ where 139 kg K ha^{-1} was applied (Table 4). Further increase in K application rate (T$_5$ to T$_7$) could not bring about any significant change in potato tuber yield. The response of the crop to K application was due to low available status of K in soil before planting (Table 1).

The results showed that tuber dry matter contents increased with increase in potassium fertilizer rates up to a certain level and then decreased. The maximum dry matter contents (19.80 %) were observed in the treatment with K fertilizer rate of 183 kg ha^{-1} followed by18.14 % in the treatment with 273 kg K ha^{-1} (Table 4). In the present study, the increasing potassium levels increased the starch contents in tubers up to 81.20 % in the treatment T$_5$ with 228 kg ha^{-1} of potassium (Table 5). This

increase was of 9.21 % when compared with control plots. A decrease in starch contents was observed with further increase in potassium rate.

Effect of K fertilizer on quality parameters

The results showed that protein contents of potato tubers increased with increase in potassium fertilizer rates (Table 5). This increase ranged from 11.67 % in T$_0$ to the maximum values of 13.28% in the treatment when K fertilizer was applied at 139 kg ha^{-1} i.e. 15 mg K L^{-1} of soil solution. In the present study, the reducing sugars concentration was maximum (42 mg per 100g fresh weight) in control treatment that decreased with increase in K application rate. The minimum concentration (26 mg per 100g fresh weight) was observed in the treatment where 273 kg K ha^{-1} was applied (Table 5).

The data showed that vitamin C contents in potato tubers increased with increasing level of K fertilizer and maximum (18.64 mg per 100g fresh weight) vitamin C contents were observed in the treatment with 228 kg K ha^{-1} (Table 5). This increase was statistically significant when compared with control treatment.

Evaluation of yield response models to predict economic optimum K rate for maximum tuber yield

The optimum potassium fertilizer rates for potato tuber yield predicted by the models tested i.e. square root, linear plus plateau, quadratic plus plateau, exponential and quadratic model were 67, 100, 100, 32 and 179 kg K ha^{-1}, respectively (Table 6). The five tested models showed minor difference in R^2 values however, they showed large variation in calculated optimum K fertilizer rates with similar R^2 values. This is illustrated by present study which exhibited variations in K rates between 32 and 179 kg K ha^{-1} (Table 6). The linear plus plateau models fitted the data with less bias on the basis of R^2 and S.E than the other models (Table 6).

Evaluation of yield response models to predict economic optimum K rate for quality parameters

The optimum potassium fertilizer rates for dry matter predicted by the models tested i.e. square root, linear plus plateau, quadratic plus plateau, exponential and quadratic model were 74, 103, 103, 428 and 144 kg K ha^{-1}, respectively (Table 7). In the present study, the optimum K fertilizer rate for potato dry matter predicted by quadratic model was lower than exponential model. It has been observed that the optimum K fertilizer rate predicted by exponential model was very high i.e. 428 kg K ha^{-1} than the applied K rates.

The five tested models showed little difference in R^2 values however, they showed large variation in calculated optimum K fertilizer rates with similar R^2 values. The optimum K rate calculated by both quadratic plus plateau and linear plus plateau was (103 kg K ha^{-1}). Optimum K rate for protein predicted by square root, quadratic plus plateau, linear plus plateau, exponential and quadratic models were 74, 140, 180, 179 and 118 kg K ha^{-1}, respectively (Table 8). Optimum K rates for starch predicted by square root, quadratic plus plateau, linear plus plateau, exponential and quadratic models were 77, 222, 203, 561 and 150 kg K ha^{-1}, respectively (Table 9). Optimum K rates for vitamin C predicted by square root, quadratic plus plateau,

linear plus plateau, exponential and quadratic models were 74, 103, 200, 430 and 84 kg K ha^{-1}, respectively (Table 10).

The optimum K rate for protein, starch and vitamin C calculated by the quadratic model was very less (0.283, 0.00142 and 0.00142 kg K ha^{-1} respectively) than the applied K rates. The standard error of the estimate also varied greatly among models. The linear plus plateau models fitted the data with less bias on the basis of R^2 and S.E than the other models.

External K requirements of potato crop

Optimum K requirement for maximum potato tuber yield was 100 kg K ha^{-1} (Table 11). Regarding potato quality parameters K fertilizer was 180, 230 and 200 kg ha^{-1} for protein, starch and vitamin C content respectively.

Table 1: Physical and chemical properties of the soil used for experiment

Determinant	Unit	Value
EC$_e$	dS m^{-1}	0.75
pH$_s$	-	7.81
Organic matter	%	0.72
Total N	%	0.04
Available P	mg kg^{-1}	5.80
Available K	mg kg^{-1}	71.00
Cation exchange capacity	cmol^c kg^{-1}	9.26
CaCO$_3$	%	8.71
Sand	%	43.00
Silt	%	35.00
Clay	%	22.00
Textural Class	-	Loam (Typic Ustochrept)

Table 2: Freundlich model based K rate applied to potato crop in the field experiment

Treatment	T$_0$	T$_1$	T$_2$	T$_3$	T$_4$	T$_5$	T$_6$
Adjusted soil solution K levels (mg L^{-1})	0	5	10	15	20	25	30
K rate (kg ha^{-1})	0	49	94	139	183	228	273

Table 3:Effect of potassium on plant height, leaf area and chlorophyll contents of potato crop under field conditions (Average of three replicates)

Treatment	Adjusted soil solution K levels (mg L⁻¹)	K rate (kg ha⁻¹)	Plant height (cm)	Leaf area (cm⁻² plant⁻¹)	Chlorophyll (%)
T$_0$	0	0	40.67c	2081d	35.00d
T$_1$	5	49	44.00c	2358c	36.00cd
T2	10	94	48.67b	2514c	39.33bc
T3	15	139	56.33a	2924b	39.00bc
T4	20	183	56.00a	3245a	41.00ab
T5	25	228	55.33a	3438a	42.67ab
T6	30	273	55.67a	3438a	43.67a

The means sharing same letters are not statistically different at 5% level of probability

Table 4: Effect of potassium on potato tuber yield, dry matter percentage and starch contents of potato crop under field conditions (Average of three replicates)

Treatment	Adjusted soil solution K levels (mg L⁻¹)	K rate (kg ha⁻¹)	Potato yield (T ha⁻¹)	Dry matter (%)	Starch content (%)
T$_0$	0	0	20.04c	12.62f	73.56c
T$_1$	5	49	25.97b	16.23e	74.44c
T2	10	94	31.30a	17.75d	75.80bc
T3	15	139	34.05a	18.80b	75.76bc
T4	20	183	30.84ab	19.80a	78.20ab
T5	25	228	33.20a	18.39c	81.20a
T6	30	273	30.09ab	18.14c	79.56a

The means sharing same letters are not statistically different at 5% level of probability

Table 5: Effect of potassium on protein, reducing sugar, vitamin C contents of potato crop under field conditions (Average of three replicates)

Treatment	Adjusted soil solution K levels (mg L⁻¹)	K rate (kg ha⁻¹)	Protein (%)	Reducing sugar mg (100g FW)⁻¹	Vitamin C mg (100g FW)⁻¹
T$_0$	0	0	11.67c	42.00a	14.98c
T$_1$	5	49	11.80bc	40.33ab	15.40c
T2	10	94	12.11b	41.33ab	15.94bc
T3	15	139	13.28a	37.33ab	17.17bc
T4	20	183	13.16a	36.00bc	18.49ab
T5	25	228	13.27a	31.00cd	18.64a
T6	30	273	13.27a	26.00d	18.35a

The means sharing same letters are not statistically different at 5% level of probability

Table 6: Optimum rates of K fertilization predicted for potato yield by each model along with their coefficients of determination (R^2) and standard error of estimate values

Model	Optimum K rate (kg ha^{-1})	Coefficien t of determination (R_2)	Coefficients of equations		
			A	b	C
Square root	67	0.86**	2.16(19.46) *	0.5375(1.775)	0.03108(0.06043)
Quadratic plus plateau	100	0.98**	11.95(0.821)	0.091(0.0389)	0.00062(0.00037) Plateau value=32.70 (0.394)
Linear plus plateau	100	0.99**	10.83(0.776)	0.1639(0.01012)	Plateau value=32.71 (0.545)
Exponential / Mitscherlich	32	0.83**	-	25.49(9.468)	0.003737(0.00197)
Quadratic	179	0.92**	1.395(20.19)	0.02314(0.1470)	8.288(0.0004086)

*Values in parenthesis are standard error of estimate ** = (p = 0.05)

Table 7: Optimum rates of K fertilization predicted for dry matter by each model along with their coefficients of determination (R^2) and standard error of estimate values

Model	Optimum K rate (kg ha^{-1})	Coefficient of determination (R_2)	Coefficients of equations		
			A	b	c
Square root	74	0.92**	0.7592(12.42)	0.1871(0.844)	0.0585(0.02883)
Quadratic plus plateau	103	0.98**	0.7309(12.02)	0.03946(0.06944)	0.00041(0.000424) Plateau value=32.70 (0.394)
Linear plus plateau	103	0.99**	0.01112(0.05486)	0.6805(12.92)	0.3696(18.78)
Exponential / Mitscherlich	428**	0.88**	-	23.40(81.44)	0.002725(0.01270)
Quadratic	144	0.92**	1.4113(12.83)	0.006962(0.07125)	2.44e-005(0.0001936)

*Values in parenthesis are standard error of estimate ** = (p = 0.05)
**optimum rate is very high than applied rates

Table 8: Optimum rates of K fertilization predicted for protein content by each model along with their coefficients of determination (R^2) and standard error of estimate values

Model	Optimum K rate (kg ha^{-1})	Coefficient of determination (R_2)	Coefficients of equations		
			A	b	c
Square root	74	0.82**	0.3849 (11.46)	0.09484 (0.04050)	0.05484 (0.004711)
Quadratic plus plateau	140	0.98**	0.05686 (11.67)	0.003070 (0.0004450)	3.159e-005(4.50e-005)
Linear plus plateau	180	0.99**	0.06247(0.0114)	0.2401 (11.44)	Plateau value=13.23 (0.1639)
Exponential / Mitscherlich	179	0.80**	-	0.003352 (0.01008)	77.04 (196.8)
Quadratic	118	0.81**	0.2969 (11.45)	0.005027(0.01315)	1.76e-005(2026e-005)

*Values in parenthesis are standard error of estimate ** = (p= 0.05)

Table 9: Optimum rates of K fertilization predicted starch content by each model along with their coefficients of determination (R²) and standard error of estimate values

Model	Optimum K rate (kg ha⁻¹)	Coefficient of determination (R₂)	Coefficients of equations		
			A	b	c
Square root	77	0.90**	1.053 (73.59)	0.2595(0.08634)	0.015(0.03190)
Quadratic plus plateau	222	0.97**	0.6680 (73.68)	0.1707(0.1220)	8.942e-005(6.979e-005)
Linear plus plateau	230	0.99**	0.00437(0.03144)	0.6087(72.98)	0.8304(79.69)
Exponential / Mitscherlich	561**	0.83**	-	0.001444 (0.005986)	106.9 (38.14)
Quadratic	150	0.89**	0.9575 (73.30)	0.001612(0.0276)	5.69e-005(2.51e-006)

*Values in parenthesis are standard error of estimate ** = (P= 0.05)
**optimum rate is very high than applied rates

Table 10: Optimum rates of K fertilization predicted for vitamin C by each model along with their coefficients of determination (R²) and standard error of estimate values

Model	Optimum K rate (kg ha⁻¹)	Coefficient of determination (R₂)	Coefficients of equations		
			A	b	c
Square root	74	0.89**	0.5945(14.86)	0.1465(0.01918)	0.08469(0.01405)
Quadratic plus plateau	180	0.96**	0.1360(15.01)	0.004653(0.001454)	3.215e-005(18.46) Plateau value=32.70 (0.394)
Linear plus plateau	200	0.99**	0.002625 (0.01919)	0.2971(14.61)	0.2677(18.50)
Exponential / Mitscherlich	430**	0.65**	-	0.008223(0.002060)	183.6(58.24)
Quadratic	84	0.54**	0.4814(14.62)	0.00815(0.022631)	2.860e-005(2.749e-005)

*Values in parenthesis are standard error of estimate ** = (P= 0.05)
**optimum rate is very high than applied rates

Table 11: Economic optimum K rate predicted by Linear plus plateau model

Variable	Optimum K rate (kg ha⁻¹)
Potato tuber yield (Mg ha⁻¹)	100
Dry matter (%)	103
Protein (%)	180
Starch (%)	230
Vitamin C mg (100g FW)⁻¹	200

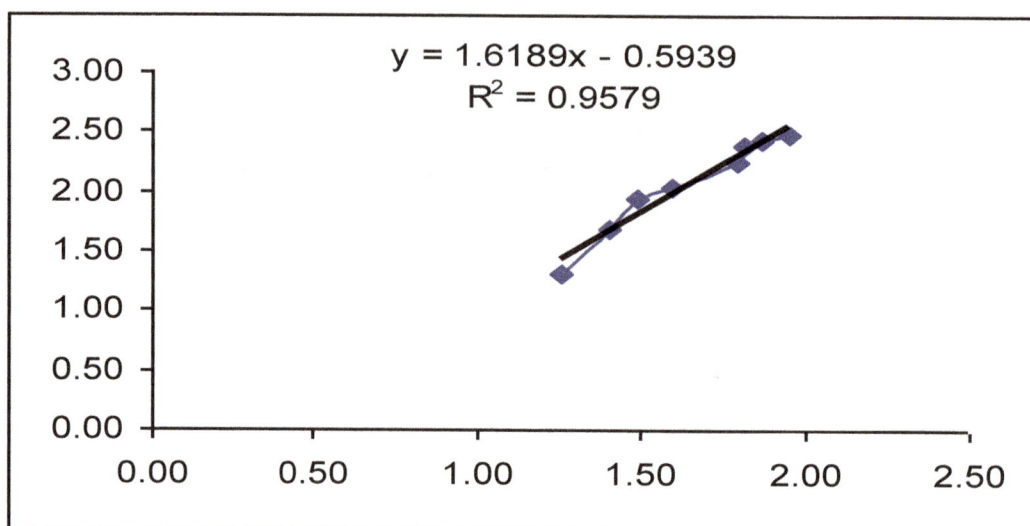

$$y = 1.6189x - 0.5939$$
$$R^2 = 0.9579$$

Figure 1. Freundlich adsorption isotherm for the selected soil

Discussion

Field studies were conducted to see the effect of model based potassium fertilizer rate using adsorption isotherm technique for potato crop. The fate of nutrients added in the soil depends upon the initial nutrient level in the soil while the soil solution concentration depends upon the rate of nutrient removal by the plants. It also depends desorption rate of nutrient from solid phase. The adsorption based K equilibrium solution level serves as an index of K availability. So, it has been reported that equilibrium K concentration using adsorption isotherm technique provides a better index of fertility of the soil (Singh and Jones, 1975). In the present study, Freundlich isotherm of equilibrium K concentration against adsorbed K gave a highly significant linear relationship (Fig. 1). It might be due assumption of unlimited sorption sites for heterogeneous medium in Freundlich model which in turn gave better correlation in soil with illite as the dominant clay mineral. The K fertilizer doses were calculated against these specific soil solution levels to see their effect on growth, yield and quality of potato crop. All potassium fertilizer rates were applied as basal dose during potato planting.

The results of the present study showed that plant height increased significantly with the application of K fertilizer. Although plant height is a genetic factor but it can be used as an indicator of crop performance and can be improved through balanced nutrition.

Different growth, physiological and yield parameters contribute to the yield of the cop. For potato crop, the plant height, leaf area and chlorophyll contents are considered as important determinations used to describe crop performance (Taiz and Zeiger, 2006). These indices indicate the carbon assimilation rate of the potato plant system and its ultimate conversion in to sink (tubers).

The results of our study showed that potassium fertilization improved leaf area of potato crop. Leaf area is used an indicator for photosynthetic efficiency of the plants as it captures light thus an increase in leaf area results an improvement in photosynthetic rate. It has been observed that leaf area has critical role in studies of plant competition, plant nutrition, plant protection measures, plant soil-water relations and crop ecosystems (Mohsenin, 1986). Further increase in leaf area with increasing K rate was statistically non-significant. Similarly, Al-Moshileh et al. (2005) reported significant the improvement in growth parameters such as plant height, and leaf area of potato crop with increasing level of potassium fertilizer.

Chlorophyll is the real plant factory for manufacturing food. Adequate supply of K improved chlorophyll contents of potato plant significantly. The improvement in plant height, leaf area and chlorophyll contents (%) might be due to increase in photosynthetic rate which in turn related to improved stomatal conductance and higher ribulase bisphosphate carboxylase activity resulting in rapid rate of CO_2 fixation (Cakmak and Engels, 1999).

Results of our study showed that tuber yield was increased with increasing rate of potassium fertilizer. The response of the crop to K application might be due to low available status of K in soil before planting (Table 1). The farmers, in general depend on the native sources of K in irrigation water and soil minerals. The K requirement of potato crop is more that is required for carbohydrate metabolism and other physiological functions (Singh and Trehan, 1998; Incrocci et al., 2017) along with conversion of N and P in plants (Mengel and Kirkby, 1987). Potassium is known to facilitate the efficient translocation of photosynthates to the developing tubers (Beringer, 1978; Kavvadias et al., 2012). This fact is evident from the present study, that there was a progressive increase in tuber yield with each incremental level of added K up to 155 kg ha^{-1}. These results are in line with those of Hannan et al. (2011) who reported an increase in potato tuber yield with increasing K rate using isotherm technique. These might be due to translocation of more photosynthates to the tubers. In fact, potassium has critical role in the translocation photosynthates to storage organs (Romheld and Kirkby, 2010).

The results showed that tuber dry matter contents increased with increase in potassium fertilizer rates up to a certain level and then decreased. The decrease in dry matter contents with higher levels of potassium might be due to increase in water contents to maintain cell turgor pressure (Hannan et al., 2011). The results are in line with the previous studies conducted by Kumar et al. (2004) who have reported a decrease in dry matter contents with increasing K fertilizer rates. The potassium affects the water contents of the plasma thus increases the water contents of tubers. The reduction in dry matter contents with increase in potassium rate might also be due to translocation of more photosynthates and water to the tubers which in turn decreased the dry matter contents (Kavvadias et al., 2012).

Potassium stimulates the activity of different important enzymes in the plants such as starch synthatase that catalyzes simple sugars into complex polysaccharide i.e. starch (Mengel and Kirkby, 1987). In the present study, the increasing potassium levels increased the starch contents in tubers. A decrease in starch contents was observed in the present study that might be due to increase in water uptake through potassium application that restricted the number of amyloplasts in cytoplasm (Perrenoud,

1993). The same results have also been reported in previous work conducted by Hannan et al. (2011).

The results showed that protein contents of potato tubers increased with increase in potassium fertilizer rates in our study. These results are supported with the work of Khan et al. (2012 who reported that protein contents were improved with the application of potassium and these increased with increasing rates. This improvement in protein contents might be due to more conversion of photosynthates to protein and other storage components such as starch.

The concentration of reducing sugars in tubers is an important quality indicator for potato processing industries. In the present study, the reducing sugars concentration decreased with increase in K application rate. It appeared that still higher K levels need to be tested for finding out optimum K fertilizer rate for minimum reducing sugars in the potato tubers. Similarly, Hannan et al. (2011) reported a decrease in reducing sugars contents with increasing K fertilizer levels.

The data showed that vitamin C contents in potato tubers increased with increasing level of K fertilizer. This increase was statistically significant when compared with control treatment. It has been reported that potassium triggers the photosynthates production their translocation to tubers thus improves the conversion of photosynthates in to secondary metabolites such as vitamins (Mengel and Kirkby, 1987: Khan et al., 2012). Potassium has been reported to be engaged in the conversion of radiant energy into chemical energy. This metabolic energy is required for plant metabolism that results in increased production of proteins and starch, and decreases the reducing sugars contents thus improving the quality of potato tubers (Incrocci et al., 2017). The higher energy status resulting with optimum K supply promotes the synthesis of Vitamin C and other secondary metabolites.

In the present study, the optimum K fertilizer rate for tuber yield predicted by quadratic model was higher than exponential model while the optimum K fertilizer rate for potato dry matter predicted by quadratic model was lower than exponential model. These results are supported by the work of Hannan et al. (2011) where they reported the large variation in optimum K rate due to inappropriate model selection. They reported that Linear plus plateau was the most suitable model and best option for optimum K fertilizer recommendations as observed in these studies. The optimum K fertilizer rates predicted by different models varied greatly (Neeteson and Wadman, 1987; Incrocci et al., 2017)).

It has been observed that coefficient of determination is a poor criterion to select a model for optimization of K fertilizer rate. The same results have been reported in previous studies (Colwell, 1994; Hannan et al., 2011). The standard error of the estimate also varied greatly among models. So, it can be concluded from this study that the linear plus plateau model is best suited to describe the yield response of potato to K fertilizer and to predict the economic optimum K rates.

Conclusion

The results showed that growth parameters like plant height, leaf area and chlorophyll contents contribute significantly to potato tuber yield. The quality parameters like protein, starch and vitamin C contents were significantly affected by increasing K rates. Out of the tested models, linear plus plateau and Quadratic plus plateau predicted (equally well) the optimum fertilizer K rate for both for yield and quality attributes of potato crop. It was observed that the optimum fertilizer K rates (at 95 % relative yield) for potato tuber yield, protein, starch and vitamin C contents were 100, 180, 230 and 200 kg K ha^{-1}, respectively. So, it is suggested that adsorption isotherm technique should be used to calculate site specific and crop specific fertilizer requirements of crops. It can overcome the problem of sporadic responses to fertilizer as well. The K fertilizer application would increase the quality of crop that would support the quality based marketing system in Pakistan leading to improved profitability for the farmers.

Acknowledgement

The authors acknowledge the support of technical staff during the laboratory analysis. The first author also acknowledges the guidance from Dr. Abdul Hannan during the conduct and write-up of the work.

References

Al-Moshileh A, Errebhi MMA and Motawei MI, 2005. Effect of various potassium and nitrogen rates and splitting methods on potato under sandy soil and arid environmental conditions. Emir. J. Agric. Sci. 17:01-09.

AOAC, 2005. Official Methods of Analysis. 18th edn. Association of Official Analytical Chemists, Arlington, VA, USA.

Beringer H, 1978. Functions of potassium in plant metabolism with particular reference to yield, pp. 185-202. in Sekhon G.S (ed.), Potassium in soils and crops. Gurgaon, Haryana, India: Potash Research Institute of India.

Blankensh SM, Ellsuworth DD and Powell RL, 1993. A ripening index for banana fruit based on starch content. J. Hort. Technol. 3:338-339.

Cakmak I and Engels C, 1999. Role of mineral nutrition in photosynthesis and yield formation, pp. 141-168.in Rengel Z (ed.), Mineral nutrition of crops: mechanisms and implications. The Haworth Ptress, New York, USA.

Chapman HD and Parker F, 1961. Determination of NPK, pp. 150-179. In methods of analysis for soil, plant and water. Division of Agriculture, University of California, USA.

Colwell JD, 1994. Estimating fertilizer requirements: A quantitative approach. CAB International, Wallingford, UK.

FAO, 2016. Fertilizer use by crop in Pakistan. Natural Resources Management and Environment Department, Food and Agriculture Organization of the United Nations, USA. Available at: http://www.fao.org/docrep/007/y5460e/y5460e0. htm. Accessed on 22-08-2016.

Hannan A, Arif M, Ranjha AM, Abid A, Fan XH and Li YC, 2011. Response models to determine potassium fertilizer rates for potato crop on a calcareous soil in Pakistan. Comm. Soil Sci. Plant Anal. 42(6): 645-655.

Hortwitz W, 1960. Official methods of analysis. Association of Official Agric. Chemists, Washington, DC.9:314-320.

Incrocci L, Massa D and Pardossi A, 2017. New trends in the fertigation management of irrigated vegetable crops. Horticulturae 3, 37; doi:10.3390/horticulturae3020037.

Kavvadias V, Paschalidis C, Akrivos G and Petropoulos D, 2012. Nitrogen and potassium fertilization responses of potato (*Solanum tuberosum*) cv. Spunta. Comm. Soil Sci. Plant Anal. 43(1-2): 176-189.

Kenyanya O, Mbuvi HM, Muthengia JM and Omayo EK, 2014. Use of adsorption isotherm models to determine potassium fertilizer acreage doses for optimum maize growth and yields in Nyamira County, Kenya. Intl. J. Agri. Crop Sci. 8 (15): 1525-1536.

Khan MZ, Akhtar ME, Mahmood-ul-Hassan M, Mahmood MM and Safdar MN, 2012. Potato tuber yield and quality as affected by rates and sources of potassium fertilizer. J. Plant Nutr. 35(5): 664-677.

Kumar D, Singh BP and Kumar P, 2004. An overview of the factors affecting sugar content of potatoes. Ann. Appl. Biol. 145: 247-256.

Louison L, Omrane A, Ozier-Lafontaine H and Picart D, 2015. Modeling plant nutrient uptake: Mathematical analysis and optimal control. Lect. Notes Pure Appl. 4: 193–203.

Mengel K and Kirkby EA, 1987. Principles of plant nutrition, 4th Ed. International Potash Institute, Worblaufen-Bern, Switzerland.

Mengel K, Rahmatullah and Dou H, 1998. Release of potassium from the silt and sand fraction of loss-derived soils. Soil Sci. 163: 805-813.

Mohsenin N, 1986. Physical properties of plant and animal materials, Gordon and Breach Science Publishers, New York, USA.

Neeteson JJ and Wadman WP, 1987. Assessment of economically optimum application rates of fertilizer N on the basis of response curves. Fert. Res. 12: 37-52.

Perrenoud S, 1993. Fertilizing for higher yield potato, IPI Bull. 8, 2nd Ed. International Potash Institute, Worblaufen-Bern, Switzerland.

Romheld V and Kirkby EA, 2010. Research on potassium in agriculture: Needs and prospects. Plant Soil 335: 155-180.

Ruck JA, 1961. Chemical methods for analysis of fruits and vegetable products. Res. St. Summer Land. Res. Branch Canada. Deptt. of Agric. 1154.

Ryan J, Estefan G and Rashid A, 2001. Soil and plant analysis laboratory manual, 2nd edition, International Center for Agriculture in Dry Areas (ICARDA), Syria, 172p.

Samadi A, 2003. Predicting phosphate requirement using sorption isotherms in selected calcareous soils of western Azarbaijan province, Iran. Comm. Soil Sci. Plant Anal. 34(19-20): 2885-2899.

Samadi A, 2006. Potassium exchange isotherms as a plant availability index in selected calcareous soils of Western Azarbaijan Province, Iran.Turk. J. Agric. 30: 213-222.

Singh BB and Jones JP, 1975. Use of sorption-isotherms for evaluating potassium requirement of plants. Soil Sci. Soc. Am. J. 39:881-886.

Singh JP and Trehan SP, 1998. Balanced fertilization to increase the yield of potato. in Barar MS and Bansal SK (eds.), Proceedings of the IPI-PRII-PAU Workshop on Balanced Fertilization in Punjab Agriculture, International Potash Institute, Worblaufen-Bern, Switzerland.

Steel RGD, Torrie JH and Dicky DA, 1997. Principles and procedures of statistics a biometrical approach. (3rd Ed.). pp. 204-227. McGraw Hill Book International Co., Singapore.

Taiz L and Zeiger E, 2006. Plant defenses: Surface protectants and secondary metabolites, pp. 283-308. In Taiz L and Zeiger E (ed.), Plant Physiology, Sinauer Associates, Sunderland, MA.

Thompson RB, Incrocci L, Voogt W Pardossi A and Magán JJ, 2017. Sustainable irrigation and nitrogen management of fertigated vegetable crops. Acta Hortic. 1150: 363–378.

Wang C, Boithias L, Ning Z, Han Y, Sauvage S, Sánchez-Pérez JM, Kuramochi K and Hatano R, 2016. Comparison of Langmuir and Freundlich adsorption equations within the SWAT-K model for assessing potassium environmental losses at basin. Agric. Water Manage. doi.org/10.1016/j.agwat.2016.08.001.

14

Morpho-physiological responses of rice genotypes and its clustering under hydroponic iron toxicity conditions

Turhadi Turhadi[1]*, Hamim Hamim[1], Munif Ghulamahdi[2], Miftahudin Miftahudin[1]

[1]Department of Biology, Faculty of Mathematics and Natural Sciences, Bogor Agricultural University, Kampus IPB Darmaga, 16680 Bogor, Indonesia

[2]Department of Agronomy and Horticulture, Faculty of Agriculture, Bogor Agricultural University, Kampus IPB Darmaga, 16680 Bogor, Indonesia

Abstract

The acid soil area covers major topics land where Iron (Fe) toxicity is one of a limiting factor for rice production which can be overcome by planting the tolerant variety. The information of morpho-physiological characters and the genetic variation of tolerant genotypes is very important. Here we study the variation of root and shoot growth as well as physiological responses to iron toxicity between ten rice genotypes under hydroponic conditions with agar addition. Growth parameters, leaf bronzing score, Fe content in the shoot and root plaques, total chlorophyll, carotenoids, and malondialdehyde (MDA) were observed in this study. Based on morpho-physiological data related to iron toxicity, ten rice genotypes were clustered into three groups which the best performance genotypes were Pokkali and Hawara Bunar. Leaf bronzing score showed correlated with Fe content in the shoot, but tolerant and sensitive genotypes could be differentiate based on this character because it showed non significant Fe content between those two groups. Our study found that the pattern in morpho-physiological characters variation could be useful for selection of desirable genotypes for Fe tolerant rice.

Keywords: Iron toxicity, Leaf bronzing score, MDA, Morpho-physiological characters, Tolerance

Corresponding author email:
miftahudin@ipb.ac.id

Introduction

Iron (Fe) is a micronutrient that participates in photosynthesis, the respiratory chain (in cytochromes), and as a cofactor in various enzymes (Marschner, 1995). However, excess of Fe can be toxic to plants, its abundance inside the cell must be tightly controlled. The critical Fe concentration that causes toxicity for rice is about 250-500 ppm (Yoshida et al., 1976).

Iron toxicity is a symptom associated with high Fe concentration inside the cell, which is dangerous for plants because it leads to oxidative stress (Kampfenkel et al., 1995). The general symptom associated with iron toxicity in the plant is brown spot (bronzing) developed in leaf blades (Takehisa and Sato, 2007). Iron toxicity causes a decrease in growth of some crop rice (Audebert and Fofana, 2009; Nugraha et al., 2016a), Australian hexaploid wheat (Khabaz-Saberi et al., 2010), and tobacco (Nicotiana plumbaginifolia) (Kampfenkel et al., 1995).

Iron toxicity in rice affects the regulation of Fe homeostasis through protein transporters, ROS generation, carbohydrates, hormones, and secondary metabolisms (Quinet et al., 2012). Iron excess in the plant cells generates reactive oxygen species (ROS), which cause lipid peroxidation of the cell membranes and it indicated with malondialdehyde (MDA) production (Fang et al., 2001; Polit 2007; Hamim et al., 2017). ROS production in Iron excess condition through two chemical reactions, named oxidation and Fenton reaction. In the oxidation reaction, ferrous (Fe^{2+}) can react with oxygen then produce superoxide radical. Meanwhile, in the Fenton reaction, ferrous (Fe^{2+}) can react with hydrogen peroxide then produce hydroxyl radical (Marschner, 1995).

There are many information on hypothetic tolerance strategies in plant when it is exposed to iron excess (Engel et al., 2012; Nugraha and Rumanti, 2017), but it can be simplified in the exclusion and inclusion strategies that involve complex physiological processes. For excluder plants, roots improve the oxidation power in the rhizosphere area to oxidize Fe^{2+} to become Fe^{3+} in the root surface layer, which causes plaque accumulation (Nugraha et al., 2016b). For the inclusion strategies the plant may accumulate the Fe inside the cell with a compartmentation strategy in the vacuole and produce ferritin molecules (Onaga et al., 2016) or through ROS detoxification, which involves enzymes and metabolite antioxidants (da Silveira et al., 2009; Kang et al., 2011; Kabir et al., 2016). However, the tolerance level of plant to iron toxicity stress depends on plant development stage, stress intensity, stress duration, and climatic conditions (Engel, 2009).

Many traits in plants has been known affected by iron toxicity, these include LBS and morpho-physiological characters. LBS is a key secondary trait to differentiate the tolerance level of plants to iron toxicity, especially in rice (Sikirou et al., 2015). However, the LBS did not always correlate with the tolerance level of rice (Becker and Asch, 2005; de Dorlodot et al., 2005; Nugraha et al., 2015). Enhanced tolerance to iron toxicity in breeding program of rice needs the information of morpho-physiological characters (Nugraha et al., 2016c).

Morpho-physiological of some Indonesian rice varieties in responses to iron toxicity may varied that could be tolerant-excluder type or tolerant-includer type. Morpho-physiological analysis has been success to determinate the adaptation pattern of durum wheat (Annicchiarico et al., 2008), genetic diversity on heat tolerance of tall fescue (Festuca arundinacea Schreb.) accessions (Sun et al., 2015) and tolerance level to Aluminum toxicity in rice varieties of North East India (Awasthi et al., 2017). Based on those some previous papers, morpho-physiological characters might be also used in the determination of iron toxicity tolerance level in rice. In this early investigation, we use morpho-physiological characters from two sensitive and eight tolerant genotypes to iron excess to investigate and classify their tolerance type. This paper reports the variations in rice root and shoot growth as well as physiological responses to iron toxicity in ten rice genotypes.

Material and Methods

Plant materials

The plants used in this research were Fe-sensitive (IR64 and Inpara 5) and Fe-tolerant rice genotypes (Mahsuri, Pokkali, Inpara 2, Inpara 6, Danau Gaung, Indragiri, Hawara Bunar, IRH108) provided by Indonesian Center for Rice Research and Plant Physiology and Molecular Biology Laboratory, Department of Biology, Bogor Agricultural University (Table 1).

Analysis of growth and physiological responses of ten rice genotypes to iron toxicity

This experiment aimed to evaluate the variation of tolerance-type of 10 rice genotypes to iron toxicity. The seeds were surface sterilized using sodium hypochlorite 1 % (v/v) for 15 minutes and rinsed with sterile distillate water. The seeds were germinated in the incubator (27 °C) for three days. Uniform seedlings were transferred to a sheet-holed styrofoam floating on 9 L plastic trays (35 x 28.5 x 12) cm^3 filled with 8 L half-strength Yoshida's solution until the plants reach two-weeks of age with pre-culture solution renewal every 7 days. Two-weeks old seedlings were then grown on 800 ml plastic pot (∅: 8.5 cm; height: 15 cm) containing 750 ml HSYA or nutrient culture solution were prepared by dissolving 0.2% agar powder in Yoshida's solution (Yoshida et al., 1976). The Fe treatment followed the procedure conducted by Nugraha et al. (2015). The Fe was added to every nutrient culture solution with two different Fe levels of concentration i.e. 0 (control) and 400 ppm in the form of $FeSO_4 \cdot 7H_2O$. The experiment was arranged as complete randomized design with three replications. Leaf bronzing score at the 10th DAS (LBS10), maximum root length, shoot height, root- and shoot elongation rate, relative growth rate, iron plaque content, and shoot iron content were measured in this experiment.

The LBS10 were determined at 2nd, 3rd, and 4th leaves (Shimizu et al., 2005) with scoring index scale from 1 (no bronzing symptom on the leaves) to 10 (the plant died) according to IRRI (2013).

Root- and shoot elongation rates (RER and SER) were measured on two plants in each pot before and after 10 days of treatment. The dry biomass of root and shoot of both stressed and control plants were determined after being dried at 80 °C for 72 hours. The dry biomass of root and shoot used for relative growth rate (RGR) determination.

Fe content of root plaque and shoot tissues were determined according to Nugraha et al. (2015) and Engel et al. (2012) at 10th DAS respectively. Fe content was analyzed using atomic absorption spectrometry (AAS) (Agilent 200 Series AA Systems, Agilent Technologies, Inc, USA).

Evaluation of physiological responses between tolerance-type to iron toxicity

This experiment aimed to analyze the difference of physiological responses between tolerance-type to iron toxicity. The condition of the experiment was similar to the previous experiment as described above. The experiment was arranged as a complete randomized design with three replications. Three rice genotypes were used in this experiment, IR64, Inpara 2, and Pokkali. The Fe was added by two different Fe concentrations, 0 (control) and 400 ppm in HSYA solution.

Total chlorophyll (Chl) and carotenoid extraction were determined according to Quinet et al. (2012) and were calculated according to Lichtenthaler (1987). The lipid peroxidation level was detected using malondialdehyde (MDA) quantification in the root and leaf organs. MDA extraction was carried out according to Muller et al. (2015) with a small modification. The MDA concentration was calculated according to Wang et al. (2013).

Statistical analysis

The collected data were analyzed by Duncan's Multiple Range Test (α=5%) for comparison among means of morpho-physiological data using SPSS version 16. The average data of morpho-physiological data were subjected by correlation analysis using SPSS version 16 and principal component analysis (PCA) and clustering analysis using PAST version 3.06 (Hammer et al., 2001). The unstandardized squared Euclidean distance and the unweighted pair group arithmatic averaging (UPGMA) were imputed in a cluster analysis to classify the genotypes based on

their tolerance type to iron toxicity according to Annicchiarico et al., (2008).

Results

Ten rice genotypes shows different reaction of morpho-physiological under iron toxicity

Iron toxicity reduced both shoot and root growth of 2-week-old rice seedlings in both Fe-tolerant and -sensitive genotypes. Shoot height of Fe-tolerant genotypes was reduced between 16.0-22.7 %, while the shoot height of Fe-sensitive genotypes was reduced between 22.5-22.7 % (Table 2). Shoot elongation rate was also reduced in both Fe-tolerant and -sensitive genotypes with reduction ranges of 41.8-61.6 % and 80.6-87.3 %, respectively (Table 2). Iron toxicity also decreased shoot relative growth rate (SRGR) between 9.5-25.3 % in all genotypes (Table 2).

Iron toxicity inhibited the root growth, which the highest growth inhibition was shown by Inpara 5, and the lowest one was shown by Inpara 6. The percentage of maximum root length was reduced between 12.7-26.5 % (Table 2). Root elongation rate was also inhibited between 72.2-75.9 % and 32.8-54.4 % in sensitive tolerant genotypes, respectively (Table 2). Root relative growth rate (RRGR) of 10 rice genotypes were decreased significantly (23.8-61.3 %) in both sensitive and tolerant genotypes under iron toxicity (Table 2).

When the rice plants were exposed to Fe treatment, almost all the plants showed bronzing symptom in the leaf blades. There was a variation in bronzing symptom among the genotypes. LBS10 (leaf bronzing score at 10th DAS) showed that score in sensitive genotypes was higher than that of the tolerant genotypes. LBS10 data showed that the highest bronzing score occurred in Inpara 5 and IR64 (10), while the lowest score was in Mahsuri (3.7) (Figure 1). Fe content in shoot tissue and root plaque using atomic absorption spectrometry (AAS) showed that in shoot tissue ranged 8.8-11.1 mg.g⁻¹ DW (Table 3). The Fe content in shoot tissue of sensitive genotypes was higher than that in Fe-tolerant genotypes. AAS data in this research also showed that the Fe content in root plaque was varied among genotypes, but Fe content in root plaque of tolerant genotypes was not always higher than that of sensitive genotypes. Fe content in root plaque of tolerant genotypes was 3.6-5.3 mg.g⁻¹ DW, while in sensitive genotypes it was 4.9-5.1 mg.g⁻¹ DW (Table 3). This study also showed the variation of Fe content in both shoot tissues and root plaques (Table 3). Average Fe content in the shoot of sensitive genotypes (10.2 mg.g^{-1} DW) was higher than that of tolerant genotype (9.9 mg.g^{-1} DW) (Table 3). The highest Fe content in shoot is showed by Hawara Bunar (11.1 mg.g^{-1} DW) and the lowest one is showed by Inpara 2, (8.8 mg.g^{-1} DW).

Physiological responses between rice genotypes and tolerance type to iron toxicity

Iron toxicity caused a significant decrease in total chlorophyll content in leaves of all genotypes in comparison with control plants, which decreased between 27.3-34.4 % (Table 4). In addition, total carotenoid content in leaves of three genotypes were also significantly reduced under iron toxicity between 23.5-28.6 % (Table 4). There was no particular pattern among genotypes regarding chlorophyll and carotenoid content.

Malondialdehyde data indicated that lipid peroxidation occurred both in roots and leaves after being exposed to iron toxicity for 10 days. MDA content significantly increased in leaves of IR64, Inpara 2, and Pokkali in comparison with the control plants (Table 4). The increase of leaves MDA content in sensitive genotype (IR64) was higher than that of both tolerant genotypes (Inpara 2 and Pokkali). However, MDA content did not always significantly increase in roots of those genotypes in response to iron toxicity. The increase of MDA content in roots of both tolerant genotypes (Inpara 2 and Pokkali) did not significantly increase while MDA content in IR64 roots showed significant increased under iron toxicity (Table 4).

The relationship among morpho-physiological characters in rice were demonstrated using Pearson's correlation. A significant positive correlation was found between MDA content in leaves and LBS10, while significant negative correlations were found between MDA content in roots, shoot and root elongation rate to LBS10 (Table 5).

Clustering of tolerance in ten genotypes based on morpho-physiological characters

To know the variation among morpho-physiological characters on ten rice genotypes, the PCA was carried out. In this study, out of total nine principal components (PC), two PC were have Eigen value >1 and it contributed 78.8% of total variation among the rice genotypes observed for nine morpho-physiological characters (Table 6). The PC1 has

highest contributed to the variation (59.9%) followed by PC2 (18.9%). PC1 was positively and strongly associated with LBS10, shoot height, shoot extention rate, root length, and root extention rate. Shoot relative growth rate, Fe content in shoot tissues, and Fe content in root plaques have most important contribution in PC2 (Table 6). PCA result showed clear differentiation between sensitive- and tolerant genotypes (Figure 2).

The cluster analysis demonstrated using UPGMA method to classify ten rice genotypes based on their morpho-physiological characters to iron toxicity. Using nine morpho-physiological characters, the classification results showed that cluster 1 comprised of 2 genotypes, cluster 2 comprised of 6 genotypes, and cluster 3 comprised of 2 genotypes (Figure 3). The genotypes in cluster 1 have high inhibition on root relative growth rate, high Fe in shoot and root plaque. The genotypes in cluster 2 have low inhibition on root relative growth rate. The genotypes in cluster 3 have high inhibition in all morpho-physiological characters observed in this study. This study also demonstrated that three belong to the total clorophyll, carotenoid, and malondialdehyde (MDA) content of the rice leaves of the representative genotypes were analyzed in this study. IR64, Inpara 2, and Pokkali were the representative genotypes of Cluster 1, Cluster 2, and Cluster 3, respectively.

Discussion

Varied response demonstrated in the tolerant genotypes according their tolerance level to iron toxicity (Table 2). Iron toxicity in previous studies reduced shoot height between 38-62% under 300 ppm iron stress for 4 weeks (Suryadi, 2012). We noted in this study that the root elongation rate and maximum root length inhibited more than shoot elongation rate and shoot height. The data showed that the root relative growth rate inhibited more than the shoot, and it was consistent between Fe-sensitive and -tolerant genotypes (Table 2). According to Li et al. (2016) iron toxicity decreases both root cell elongation and division and inhibits lateral root initiation process because of direct contact between root tips and Fe.

The variation of Fe content in both shoot tissues and root plaques also showed in this study (Table 3). The small amount of Fe content in shoot tissues of tolerant genotypes such Inpara2 is an indication of the Fe exclusion strategy through oxidation of Fe in the rhizosphere, which is an efficient strategy of Fe tolerance mechanism (Kang et al., 2013). In contrast to the Fe in the shoot, we also observed that the Fe content in root plaques showed inconsistent pattern between Fe-sensitive and -tolerant genotypes (Table 3). Several tolerant rice genotypes include Mahsuri, Indragiri, Inpara 6, Inpara 2, and IRH108, showed less Fe content in root plaques than that of sensitive genotypes (Inpara 5 and IR64), while the other three tolerant genotypes (Hawara Bunar, Pokkali, and Danau Gaung) showed Fe content in root plaques higher than that of sensitive genotypes. This research noted that the Hawara bunar has the highest capability to accumulate Fe in root plaques, while Indragiri has the lowest one.

The tolerant genotypes to iron toxicity have the ability to oxidize Fe^{2+} on the surface of the roots. Plants with high ability to develop aerenchyme cells are able to receive more O^2 in the roots area. However, highly reduced Fe in roots area will react with O^2, which leads to the Fe plaque formation on the root surface (Harahap et al., 2014). Based on Fe content in root plaques and leaf bronzing score, Pokkali and Hawara Bunar have a strategy to tolerate iron toxicity through Fe exclusion, which is in agreement with the previous research that suggests Pokkali is an excluder type (Engel et al., 2012). In addition, this research also showed that Pokkali and Hawara Bunar genotypes have inclusion strategy based on their Fe content in shoots (Table 3). We suggested Pokkali and Hawara Bunar have double mechanisms to tolerate iron toxicity, which involved both exclusion and inclusion strategies. Pokkali and Hawara Bunar indicated the best performance in tolerate to iron excess condition compared with other tolerance genotypes. Both genotypes have good strategy to accumulate Fe in the shoot and root plaque and their growth performance was not affected significantly by the iron stress condition.

The PCA and cluster analysis were performed for grouping ten rice genotypes based on their morpho-physiological responses to iron toxicity. Based on both analyzes, the sensitive genotypes showed clearly separate from the tolerant genotypes (Figure 2 and 3). There are two sub-group in tolerant genotypes, which is illustrated by Cluster 1 and 2. Both clusters were differentiated by root relative growth rate (RRGR) character. The genotypes in Cluster 1 have low root growth activity, while the genotype in Cluster 2 have high growth activity (Table 2; Figure 3).

Cluster analysis grouped ten rice genotypes into 3 group tolerance-type. Cluster 1 was categorized as tolerance group, Cluster 2 was categorized as moderate tolerance group, and Cluster 3 grouped sensitive genotypes. Several previous classifications of tolerance-type was carried out by Engel et al., (2012), Harahap et al., (2014), Nugraha et al., (2016b) rice, but they grouped the rice genotypes only based on the LBS and iron fate in the tissues. In this study, we classify tolerance-type to iron toxicity in rice based on their morpho-physiological characters. In this research, we showed Mahsuri, IRH108, Danau Gaung, Indragiri, Inpara 2, and Inpara 6 categorized as moderate tolerance group (Figure 3). This group is separated from the other two groups based only on root relative growth rate.

Interestingly, LBS10 did not show correlation with Fe content in the roots and shoot (Table 5). This corresponds to the reports of Becker and Asch (2005) and de Dorlodot et al. (2005) who demonstrated LBS did not always correlate with the tolerance level of rice. Based on our findings, rice genotype selection under Fe excess condition could be based on not only LBS but also based on root relative growth rate character (Table 2).

To know the physiological responses among tolerance-type of rice genotypes to iron toxicity, the total chlorophyll, carotenoid, and MDA content were analyzed in the representative genotypes of sensitive (IR64), moderate (Inpara2), and tolerant (Pokkali) to iron toxicity. Decrease in the chlorophyll content under iron toxicity suggested with the decrease of photosynthesis rate and chlorophyll biosynthesis (Quinet et al., 2012).

The study also showed different pattern of lipid peroxidation between roots and shoots under iron toxicity based on MDA content (Table 4). IR64 has the highest increase in MDA content under iron toxicity indicated that IR64 experienced highest stress than that of the other two genotypes. In this present study, Inpara 2 and Pokkali had differ MDA content. Those two genotypes suggested had differ in their tolerance-type to iron toxicity. Inpara 2 classified as the tolerant-includer type, meanwhile Pokkali as the tolerant-excluder type (Engel et al., 2012; Nugraha et al., 2016b). The MDA content indicates the level of lipid peroxidation of the cell membranes caused by ROS and oxidative stress (Polit, 2007). Iron toxicity could produce ROS in many organelles and parts of the cell, such as chloroplast, endoplasmic reticulum, peroxisomes, mitochondria, plasma membrane, cell wall, and apoplast (Sharma et al., 2012). ROS molecules also damage essential molecules inside the cell, i.e.: DNA, RNA, and protein (Mittler, 2016). The tolerant genotypes have some strategies to tolerate iron toxicity through increasing oxidation power in the rhizosphere area, ROS detoxification, and compartmentation in the apoplasm, chloroplas and vacuole (Onaga et al., 2016).

ROS detoxification involves both enzymatic and non-enzymatic mechanisms. Inpara 2 and Pokkali have significantly decrease the carotenoid content compared to the control plant (Table 4), but the Pokkali has a small decrease in carotenoid content compared to IR64, which is a sensitive genotype to iron toxicity. Carotenoids will protect the chloroplasts as photosynthetic organelles by inhibiting the formation of triplet chlorophyll (3Chl*) and excited chlorophyll (Chl*) to prevent the production of $^1O_2$ (Sharma et al., 2012; Puthur, 2016). According to Kabir et al., (2016) increased catalase enzyme activity (CAT), peroxidase (POD), glutathione reductase (GR), and superoxide dismutase (SOD) as well as increased ascorbic acid content, glutathione and amino acids (cysteine, methionine, proline) in the Pokkali are suggested to be strong antioxidant defenses of tolerant plant when iron toxicity occurs. The study demonstrated total chlorophyll and carotenoid content did not significant correlation with MDA content in the shoots. However, chlorophyll content had significant correlation with Fe content in root plaque, while the carotenoid content did not significant correlation with Fe content in root plaque. Based on our findings it was suggested that leaf chlorophyll and carotenoid content did not directly relate with tolerance strategy to iron toxicity. The decrease in total

chlorophyll suggested as an impact of stress condition in leaves, which is demonstrated with the high MDA content. On the whole results of our study suggested that the morpho-physiological characters could be used to predict the tolerance level to iron toxicity in rice. This study supports the genetic improvement strategies in rice to obtain rice genotype tolerant to iron toxicity. Further analysis on basic genetic, molecular, and physiological processes underlying the tolerance mechanism of rice to iron toxicity is still required to be investigated.

Table 1: The tolerance level of ten rice genotypes in this present study

Geno types	Tolerance level to iron toxicity	Reference
Inpara 5	Sensitive	Nugraha and Rumanti (2017)
IR64	Sensitive	Nugraha and Rumanti (2017)
Mahsuri	Tolerant	Nugraha and Rumanti (2017)
Pokkali	Tolerant	Engel et al., (2012)
Inpara 2	Tolerant	Suprihatno et al., (2010)
Inpara 6	Tolerant	Suprihatno et al., (2010)
Danau Gaung	Tolerant	Suprihatno et al., (2010)
Indragiri	Tolerant	Suprihatno et al., (2010)
Hawara Bunar	Tolerant	Amnal (2009)
IRH108	Tolerant	Kolaka (2016)

Table 2. Shoot and root growth of 10 rice genotypes after exposed to 400 ppm Fe for 10 days.

Genotypes	Percentage of growth decreasing (%)					
	SH	SER	RRGR	RL	RER	RRGR
Inpara 5	22.5[bc]	80.6[c]	18.5[bc]	26.5[d]	72.2[d]	50.9[c-e]
IR64	22.7[c]	87.3[c]	25.3[c]	25.1[d]	75.9[d]	61.3[e]
Hawara Bunar	18.8[a-c]	44.2[ab]	9.5[a]	17.2[bc]	40.3[ab]	51.6[c-e]
Pokkali	16.2[a]	61.6[b]	15.4[ab]	16.4[a-c]	35.9[ab]	57.6[de]
Mahsuri	16.0[a]	53.8[ab]	14.5[ab]	17.1[bc]	42.8[ab]	23.8[a]
Danau Gaung	17.9[a-c]	47.0[ab]	14.1[ab]	14.5[ab]	37.7[ab]	36.8[a-c]
Indragiri	16.2[a]	41.8[a]	18.5[bc]	18.7[bc]	44.1[a-c]	45.0[b-d]
Inpara 6	19.1[a-c]	49.6[ab]	16.4[b]	12.7[a]	32.8[a]	34.7[ab]
Inpara 2	17.3[a-c]	43.8[ab]	16.8[b]	18.6[bc]	44.5[bc]	36.1[a-c]
IRH108	17.2[ab]	59.9[ab]	13.8[ab]	19.3[c]	54.4[c]	28.9[a]

[a-e]Different letters in the same column show significant differences (p<0.05) according to Duncan's Multiple Range Test. LBS10=Leaf bronzing score at 10th DAS; SH=Shoot height; SER=Shoot extension rate; SRGR=Shoot relative growth rate; RL=Root length; RER=Root extension rate; RRGR=Root relative growth rate.

Table 3: Fe concentration in shoot tissues and root plaques of 10 genotypes under iron toxicity. Data were taken on 10th day after stress.

Genotypes	Fe content in shoot tissues (mg.g^{-1} DW)	Fe content in root plaques (mg.g^{-1} DW)
Inpara 5	10.4[a]	5.1[de]
IR64	10.0[a]	4.9[de]
Hawara Bunar	11.1[a]	5.3[e]
Pokkali	9.9[a]	5.0[de]
Mahsuri	9.5[a]	4.3[c]
Danau Gaung	9.4[a]	5.1[de]
Indragiri	9.7[a]	3.6[a]
Inpara 6	10.9[a]	4.1[bc]
Inpara 2	8.8[a]	3.8[ab]
IRH108	9.6[a]	4.7[d]

[a-d]Different letters in the same column show significant differences (p<0.05) according to Duncan's Multiple Range Test.

Table 4: Physiological responses of IR64, Inpara 2, and Pokkali were grown under control and iron toxicity hydroponic conditions.

Characters	Genotypes					
	IR64		Inpara 2		Pokkali	
	C	++Fe	C	++Fe	C	++Fe
Total chlorophyll content (mg.g^{-1} FW)	306.5[c]	215.2[ab]	347.2[d]	185.2[a]	340.8[cd]	246.1[b]
Carotenoid content (mg.g^{-1} FW)	84.9[c]	62.9[ab]	95.4[c]	55.3[a]	92.2[c]	70.8[b]
MDA content in leaves (μmol.g^{-1} FW)	0.2[a]	1.8[c]	0.2[a]	1.5[ab]	0.2[a]	1.3[b]
MDA content in roots (μmol.g^{-1} FW)	0.3[b]	0.7[c]	0.1[a]	0.1[b]	0.2[ab]	0.2[ab]

[a-d]Different letters in the same row show significant differences (p<0.05) according to Duncan's Multiple Range Test. Control=0 ppm FeSO$_4$·7H$_2$O; ++Fe= 400 ppm FeSO$_4$·7H$_2$O.

Table 5. Correlation analysis among morpho-physiological characters under iron toxicity of ten rice genotypes

	LBS10	SRGR	RRGR	FeS	FeR	CHL	CAR	SMDA	RMDA	SER	RER
LBS10	1.000	-0.515	-0.377	0.123	0.195	-0.249	-0.207	0.716*	0.857**	-0.816**	-0.876**
SRGR		1.000	0.816**	-0.226	-0.734*	-0.231	-0.221	-0.250	-0.598	0.768*	0.382
RRGR			1.000	-0.111	-0.813**	-0.495	-0.511	-0.219	-0.381	0.526	0.155
FeS				1.000	0.329	0.083	-0.240	0.189	0.531	-0.231	-0.222
FeR					1.000	0.768*	0.671*	-0.026	0.379	-0.477	-0.137
CHL						1.000	0.942**	-0.514	-0.035	-0.051	0.247
CAR							1.000	-0.518	-0.137	-0.054	0.258
SMDA								1.000	0.502	-0.393	-0.687*
RMDA									1.000	0.285	0.717*
SER										1.000	0.668*
RER											1.000

**=Correlation is significant at the 0.01 level (2-tailed); *=Correlation is significant at the 0.05 level (2-tailed). . LBS10=Leaf bronzing score at 10th day after stress; SRGR=Shoot relative growth rate; RRGR=Root relative growth rate; FeS=Fe content in shoot tissues; FeR=Fe content in root plaques; CHL=Total chlorophyll content; CAR=Carotenoid content; SMDA=MDA content in leaves; RMDA=MDA content in roots; SER=Shoot elongation rate; RER=Root elongation rate.

Table 6: Loading score of two principal components, Eigen value, % variance and % cumulative variance on inhibition percentage (%) of morpho-physiological characters of genotypes evaluated in this study

Characters	Loading score of principal component	
	PC1	PC2
LBS10	0.42	-0.06
Shoot height	0.38	0.16
Shoot elongation rate	0.39	-0.01
Shoot relative growth rate	0.31	-0.41
Root length	0.38	-0.17
Root elongation rate	0.40	-0.17
Root relative growth rate	0.28	0.27
Fe content in shoot tissues	0.10	0.61
Fe content in root plaques	0.17	0.55
Eigen value observed	5.39	1.70
Proportion of total variance (%)	59.9	18.9
Cumulative proportion of total variance (%)	59.9	78.8

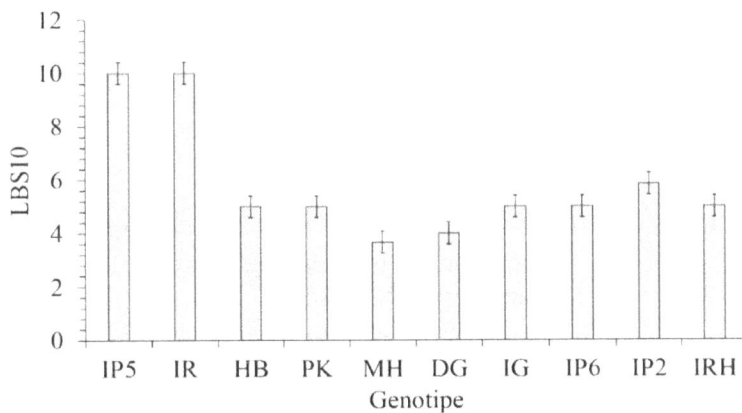

Figure 1: Leaf bronzing score at 10th DAS (LBS10) of ten rice genotypes under iron toxicity. Data were taken on 10th DAS. IP5=Inpara 5; IR=IR64; HB=Hawara Bunar; PK=Pokkali; MH=Mahsuri; DG=Danau Gaung; IG=Indragiri; IP6=Inpara 6; IP2=Inpara 2; IRH= IRH108. Bars indicate the standard error.

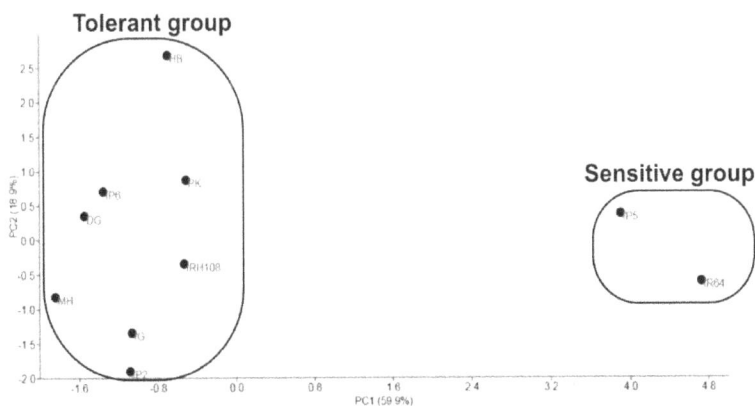

Figure 2: Principal component analysis (PCA) of 10 rice genotypes under iron toxicity. IP5=Inpara 5; IR=IR64; HB=Hawara Bunar; PK=Pokkali; MH=Mahsuri; DG=Danau Gaung; IG=Indragiri; IP6=Inpara 6; IP2=Inpara 2; IRH= IRH108.

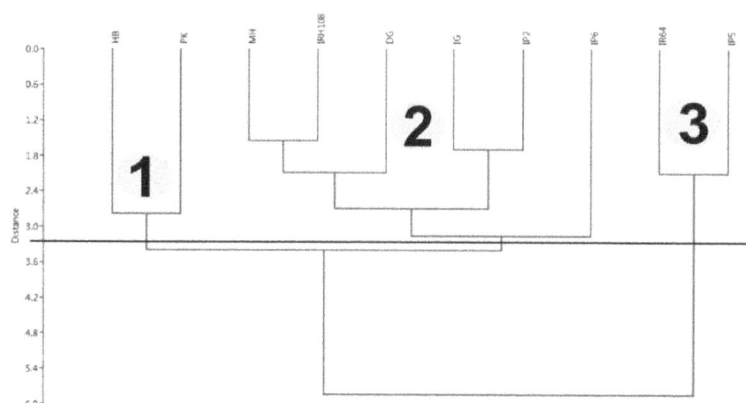

Figure 3: Cluster analysis of ten rice genotypes under iron toxicity. IP5=Inpara 5; IR=IR64; HB=Hawara Bunar; PK=Pokkali; MH=Mahsuri; DG=Danau Gaung; IG=Indragiri; IP6=Inpara 6; IP2=Inpara 2; IRH= IRH108.

Conclusion

The present study concluded that iron toxicity significantly affect morphological and physiological characters on rice genotypes. PCA and cluster analysis using those characters demonstrated there are three groups in related to iron toxicity tolerance level. The sensitive group consisted of IR64 and Inpara 5. The moderate group consisted of Mahsuri, IRH108, Danau Gaung, Indragiri, Inpara 2, and Inpara 6. The tolerant group consisted of Hawara Bunar and Pokkali. In addition, iron toxicity decreased chlorophyll and carotenoid concentration and increased the lipid peroxidation level in both the root and shoot. Rice var. Pokkali and Hawara Bunar showed the best growth performance under the iron toxicity condition in comparison with other genotypes. This valuable finding will be useful to improve iron toxicity tolerance in rice through selection strategy in breeding programs.

Acknowledgment

The authors would like to thank PMDSU grant from Ministry of Research, Technology and Higher Education, the Republic of Indonesia F.Y. 2016 (No. 330/SP2H/LT/DRPM/IX/2016) granted to Dr. Miftahudin for the financial support.

References

Amnal, 2009. Physiological Responses of Several Rice Varieties to Iron Stress. M.S. thesis, Bogor Agricultural University, Indonesia.

Annicchiarico P, Royo C, Bellah F and Moragues M, 2008. Relationships among adaptation patterns, morphophysiological traits and molecular markers in durum wheat. Plant Breeding. 128: 164-171.

Audebert A and Fofana M, 2009. Rice yield gap due to iron toxicity in West Africa. J. Agron. Crop. Sci. 195: 66-76.

Awasthi JP, Saha B, Regon P, Sahoo S, Chowra U, Pradhan A, Roy A and Panda SK, 2017. Morpho-physiological analysis of tolerance to aluminum toxicity in rice varieties of North East India. PLoS ONE. 12(4): e0176357.

Becker M and Asch F, 2005. Iron toxicity in rice-conditions and management concepts. J. Soil. Sci. Plant. Nutr. 168: 558-573.

Da Silveira VC, Fadanelli C, Sperotto RA, Stein RJ, Basso LA, Santos DS, Junior ISV, Dias JF and Fett JP, 2009. Role of ferritin in the rice tolerance to iron overload. Sci. Agric. (Piracicaba, Braz.) 66(4): 549-555.

De Dorlodot S, Lutts S and Bertin P, 2005. Effects of ferrous iron toxicity on the growth and mineral composition of an interspecific rice. J. Plant. Nutr. 28: 1-20.

Engel K, Asch F and Becker F, 2012. Classification of rice genotypes based on their mechanisms of adaptation to iron toxicity. J. Plant. Nutr. Soil. Sci. 175: 871-881.

Engel, K, 2009. Efficiency of adaptation mechanisms of rice to diverse conditions iron toxicity. http://escholarship.org/uc/item/3cz2f4rj.

Fang WC, Wang JW, Lin CC and Kao CH, 2001. Iron induction of lipid peroxidation and effects on antioxidative enzyme activities in rice leaves. Plant. Growth. Regul. 35: 75-80.

Hamim H, Violita V, Triadiati T and Miftahudin M, 2017. Oxidative stress and photosynthesis reduction of cultivated (*Glycine max* L.) and wild soybean (*G. tomentella* L.) exposed to drought and paraquat. Asian. J. Plant. Sci. 16: 65-77.

Hammer O, Harper DAT, Ryan PD, 2001. PAST: palaentological statictics software package for education and data analysis. Palaentologia Electronica 4(1): 1-9.

Harahap SM, Ghulamahdi M, Aziz SA, Sutandi A and Miftahudin, 2014. Relationship of ethylene production and aerenchyme formation on oxidation ability and root surfaced-iron (Fe^{2+}) accumulation under different iron concentrations and rice genotypes. IJAST 4(1): 186-194.

International Rice Research Institute (IRRI), 2013. Standard Evaluation System for Rice. IRRI, Manila.

Kabir AH, Begum MC, Haque A, Amin R, Swaraz AM, Haider SA, Paul NK and Hossain MM, 2016. Genetic variation in Fe toxicity tolerance is associated with the regulation of translocation and chelation of iron along with antioxidant defence in shoots of rice. Func. Plant. Biol. 43(11): 1070-1081.

Kampfenkel K, Montagu MV and Inze D, 1995. Effects of iron excess on *Nicotiana plumbaginifolia* plants: implication to oxidation stress. Plant. Physiol. 107: 725-735.

Kang DJ, Futakuchi K, Seo YJ, Vijarnsorn P and Ishii R, 2011. Relationship of Fe-tolerance to morphological changes in roots in upland NERICA lines under fe-treated hydroponic conditions. JCSB 14(4): 311-315.

Khabaz-Saberi H, Rengel Z, Wilson R and Setter TL, 2010. Variation for tolerance to high concentration of ferrous iron (Fe^{2+}) in Australian hexaploid wheat. Euphytica 172: 275-283.

Kolaka L. 2016. Growth and production of Fe tolerant rice lines derived from a cross between Var. IR64 x Hawara Bunar in tidal soil, Banyuasin, South Sumatera. M.S. thesis, Bogor Agricultural University, Indonesia.

Li G, Kronzucker HJ and Shi W, 2016. The response of the root apex in plant adaptation to iron heterogeneity in soil. Front. Plant. Sci. 7: 1-7.

Lichtenthaler HK, 1987. Chlorophylls and carotenoid: pigments of photosynthetic biomembranes. Methods Enzymol. 148: 350-382.

Marschner H, 1995. Mineral nutrition of higher plants, 2nd ed. Academic Press, San Diego.

Mittler R, 2016. ROS are good. Trends. Plant. Sci. 22(1): 11-19.

Müller C, Kuki KN, Pinheiro DT, de Souza LR, Silva AIS, Loureiro ME, Oliva MA and Almeida AM, 2015. Differential physiological responses in rice upon exposure to excess distinct iron forms. Plant Soil. 391: 123-138.

Nugraha Y, Ardie SW, Ghulammahdi M, Suwarno, Aswidinnoor H. 2015. Nutrient culture media with agar is effective for early- and rapid-screening of iron toxicity tolerant in rice. J. Crop Sci. Biotech. 19(1):61-70.

Nugraha Y, Ardie SW, Suwarno, Ghulammahdi M, Aswidinnoor H. 2016a. Implication of gene action and heritability under stress and control conditions for selection iron toxicity tolerant in rice. Agrivita 38(3): 282-295.

Nugraha Y, Rumanti IA, Guswara A, Ardie SW, Ghulammahdi M, Suwarno, Aswidinnoor H. 2016b. Response of selected rice varieties under excess iron condition. J. Pen. Pert. Tan. Pangan 35(3): 181-190.

Nugraha Y, Ardie SW, Ghulammahdi M, Suwarno, Aswidinnoor H. 2016c. Generation mean analysis of leaf bronzing associated with iron toxicity in rice seedlings using digital imaging methods. SABRAO J. 48(4): 453-464.

Nugraha Y and Rumanti IA. 2017. Breeding for rice variety tolerant to iron toxicity. Iptek Tan. Pangan. 12(1): 9-24.

Onaga G, Dramé KN and Ismail AM, 2016. Understanding the regulation of iron nutrition: can it contribute to improving iron toxicity tolerance in rice?. Func. Plant. Biol. 43(8): 709-726.

Polit ES, 2007. Lipid peroxidation in plant cells, its physiological role and changes under heavy metal stress. Acta. Soc. Bot. Pol. Pol. Tow. Bot. 76(1): 49-54.

Puthur JT, 2016. Antioxidants and cellular antioxidation mechanism in plants. SIJBS 2(1): 14-17.

Quinet M, Vromman D, Clippe A, Bertin P, Lequeux H, Dufey I, Lutts S and Lefèvre I, 2012. Combined transcriptomic and physiological approaches reveal strong differences between short- and long-term response of rice (*Oryza sativa*) to iron toxicity. Plant. Cell. Environ. 35(10): 1837-1859.

Sharma P, Jha AB, Dubey RS and Pessarakli M, 2012. Reactive oxygen species, oxidative damage, and antioxidative defense mechanism in plants under stressful conditions. J. Bot. 2012: 1-26.

Shimizu A, Guerta CQ, Gregorio GB and Ikehashi H, 2005. Improved mass screening of tolerance to iron toxicity in rice by lowering temperature of culture solution. J. Plant. Nutr. 28: 1481-1493.

Sikirou M, Saito K, Achigan-Dako EG., Dramé KN, Ahanchédé A and Venuprasad R, 2015. Genetic improvement of iron toxicity tolerance in rice - progress, challenges and prospects in West Africa. Plant. Prod. Sci. 18(4): 423-434.

Suprihatno B, Daradjat AA, Satoto, Baehaki, Surprihanto, Setyono A, Indrasari SD, Wardana IP, Sembiring H, 2010. Description of rice varieties. Indonesian Center for Rice Research, Indoenesia.

Sun X, Xie Y, Bi Y, Liu J, Amombo E, Hu T and Fu J, 2015. Comparative study of diversity based on heat tolerant-related morpho-physiological traits and molecular markers in tall fescue accessions. Sci. Rep. 5: 18213.

Suryadi D, 2012. Screening of F7 rice lines derived from a cross between IR64 and Hawara Bunar under iron stress. M.S. Thesis. Bogor Agricultural University, Bogor Indonesia.

Tekehisa H and Sato T, 2007. Stress, physiological and genetic factors of rice leaf bronzing in paddy fields. Proc. Crop. Sci. Soc. Jpn. 1(2): 63-68.

Wang YS, Ding DM, Gu XG, Wang JL, Yunli P, Gao LP and Xia T, 2013. Analysis of interfering substances in the measurement of malondialdehid content in plant leaves. Am. J. Biochem. Biotechnol. 9(3): 235-242.

Yoshida S, Forno DA, Cock JH and Gomez KA, 1976. Laboratory manual for physiological studies of rice. IRRI, Manila.

15

Salicylic acid improves physiological traits of *Zea mays* L. seedlings under copper contamination

Nosheen Elahi[1], Muhammad Ishaq Asif Rehmani*[2], Abdul Majeed[1], Muhammad Ahmad[3]
[1]Institute of Pure and Applied Biology, B. Z. University, Multan, Pakistan
[2]Department of Agronomy, Ghazi University, Dera Ghazi Khan, Pakistan
[3]Agricultural Extension Department, Govt. of Punjab, Sanghoie, Jhelum, Pakistan

Abstract

A pot sand culture experiment was conducted to study the effect of copper (Cu) on maize plant growth with or without salicylic acid (SA). Nutrient medium (i.e. half strength Hoagland) along with sixteen different concentrations of Cu and SA were applied as rooting medium twice a week. The lower level of salicylic acid (0.1 mM) increased the biomass production, length of shoot and root, number and area of leaves. While higher level of salicylic acid (10.0 Mm) reduced all the growth parameters. The excess copper (10.0 mM) reduced the shoot and root length, biomass production, number and area of leaves. In both harvests copper accumulations in root had highest value by treating with 10 mM SA and 5 μM copper contamination and minimum value by applying the same concentration of SA but lowest concentration of copper 0.5 μM. While copper accumulation in shoot is not effected by addition of different levels of copper and salicylic acid at both harvest levels. Results indicate that SA application may be one approach to improve growth of this crop under copper contamination but high concentration of SA can decrease the crop growth and Cu accumulation in roots increased with increased Cu contamination.

Corresponding author email:
dr.rehmani.mia@hotmail.com

Keywords: Plant hormone; Cu accumulation, Plant growth

Introduction

Salicylic acid (SA) is an endogenously produced phenolic compound, which is regarded as an important plant growth regulator (Shakirova et al., 2003) and a signaling molecule and plays vital role in regulating various physiological processes in plant systems(Hayat et al., 2010; Rivas-San Vicente and Plasencia, 2011; Wani et al., 2017). Under adverse environmental conditions, itis actively involved in conferring plants resistance against various biotic (Hussain et al., 2008) and abiotic stresses (Noreen and Ashraf, 2010) including the heavy metal stress (Liu et al., 2016; Shakirova et al., 2016; Wani et al., 2017). Heavy metal contaminated soil soriginating from natural depositions and anthropogenic activities have posed serious environmental, nutritional and human health implications (Zhao et al., 2013; Najam et al., 2015; Clemens and Ma, 2016; Shakoor et al., 2017). Expensive reclamation process has alternatively forced to go for phytoremediation technologies, which are considered to be innovative, cost effective and environment friendly solution for remediation of soils contaminated with heavy metals (Bennett et al., 2003; O'connor et al., 2003; Wang et al., 2003; Ehsan et al., 2016).Plants plays vital role in remediation of contaminated soils; however, they must have intrinsic tolerance against higher concentration of heavy metal contamination (Krämer, 2010). Certain plants have been reported to have ability to tolerate heavy metal toxicity including *Zea mays* L. (Lombi et al., 2001), *Brassica juncea* L. (Bennett et al., 2003) *Polygonum hydropiper* L. (Wang et al., 2003). Copper is one of the most destructive heavy metals and may induce growth abnormalities in maize (Brennan et al., 2014). High concentrations of copper are toxic for plants and have direct or indirect impacts on plant growth and development due to interference with various physiological processes (Liu et al., 2014; Adrees et al., 2015; Rizwan et al., 2017), restrict root development and ultimately results in inhibition of plant growth (Weckx and Clijsters, 1996; Madejonb et al., 2009; Kopliku and Mesi, 2012).Plenty of scientific literature is available to highlight copper toxicity in plants and its negative impacts on growth parameters, mineral composition and chlorophyll contents, and activities of several key enzymes (Mocquot et al., 1996; Chen et al., 2000; Brun et al., 2003; Ouzounidou and Ilias, 2005; Maksymiec and Krupa, 2006; Sağlam et al., 2016). Maize seedling exposed to copper solution lead to significant inhibition of root growth, although its impact on shoot growth remained non-significant (Bashmakov et al., 2005). Plant roots have strong affinity to absorb copper ions thus hindering the development of fine roots and uptake of trace element (particularly Fe) (Lexmond and Van der Vorm, 1981; Minnich et al., 1987). Macnicol and Beckett (1985) reported that in maize plants copper toxicity (21 mg/kg) in shoots can lead to significant yield reduction (10% yield reduction). Subsequent studies reconfirmed 20 mg/kg copper concentration in maize leaf or shoot as critical thresholds of Cu toxicity indicator (Mocquot et al., 1996; Borkert et al., 1998). SA improved antioxidant capacity and influences adaptability to the toxicity of Hg and Pb by preventing oxidative stress (Zhou et al., 2009; Wang et al., 2011). Further, the role of SA against Cu stress was investigated in tobacco and cucumber (Strobel and Kuc, 1995). This study was also aimed to investigate the potential effectiveness of salicylic acid in protecting maize (*Zea mays* L.) against copper toxicity.

Materials and Methods

Seeds of maize hybrid 30Y87 were sown in petri dishes under natural light condition (with average temperature and relative humidity ranges from 35°C to 60%, respectively) in a glasshouse at Bahauddin Zakariya University, Multan, Pakistan. The ceiling of glasshouse was covered with a green sheet. The seeds were kept moist with distilled water. After germination, five seedlings were transplanted in each

of 128 plastic pots of 28 cm diameter, containing 3.5 kg of thoroughly washed dried sand, which were irrigated with 250ml water containing half strength of Hoagland Solution(as described in Table A).

Table – A: Composition of nutrient solution

Compound	Concentration of Stock Solution g/L	½ strength Hoagland soln. mL Stock Solution/1L
Macronutrients		
KNO_3	101.10	3
$Ca(NO_3)_2 \cdot 4H_2O$	236.16	2
$NH_4H_2PO_4$	115.08	1
$MgSO_4 \cdot 7H_2O$	246.48	0.5
Micronutrients		
H_3BO_3	0.773	1
$MnSO4 \cdot H_2O$	0.169	1
$ZnSO_4 \cdot 7H_2O$	0.288	1
$CuSO_4 \cdot 5H_2O$	0.062	1
H_2MoO_4	0.04	1
Fe-EDTA	22.74	3
KCl	1.864	1

Pots were arranged in 16 sets (treatments, including control), each containing eight pots. Total 128 pots were used for this experiment, 64 pots for 1st harvest and other 64 for 2nd harvest.

Plants were exposed to Copper (Cu, copper sulphate) through nutrient media. Salicylic acid (SA) was firstly dissolved in small amount of Sodium hydroxide (NaOH) solution and then added to the mineral medium. After three days of germination plants were exposed to treatments including T1 (0.0mM SA + 0.5 µM Cu),T2 (0.1mM SA + 0.5µM Cu), T3 (1.0mM SA + 0.5µMCu), T4 (10.0mM SA + 0.5µM Cu), T5 (0.0mM SA + 2.5µM Cu), T6 (0.1mM SA + 2.5µM Cu), T7 (1.0mM SA + 2.5µM Cu), T8 (10.0mM SA + 2.5µM Cu), T9 (0.0 mM SA + 5.0µM Cu), T10 (0.1mM SA + 5.0µM Cu), T11 (1.0mM SA + 5.0µM Cu), T12 (10.0mM SA + 5.0µM Cu), T13 (0.0mM SA + 10.0µM Cu), T14 (0.1 mM SA + 10.0µM Cu), T15 (1.0mM SA + 10.0µM Cu) and T16 (10.0mM SA + 10.0µM Cu).

Nutrient medium along with corresponding copper and salicylic acid treatments were changed twice a week. The pots were slightly agitated and watered with minimum amount of tap water when required. Maize seedlings were harvested twice i.e., 15 and 30 days after sowing (DAS) during course of experimentation. Harvested seedlings were cleaned by mild shaking and plant length (cm), fresh and dry weights (g) of both shoots and roots were measured using standard procedure. Moreover, number of leaves and leaf area (cm^2), copper accumulation in shoots and roots were also recorded. Shoot and root dry weights of sampled plants were recorded from weighted freshly harvested plants, after covering each component in separate paper envelopes and dried for 48 h at 80°C. Di-acid digestion method was used to determine copper concentrations in roots and shoots by using atomic absorption spectrophotometer (Spurgeon et al., 1994).

Statistical analysis
Collected data was statistically analyzed with MSTAT-C and treatment means were compared using Duncan's Multiple Range Test (Duncan, 1955) at 5% level of probability.

Results

Shoot and Root Length
Maize plants harvest at 15 DAS showed that when treated with 0.lmM SA + 0.5 µM Cu, it produced maximum shoot and root length (Table 1). Whereas, minimum shoot length was produced when 10.0mM SA was applied in combination with5.0 µM Cu. Minimum root length was recorded in plants treated with 10.0 mM SA + 5.0 µM Cu. Maximum shoot length at 30 DAS was produced when maize plants were exposed to 0.1 mM SA and 0.5 µM Cu, however it was decreased subsequently when treated with higher concentration of coppers (0.lmM SA + 2.5 µM Cu, 0.0 mM SA + 5.0 µM Cu and 10.0mM SA + 2.5 µM Cu, respectively) (Table 2). Maize plants produced the maximum root length when treated with 0.lmM SA and 2.5 µM Cu. While, it was gradually decreased by following order as 10.0mM SA + 10.0 µM Cu, 1.0 mM SA with 5.0 µM Cu and 1.0 mM SA with 10.0 µM Cu, respectively.

Biomass Production
Highest fresh weight of shoot and root was produced when plants were treated with 0.lmM SA + 0.5 µM Cu, while lowest fresh weights of shoot and root were

observed under application of 10.0mM SA + 5.0 µM Cu (Table 3). Similarly, the maximum dry weights of shoots and roots were also produced under treatment with 0.lmM SA + 0.5 µM Cu; whereas lowest dry weights were observed when plants were treated with 10.0mM SA + 5.0 µM Cu (Table 3).Maximum fresh weights of both shoot and root at 30 DAS were recorded when plants were treated with 0.1 mM SA + 0.5 µM Cu (Table 4). However, shoot fresh weight was decreased at higher concentration of Cu as well as SA. The root fresh weight decreased as 1.0mM SA with 10.0 µM Cu, 1.0mM SA + 2.5 µM Cu, 0.0mM SA + 5.0 µM Cu and 10.0mM SA + 0.5 µM Cu respectively. Data recorded at 30 DAS indicated the maximum shoot and root dry weight at 0.1 mM SA + 0.5 µM Cu. Subsequent decline in shoot dry weight was observed in 0.1 mM SA + 5.0 µM Cu, 0.lmM SA + 2.5 µMCu, 10.0mM SA + 10.0 µMCu and 10.0mM SA + 0.5 µMCu, treatments respectively. The comparisons among treatment means indicate the maximum root dry weight being decreased by subsequent conditions as 0.1 mM SA + 5.0 µMCu, 0.0mM SA + 2.5 µMCu, 10.0 mM SA + 2.5 µM Cu and 10.0mM SA + 0.5 µMCu, respectively.

Number of leaves and leaf area
By comparing the means, treatment with 0.lmM SA + 0.5 µM Cu yielded maximum number of leaves and leaf area; while 10.0mM SA + 5.0 µM Cu produced minimum number of leaves and leaf area (Table 1). This is an outcome of 15days harvest. Plants harvest at 30 DAS indicates that the maximum numbers of leaves were recorded under 0.lmM SA + 0.5 µM Cu and minimum at I0.0mM SA + 2.5 µM. And comparisons among treatment means indicate the maximum leaf area at 0.lmM SA + 0.5 µM Cu and minimum at 10.0mM SA + 0.5 µM Cu (Table 2).

Copper Accumulation in shoots and roots
Plants exposed to different levels of copper and salicylic acid had no significant effect on copper accumulation in shoot at 15 DAS (Table 5).The accumulation of Cu in root observed at 0.0mM SA + 5.0 µM Cu was maximum.
When higher concentration of copper and salicylic acid (l0.0mM SA + 10.0 µM Cu) was applied, then no effect was recorded on copper accumulation in roots. But when higher concentration of salicylic acid and lower concentration of copper (10.0mM SA + 5.0 µM Cu) applied, the accumulation of copper in roots was

increased. The data presented indicate that the addition of different levels of copper and salicylic acid has no effect on copper accumulation in shoot at 30 DAS. The accumulation of Copper in root observed at 0.0mM SA + 5.0 µM Cu is more as compared to control (0.0mM SA + 0.5 µM Cu). When higher concentration of copper and salicylic acid (l0.0mM SA + 10.0 µM Cu) applied then no effect was calculated on copper accumulation in roots. But when higher concentration of salicylic acid and lower concentration of copper (10.0mM SA + 0.5 µM Cu and l0.0mM SA + 5.0 µM Cu) was applied, the accumulation of copper in roots increased as compared to control (0.0mM SA + 0.5 µM Cu).

Table -1: Effect of different levels of salicylic acid and copper on number of leaves, leaf area and lengths of root and shoot of maize seedlings at 15 days after sowing.

Treatment	SA (mM)	Cu (µM)	Number of Leaves	Shoot Length (cm)	Root Length (cm)	Leaf Area (cm^2)
T_1	0.0	0.5	5.5^{bcde}±0.19	46.46^{bcde}±1.57	28.61^{ab}±2.00	131.88^{ab}±10.50
T_2	0.1	0.5	6.2^{a}±0.16	52.71^{a}±1.25	31.81^{a}±1.74	145.75^{a}±9.99
T_3	1.0	0.5	5.5^{bcde}±0.19	47.99^{abed}±1.21	28.61^{be}±2.54	102.81^{be}±15.10
T_4	10.0	0.5	5.12^{de}±0.12	28.41^{g}±1.80	19.35^{d}±1.87	55.55^{e}±4.28
T_5	0.0	2.5	5.6^{bcd}±0.18	46.18^{cde}±1.43	25.18^{ab}±1.02	117.77^{abc}±17.08
T_6	0.1	2.5	5.9^{ab}±0.22	50.35^{abc}±1.17	28.39^{ab}±1.00	124.58^{abc}± 11.25
T_7	1.0	2.5	5.4^{bcde}±0.18	49.02^{abed}±1.40	26.76^{abe}±0.90	123.61^{abc}±3.26
T_8	10.0	2.5	5.4^{bcde}±0.18	33.02^{f}±0.92	25.48^{be}±2.72	71.88^{de}±6.76
T_9	0.0	5.0	5.9^{ab}±0.12	45.19^{de}±0.70	22.95^{cd}±1.49	105.51^{be} ±12.96
T_{10}	0.1	5.0	5.8^{abc}±0.16	51.10^{ab}±1.03	27.14^{abe}±1.77	120.38^{abc}±5.69
T_{11}	1.0	5.0	5.4^{bcde}±0.18	48.55^{abed}±1.88	29.55^{ab}±1.36	130.25^{ab}±2.24
T_{12}	10.0	5.0	5.4^{bcde}±0.18	35.05^{f}±2.40	25.25^{be}±1.42	96.66^{cd}±7.03
T_{13}	0.0	10.0	5.0^{e}±0.19	35.66^{f}±0.95	18.28^{d}±0.97	67.39^{e}±3.63
T_{14}	0.1	10.0	5.25^{cde} ±0.16	46.50^{bcde}±1.64	27.15^{abe}±1.01	117.32^{abc}±9.59
T_{15}	1.0	10.0	5.9^{ab} ±0.22	45.06^{de}±1.51	27.91^{abe}±2.00	118.42^{abc}±8.15
T_{16}	10.0	10.0	5.7^{abc}±0.16	42.69^{e}±1.34	22.52^{cd}±1.09	97.53^{cd}±7.36

Each value represents mean of 8 replicates ± SEM.
Means sharing the different letters are significantly different from each other's at 0.05%.
SA, Salicylic acid; Cu, copper

Table – 2: Effect of different levels of salicylic acid and copper on number of leaves, leaf area and lengths of root and shoot of maize seedlings at 30 days after sowing.

Treatment	SA (mM)	Cu (μM)	Number of Leaves	Shoot Length (cm)	Root Length (cm)	Leaf Area (cm²)
T_1	0.0	0.5	$9.12^{de}\pm0.23$	$71.84^{cd}\pm1.53$	$36.59^{cd}\pm1.27$	$338.35^{be}\pm18.8$
T_2	0.1	0.5	$10.62^{a}\pm0.30$	$81.32^{a}\pm0.61$	$44.80^{a}\pm1.30$	$445.31^{a}\pm28.8$
T_3	1.0	0.5	$8.62^{ef}\pm0.32$	$66.82^{ef}\pm2.16$	$38.82^{c}\pm1.20$	$272.90^{de}\pm23.5$
T_4	10.0	0.5	$8.00^{fg}\pm0.00$	$56.16^{g}\pm2.15$	$30.40^{e}\pm1.45$	$133.66^{f}\pm14.6$
T_5	0.0	2.5	$9.25^{cde}\pm0.31$	$72.61^{cd}\pm0.96$	$36.42^{cd}\pm1.05$	$306.58^{cd}\pm23.7$
T_6	0.1	2.5	$9.62^{bcd}\pm0.07$	$77.50^{ab}\pm1.23$	$45.35^{a}\pm0.98$	$304.58^{cd}\pm33.2$
T_7	1.0	2.5	$9.12^{de}\pm0.23$	$68.04^{def}\pm1.69$	$44.35^{a}\pm0.91$	$239.44^{e}\pm10.4$
T_8	10.0	2.5	$7.88^{fg}\pm0.13$	$57.70^{g}\pm2.77$	$30.28^{e}\pm1.54$	$161.64^{f}\pm19.3$
T_9	0.0	5.0	$8.25^{fg}\pm0.16$	$71.34^{cde}\pm1.03$	$6.65^{cd}\pm2.59$	$262.62^{de}\pm11.0$
T_{10}	0.1	5.0	$9.75^{bed}\pm0.06$	$79.26^{ab}\pm0.94$	$43.15^{ab}\pm1.13$	$263.14^{de}\pm13.1$
T_{11}	1.0	5.0	$10.00^{abc}\pm0.19$	$74.86^{be}\pm2.11$	$36.41^{cd}\pm1.31$	$364.52^{b}\pm29.6$
T_{12}	10.0	5.0	$8.00^{fg}\pm0.00$	$57.29^{g}\pm1.02$	$33.44^{de}\pm1.33$	$300.56^{cd}\pm6.9$
T_{13}	0.0	10.0	$7.75^{g}\pm0.16$	$54.64^{g}\pm1.06$	$30.66^{e}\pm0.46$	$116.69^{cd}\pm4.61$
T_{14}	0.1	10.0	$9.75^{bed}\pm0.37$	$70.56^{cdef}\pm1.82$	$38.44^{c}\pm0.75$	$322.22^{bed}\pm31.5$
T_{15}	1.0	10.0	$10.12^{ab}\pm0.18$	$78.68^{ab}\pm0.71$	$43.75^{a}\pm0.88$	$378.09^{b}\pm5.9$
T_{16}	10.0	10.0	$10.12^{ab}\pm0.30$	$66.34^{f}\pm1.47$	$39.76^{be}\pm0.93$	$337.89^{be}\pm6.5$

Each value represents mean of 8 replicates ± SEM.
Means sharing the different letters are significantly different from each other's at 0.05%.
SA, Salicylic acid; Cu, copper

Table- 3: Effect of different levels of salicylic acid and copper on fresh and dry weights of maize seedlings at 15 days after sowing.

Treatment	SA (mM)	Cu (μM)	Fresh Weight(g)		Dry Weight(g)	
			Shoot	Root	Shoot	Root
T_1	0.0	0.5	$3.34^{a}\pm0.21$	$1.21^{abc}\pm0.16$	$0.30^{ab}\pm0.02$	$0.21^{ab}\pm0.02$
T_2	0.1	0.5	$3.46^{a}\pm0.17$	$1.48^{a}\pm0.19$	$0.31^{a}\pm0.02$	$0.23^{a}\pm0.02$
T_3	1.0	0.5	$2.08^{abc}\pm0.12$	$1.25^{abc}\pm0.10$	$0.18^{def}\pm0.01$	$0.09^{fgh}\pm0.01$
T_4	10.0	0.5	$1.08^{c}\pm0.13$	$0.65^{fg}\pm0.06$	$0.11^{g}\pm0.01$	$0.08^{gh}\pm0.01$
T_5	0.0	2.5	$2.81^{ab}\pm0.18$	$0.94^{cdefg}\pm0.10$	$0.24^{bed}\pm0.02$	$0.11^{fgh}\pm0.02$
T_6	0.1	2.5	$2.87^{ab}\pm0.36$	$1.09^{bed}\pm0.17$	$0.24^{bed}\pm0.03$	$0.17^{ede}\pm0.02$
T_7	1.0	2.5	$2.83^{ab}\pm0.26$	$1.04^{bcde}\pm0.16$	$0.24^{bed}\pm0.02$	$0.13^{efg}\pm0.02$
T_8	10.0	2.5	$1.56^{be}\pm0.17$	$0.63^{fg}\pm0.05$	$0.15^{egf}\pm0.01$	$0.09^{gh}\pm0.01$
T_9	0.0	5.0	$2.24^{abe}\pm0.19$	$0.59^{g}\pm0.04$	$0.21^{cd}\pm0.02$	$0.10^{fgh}\pm0.01$
T_{10}	0.1	5.0	$2.36^{abe}\pm0.16$	$0.88^{cdefg}\pm0.04$	$0.23^{cd}\pm0.02$	$0.17^{cde}\pm0.02$
T_{11}	1.0	5.0	$2.82^{ab}\pm0.26$	$0.93^{cdefg}\pm0.08$	$0.25^{bc}\pm0.02$	$0.14^{def}\pm0.01$
T_{12}	10.0	5.0	$2.32^{abc}\pm0.20$	$0.71^{efg}\pm0.01$	$0.14^{fg}\pm0.01$	$0.11^{fgh}\pm0.02$
T_{13}	0.0	10.0	$1.62^{be}\pm0.07$	$0.57^{g}\pm0.07$	$0.11^{g}\pm0.01$	$0.07^{h}\pm0.02$
T_{14}	0.1	10.0	$2.43^{abe}\pm0.21$	$0.99^{cdef}\pm0.10$	$0.22^{cd}\pm0.02$	$0.12^{efg}\pm0.01$
T_{15}	1.0	10.0	$3.37^{a}\pm0.25$	$1.38^{ab}\pm0.20$	$0.30^{ab}\pm0.02$	$0.20^{abe}\pm0.02$
T_{16}	10.0	10.0	$2.74^{ab}\pm0.10$	$0.77^{defg}\pm0.20$	$0.20^{cde}\pm0.02$	$0.18^{bed}\pm0.01$

Each value represents mean of 8 replicates ± SEM.
Means sharing the different letters are significantly different from each other's at 0.05%.
SA, Salicylic acid; Cu, copper

Table – 4: Effect of different levels of salicylic acid and copper on fresh and dry weights of maize seedlings at 30 days after sowing.

Treatment	SA (mM)	Cu (µM)	Fresh Weight (g)		Dry Weight (g)	
			Shoot	Root	Shoot	Root
T_1	0.0	0.5	9.08[bc]±0.49	4.78[bcd]±0.13	0.94[cd]±0.06	0.46[be]±0.02
T_2	0.1	0.5	11.20[a]±0.62	6.82[a]±0.23	1.19[a]±0.08	0.62[a]±0.06
T_3	1.0	0.5	6.43[e]±0.73	4.72[bed]±0.17	0.76[def]±0.08	0.41[cd]±0.02
T_4	10.0	0.5	3.59[f]±0.58	2.79[f]±0.35	0.44[h]±0.04	0.23[e]±0.03
T_5	0.0	2.5	8.83[c]±0.42	4.05[de]±0.63	0.94[cd]±0.06	0.41[cd]±0.10
T_6	0.1	2.5	11.00[a]±0.73	5.51[b]±0.14	0.86[cdef]±0.06	0.47[be]±0.01
T_7	1.0	2.5	6.68[de]±0.52	4.17[ede]±0.20	0.73[ef]±0.06	0.31[de]±0.05
T_8	10.0	2.5	3.96[f]±0.47	2.59[f]±0.11	0.53[gh]±0.06	0.24[e]±0.02
T_9	0.0	5.0	8.54[c]±0.58	3.71[e]±0.30	0.79[cdef]±0.07	0.38[cd]±0.02
T_{10}	0.1	5.0	10.60[a]±0.51	5.26[b]±0.48	0.98[bc]±0.07	0.47[be]±0.04
T_{11}	1.0	5.0	8.50[c]±0.33	4.61[bcde]±0.39	0.92[cde]±0.07	0.42[cd]±0.04
T_{12}	10.0	5.0	6.44[e]±0.12	2.70[f]±0.26	0.86[cdef]±0.03	0.25[e]±0.03
T_{13}	0.0	10.0	5.60[e]±0.14	2.59[f]±0.09	0.41[h]±0.04	0.22[e]±0.02
T_{14}	0.1	10.0	10.30[ab]±0.27	4.09[cde]±0.37	0.72[ef]±0.04	0.44[c]±0.02
T_{15}	1.0	10.0	11.20[a]±0.17	5.12[be]±0.20	1.15[ab]±0.11	0.58[ab]±0.02
T_{16}	10.0	10.0	7.87[cd]±0.18	4.44[cde]±0.19	0.69[fg]±0.01	0.45[c]±0.02

Each value represents mean of 8 replicates ± SEM.

Means sharing the different letters are significantly different from each other's at 0.05%.

SA, Salicylic acid; Cu, copper

Table- 5: Effect of different levels of salicylic acid and copper on copper accumulation in maize seedlings.

Treatment	SA(mM)	Cu(µM)	Cu accumulation (ppm)			
			15 DAS		30 DAS	
			Shoot	Root	Shoot	Root
T_1	0.0	0.5	0.0092[a]±0.000	0.0425[h]±0.0005	0.0095[a]±0.0003	0.0535[d]±0.0003
T_2	0.1	0.5	0.0100[a]±0.000	0.0418[h]±0.0004	0.0100[a]±0.0003	0.0518[d]±0.0003
T_3	1.0	0.5	0.0101[a]±0.000	0.0410[h]±0.0004	0.0101[a]±0.0003	0.0462[d]±0.0004
T_4	10.0	0.5	0.0098[a]±0.000	0.0522[f]±0.0004	0.0098[a]±0.0002	0.0687[c]±0.0003
T_5	0.0	2.5	0.0095[a]±0.000	0.0415[h]±0.000I	0.0098[a]±0.0002	0.0462[d]±0.0004
T_6	0.1	2.5	0.0099[a]±0.000	0.0420[h]±0.0003	0.0096[a]±0.0003	0.0532[d]±0.0002
T_7	1.0	2.5	0.0095[a]±0.000	0.0415[h]±0.0004	0.0101[a]±0.0002	0.0470[d]±0.0004
T_8	10.0	2.5	0.0098[a]±0.000	0.0622[e]±0.0006	0.0102[a]±0.0002	0.0715[c]±0.0004
T_9	0.0	5.0	0.0099[a]±0.000	0.0735[d]±0.0004	0.0099[a]±0.0002	0.0818[c]±0.0004
T_{10}	0.1	5.0	0.0099[a]±0.000	0.0435[gh]±0.0003	0.0105[a]±0.0002	0.0538[d]±0.0004
T_{11}	1.0	5.0	0.0098[a]±0.000	0.0542[f]±0.0004	0.0098[a]±0.0002	0.0680[c]±0.0005
T_{12}	10.0	5.0	0.0102[a]±0.000	0.0845[c]±0.0003	0.0100[a]±0.0003	0.0968[b]±0.0004
T_{13}	0.0	10.0	0.0101[a]±0.000	0.1070[a]±0.0006	0.0099[a]±0.0002	0.1178[a]±0.0005
T_{14}	0.1	10.0	0.0101[a]±0.000	0.0910[b]±0.0005	0.0099[a]±0.0003	0.1130[a]±0.0004
T_{15}	1.0	10.0	0.0105[a]±0.000	0.0728[a]±0.0005	0.0101[a]±0.0002	0.0782[c]±0.0003
T_{16}	10.0	10.0	0.0100[a]±0.000	0.0492[fg]±0.0005	0.0105[a]±0.0002	0.0542[d]±0.0014

Each value represents mean of 8 replicates ± SEM.

Means sharing the different letters are significantly different from each other's at 0.05%.

SA, Salicylic acid; Cu, copper

Discussion

Salicylic acid is a natural plant hormone produced in response to adverse environmental condition (Hussein et al., 2007), and has also been applied exogenously to induce plants tolerance against biotic and abiotic stresses (Shakirova et al., 2003; Sawada et al., 2006; Zhou et al., 2009). Zhang et al. (2008) reported that excessive copper reduced plant root length, root dry weight, total dry weight, and root to shoot ratio and leaf area. The present study was conducted to observe the effect of copper with or without salicylic acid on maize plant growth. The higher concentration of SA (10.0 mM) reduced the growth of maize plants; while only the lower concentration (0.1 mM) was efficient in improving growth parameters. The Higher concentration of Copper (1 0.0 μM) is toxic and significantly reduces the growth of maize plants. In the present study, a negative relationship was observed between growth parameters and increasing copper concentration at both harvests (15 and 30 DAS).

The shoot length in 1.0 μM Cu alone treatment, was decreased by23.25% at15 DAS and 23.94 % at 30 DAS of the control (0.5 μM Cu). However, no significant effect was found at 2.5 μM Cu and 5.0 μM Cu. The higher concentration may have a negative effect on plant growth ultimately leading towards lower absorption of water. In a similar study, it was found that root length was decreased due to copper toxic concentration (Martín et al., 2006). The higher concentration of copper (10.0 μM) without salicylic acid showed toxic effect on root length which decreased by 36.11 % at 15 DAS. The main symptoms of excess copper can be detected in roots by a decrease in root growth (Martins and Mourato, 2006). Ouzounidou et al. (1995) also showed that excess copper in maize roots affected both root growth and ultra-structure. Shoot fresh weight and root fresh weight also decreased at higher level (10.0 μM Cu) by 51.50 % and 52.89 % respectively at 15 DAS and also decreased at higher level (10.0 μM Cu) by 38.33 % and 45.82 respectively at 30 DAS as compared of the control (0.5 μM Cu). Martins and Mourato (2006) observed that fresh weight reduced to 73 % of the control.

Shoot dry weight and root dry weight also decreased at higher level (l0.0 μMCu) by 63.33 % and 61.90 % respectively at 15 DAS and also decreased at higher level (10.0 μM Cu) by 56.38 % and 52.17 respectively at 30 DAS as compared of the control (0.5 μM Cu). The root dry weight also decreased at 5.0 μM Cu by 52.38 % and 17.59 % at 15 and 30 DAS respectively of the control. The number of leaves decreased only at 10.0 μM Cu by 9.09 % and 15.02 % at 15 and 30 DAS as compared of control. The copper accumulated in shoot has no significant result while root has great significant result. Copper is immobile and much accumulated in root as compared to shoot. The copper is highly toxic to root as compared to shoot. The shoot effect is due to root effect.

Restricted translocation of copper from root to shoot was also reported on observed in *Agrostis stolonifera* (Wu et al., 1975). Copper has no effect on copper accumulation in shoot. The copper accumulation in roots increased by 151.76 % and 120.19 % at 0.0mM SA + 1 0.0 μM Cu at 15 and 30 DAS respectively of sowing as compared to control (0.0mM SA + 0.5 μM Cu). Chen and Kao (2000) observed a progressively decreasing root length after increasing concentrations of copper from 20 to 50 μM. The accumulation of copper in root observed at 0.0 mM SA with 5.0 μM Cu 72.94 % and 52.90 % more respectively as compared to control (0.0mM SA + 0.5 μM Cu) at 15 and 30 DAS. The root has tendency to accumulate to higher quality of copper than aerial parts of the plants (Cathala and Salsac, 1975).

Salicylic acid acts as endogenous signal molecule responsible for inducing abiotic stress tolerance in plants. They emphasized that exogenous application of Salicylic acid increased plant growth (Bastam et al., 2013).

The present study showed that the application of Salicylic acid on growth of maize plant improved all growth characters i.e., shoot &root length, fresh and dry weight of both shoot and root and also area of green leaves. The highest increment was shown in dry weight of shoot and root at 30 DAS when treated with 0.lm mM SA by 26.60 % and 34.78 % respectively of the control (0.0 mM SA). Hussein et al. (2007) concluded that the highest increment was shown in stem dry weight and lowest in stem diameter when treated with Salicylic acid. The shoot length increased at 0.1 mM SA by 13 % of the control. Moreover the shoot & root length, fresh weight of shoot &root, dry weight of shoot & root and numbers of leaves similar response. The effect of Salicylic acid is reflected in higher fresh weight and dry weights in sunflower plants (El-Tayeb et al., 2006). The root length increased at 0.1 mM SA by 11.19 % at 15 DAS and 22.44 % at 30 DAS of the control (0.0 mM SA).

Shoot fresh weight and root fresh weight also increased at lower level of Salicylic acid (0.1 mM SA) by 3.59 % and 22.31 % respectively at 15 DAS and also increased at 0.1 mM SA by 23.35 % and 42.68 respectively at 30 DAS as compared of the control (0.0 mM SA). Shoot dry weight and root dry weight also increased at lower level (0.1 mM SA) by 3.33 % and 9.52 % respectively at 15 DAS and also increased at lower level (0.1 mM SA) by 26.60 % and 34.78 respectively at 30 DAS as compared of the control (0.0 mM SA). The number of leaves increased at 0.1 mM SA by 12.73 % and 16.45 % at 15 and 30 DAS as compared of control (0.0 mM SA). The area of green leaves increased at 0.1 mM SA by 10.52 % and 31.63 % at 15 and 30 DAS as compared of control (0.0 mM SA).

Salicylic acid has no effect on copper accumulation in shoot. The accumulation of copper in shoots at 15 and 30 DAS were not observed significantly different but in case of root higher concentration of salicylic acid and lower concentration of copper (10.0mM SA + 0.5 μM Cu) and (l0.0mM SA + 5.0 μM Cu) applied the accumulation of copper increased by 22.82 % and 98.82 % of the control (0.0mM SA + 0.5 μM Cu) at 15 DAS and higher concentration of salicylic acid and lower concentration of copper (l0.0mM SA + 0.5 μM Cu) and (10.0mM SA + 5.0 μM Cu) applied the accumulation of copper in roots increased by 28.41 % and 80.93 % of the control (0.0mM SA + 0.5 μM Cu) at 30 DAS. Hussein et al. (2007) reported that salicylic acid application increased number of leaves, leave area and plant dry matter. They also reported that salicylic acid improved dry matter accumulation under saline and non-saline condition. However salicylic acid was found more active in plants grown under normal condition. The interaction of salicylic acid and copper on growth plants were significant effect. The shoots length, roots length, fresh and dry weights of shoot and root, number of leaves and leaf area increased at when lower concentration of salicylic acid is applied (1.0 mM SA + 10.0 μM Cu) by 9.52 % 19.57 %, 23.35 %, 7.11 %,22.34 %, 26.09 % 10.96 % and 11.76 % respectively of the control (0.0 mM SA + 0.5 μM Cu). Sakhabutdinova et al. (2003) concluded that the salicylic acid treatment reduce the damage action of salinity and water deficit on wheat seedling growth. At Higher concentration of copper and salicylic acid (10.0mM SA + 10.0 μM Cu) no effect was observed on copper accumulation in roots at 15 and 30 DAS.

Conclusion

Contrasting results were observed by the application of different levels of salicylic acid. Application of lower concentration of salicylic acid (0.1 mM) significantly increased biomass production, shoot and root lengths, number and area of leaves, whereas application of higher level of salicylic acid (10.0 Mm) had inhibitory effects. Maize seedlings harvested at 15 and 30 DAS produced highest copper accumulations in roots when treated with 10 mM salicylic acid and 5 μM copper contamination. However, minimum copper accumulation was observed when maize seedlings were treated with 10 mM salicylic acid along with 0.5 μM copper. Copper accumulation in shoot remained uninfluenced by different levels of copper and salicylic acid. Application of salicylic acid at lower concentration can have potential to compensate reduction in plant growth under copper contaminated areas.

References

Adrees M, Ali S, Rizwan M, Ibrahim M, Abbas F, Farid M, Zia-ur-Rehman M, Irshad MK and Bharwana SA, 2015. The effect of excess copper on growth and physiology of important food crops: a review. Environ. Sci. Pollut. Res. 22(11): 8148-8162.

Bashmakov DI, Lukatkin AS, Revin VV, Duchovskis P, Brazaitytë A and Baranauskis K, 2005. Growth of maize seedlings affected by different concentrations of heavy metals. Ekologija. 3: 22-27.

Bastam N, Baninasab B and Ghobadi C, 2013. Improving salt tolerance by exogenous application of salicylic acid in seedlings of pistachio. Plant Growth Regul. 69(3): 275-284.

Bennett LE, Burkhead JL, Hale KL, Terry N, Pilon M and Pilon-Smits, 2003. Analysis of transgenic Indian mustard plants for phytoremediation of metal-contaminated mine tailings. J. Environ. Qual. 32(2): 432-440.

Borkert C, Cox F and Tucker M, 1998. Zinc and copper toxicity in peanut, soybean, rice, and corn in soil mixtures. Comm. Soil Sci. Plant Anal. 29(19-20): 2991-3005.

Brennan A, Jiménez EM, Puschenreiter M, Alburquerque JA and Switzer C, 2014. Effects of biochar amendment on root traits and contaminant availability of maize plants in a copper and arsenic impacted soil. Plant Soil. 379(1): 351-360.

Brun L, Le Corff J and Maillet J, 2003. Effects of elevated soil copper on phenology, growth and reproduction of five ruderal plant species. Environ. Pollut. 122(3): 361-368.

Cathala N and Salsac L, 1975. Absorption du cuivre par les racines de mais (*Zea mays* L.) et de tournesol (*Helianthus annuus* L.). PlantSoil. 42(1): 65-83.

Chen L-M, Lin CC and Kao CH, 2000. Copper toxicity in rice seedlings: changes in antioxidative enzyme activities, H_2O_2 level, and cell wall peroxidase activity in roots. Bot. Bull. Acad. Sinica. 41:99-103.

Clemens S and Ma JF, 2016. Toxic heavy metal and metalloid accumulation in crop plants and foods. Annu. Rev. Plant Biol. 67(1): 489-512.

Duncan DB, 1955. Multiple range and multiple F tests. Biometrics. 11(1): 1-42.

Ehsan N, Nawaz R, Ahmad S, Khan MM and Hayat J, 2016. Phytoremediation of chromium-contaminated soil by an ornamental plant, vinca (*Vinca rosea* L.). J.Environ. Agric. Sci. 7: 29-34.

El-Tayeb M, El-Enany A and Ahmed N, 2006. Salicylic acid-induced adaptive response to copper stress in sunflower (*Helianthus annuus* L.). Plant Growth Regul. 50(2-3): 191-199.

Hayat Q, Hayat S, Irfan M and Ahmad A, 2010. Effect of exogenous salicylic acid under changing environment: A review. Environ. Exp. Bot. 68(1): 14-25.

Hussain M, Malik MA, Farooq M, Ashraf MY, Cheema MA, 2008. Improving drought tolerance by exogenous application of glycinebetaine and salicylic acid in sunflower. J. Agron. Crop Sci. 194(3): 193-199.

Hussein M, Balbaa L and Gaballah M, 2007. Salicylic acid and salinity effects on growth of maize plants. Res. J. Agric. Biol. Sci. 3(4): 321-328.

Kopliku D and Mesi A, 2012. Correlative evaluation between experimental copper and lead Ion concentrations and root length of *Allium cepa* L. in some riverside points of NënShkodra lowland. J. Int. Environ. Appl. Sci. 7(5): 913.

Krämer U, 2010. Metal hyperaccumulation in plants. Annu. Rev. Plant Biol. 61(1): 517-534.

Lexmond TM, Van der Vorm P, 1981. The effect of pH on copper toxicity to hydroponically grown maize. Netherlands J. Agric. Sci. 29:217-238.

Liu JJ, Wei Z, and Li JH, 2014. Effects of copper on leaf membrane structure and root activity of maize seedling. Bot. Stud. 55(1): 47.

Liu Z, Ding Y, Wang F, Ye Y and Zhu C, 2016. Role of salicylic acid in resistance to cadmium stress in plants. Plant Cell Rep. 35(4): 719-731.

Lombi E, Zhao F, Dunham S and McGrath S, 2001. Phytoremediation of heavy metal–contaminated soils. J. Environ. Qual., 30(6): 1919-1926.

Macnicol R and Beckett P, 1985. Critical tissue concentrations of potentially toxic elements. Plant Soil. 85(1): 107-129.

Madejon P, Ramirez-benitez JE Corrales I, Barceló J and Poschenrieder C, 2009. Copper-induced oxidative damage and enhanced antioxidant defenses in the root apex of maize cultivars differing in Cu tolerance. Environ Exp Bot. 67: 415-420.

Maksymiec W and Krupa Z, 2006. The effects of short-term exposition to Cd, excess Cu ions and jasmonate on oxidative stress appearing in Arabidopsis thaliana. Environ. Exp. Bot. 57(1): 187-194.

Martín JAR, Arias ML and Corbí JMG, 2006. Heavy metals contents in agricultural topsoils in the Ebro basin (Spain). Application of the multivariate geoestatistical methods to study spatial variations. Environ. Pollut. 144(3): 1001-1012.

Martins LL and Mourato MP, 2006. Effect of excess copper on tomato plants: growth parameters, enzyme activities, chlorophyll, and mineral content. J. Plant Nutr. 29(12): 2179-2198.

Minnich M, McBride M, and Chaney R, 1987. Copper activity in soil solution: II. Relation to copper accumulation in young snapbeans. Soil Sci. Soc. Am.J. 51(3): 573-578.

Mocquot B, Vangronsveld J, Clijsters H and Mench M, 1996. Copper toxicity in young maize (*Zea mays* L.) plants: effects on growth, mineral and chlorophyll contents, and enzyme activities. Plant Soil. 182(2): 287-300.

Najam S, Nawaz R, Ahmad S, Ehsan N, Khan MM and Nawaz MH, 2015. Heavy metals contamination of soils and vegetables irrigated with municipal wastewater: A case study of Faisalabad, Pakistan. J. Environ. Agric. Sci. 4: 6-10.

Noreen S and Ashraf M, 2010. Modulation of salt (NaCl)-induced effects on oil composition and fatty acid profile of sunflower (*Helianthus annuus* L.) by exogenous application of salicylic acid. J. Sci. Food Agric. 90(15): 2608-2616.

O'connor CS, Lepp N, Edwards R and Sunderland G, 2003. The combined use of electrokinetic remediation and phytoremediation to decontaminate metal-polluted soils: a laboratory-scale feasibility study. Environ. Monit. Assess. 84(1-2): 141-158.

Ouzounidou G, Čiamporová M, Moustakas M and Karataglis S, 1995. Responses of maize (Zea mays L.) plants to copper stress—I. Growth, mineral content and ultrastructure of roots. Environ. Exp. Bot. 35(2): 167-176.

Ouzounidou G and Ilias I, 2005. Hormone-induced protection of sunflower photosynthetic apparatus against copper toxicity. Biolog. Plant. 49(2): 223-228.

Rivas-San VM and Plasencia J, 2011. Salicylic acid beyond defence: its role in plant growth and development. J. Exp. Bot. 62(10): 3321-3338.

Rizwan M, Ali S, Qayyum MF, Ok YS, Adrees M, Ibrahim M, Zia-ur-Rehman M, Farid M and Abbas F, 2017. Effect of metal and metal oxide nanoparticles on growth and physiology of globally important food crops: A critical review. J. Hazard. Mater. 322: 2-16.

Sağlam A, Yetişsin F, Demiralay M and Terzi R, 2016. Chapter 2 - Copper Stress and Responses in Plants A2 - Ahmad, Parvaiz. Plant Metal Interaction. Elsevier. pp. 21-40.

Sakhabutdinova A, Fatkhutdinova D, Bezrukova M and Shakirova F, 2003. Salicylic acid prevents the damaging action of stress factors on wheat plants. Bulg. J. Plant Physiol. 21: 314-319.

Sawada H, Shim I-S, and Usui K, 2006. Induction of benzoic acid 2-hydroxylase and salicylic acid biosynthesis—modulation by salt stress in rice seedlings. Plant Sci. 171(2): 263-270.

Shakirova FM, Allagulova CR, Maslennikova DR, Klyuchnikova EO, Avalbaev AM and Bezrukova MV, 2016. Salicylic acid-induced protection against cadmium toxicity in wheat plants. Environ. Exp. Bot. 122: 19-28.

Shakirova FM, Sakhabutdinova AR, Bezrukova MV, Fatkhutdinova RA and Fatkhutdinova DR, 2003. Changes in the hormonal status of wheat seedlings induced by salicylic acid and salinity. Plant Sci. 164(3): 317-322.

Shakoor MB, Nawaz R, Hussain F, Raza M, Ali S, Rizwan M, Oh S-E and Ahmad S, 2017. Human health implications, risk assessment and remediation of As-contaminated water: A critical review. Sci. Total Environ. 601: 756-769.

Spurgeon DJ, Hopkin SP and Jones DT, 1994. Effects of cadmium, copper, lead and zinc on growth, reproduction and survival of the earthworm *Eisenia fetida* (Savigny): Assessing the environmental impact of point-source metal contamination in terrestrial ecosystems. Environ. Pollut. 84(2): 123-130.

Steel RGD, Torrie JH and Dickey D, 1980. Principles and procedures of statistics: a biometrical approach. McGraw-Hill, New York.

Strobel NE and Kuc JA, 1995. Chemical and biological inducers of systemic resistance to pathogens protect cucumber and tobacco plants from damage caused by paraquat and CuCl2. Phytopathology. 85:1306–1310

Wang Q-R, Cui Y-S, Liu X-M, Dong Y-T and Christie P, 2003. Soil contamination and plant uptake of heavy metals at polluted sites in China. J. Environ. Sci. Health. Part A. 38(5): 823-838.

Wani AB, Chadar H, Wani AH, Singh S and Upadhyay N, 2017. Salicylic acid to decrease plant stress. Environ. Chem. Lett. 15(1): 101-123.

Wang C, Zhang S, Wang P, Hou J, Qian J, Ao Y, Lu J and Li L, 2011. Salicylic acid involved in the regulation of nutrient elements uptake and oxidative stress in Vallisneria natans (Lour.) Hara under Pb stress. Chemosphere. 84(1):136–142.

Weckx JEJ and Clijsters HMM, 1996. Oxidative damage and defense mechanisms in primary leaves of *Phaseolus vulgaris* as a result of root assimilation of toxic amounts of copper. Physiol. Plant. 96(3): 506-512.

Wu L, Thurman D and Bradshaw A, 1975. The uptake of copper and its effect upon respiratory processes of roots of copper- tolerant and non- tolerant clones of Agrostis stolonifera. New Phytol. 75(2): 225-229.

Zhang LP, Mehta, Liu ZP and Yang ZM, 2008. Copper-induced proline synthesis is associated with nitric oxide generation in *Chlamydomonas reinhardtii*. Plant Cell Physiol. 49(3): 411-419.

Zhao X-F, Chen L, Rehmani MIA, Wang Q-S, Wang S-H, Hou P-F, Li G-H and Ding Y-F, 2013. Effect of nitric oxide on alleviating cadmium toxicity in rice (*Oryza sativa* L.). J. Integ. Agric. 12(9): 1540-1550.

Zhou ZS, Guo K, Elbaz AA and Yang ZM, 2009. Salicylic acid alleviates mercury toxicity by preventing oxidative stress in roots of Medicago sativa. Environ. Exp. Bot. 65(1): 27-34.

16

Comparative and Interactive Effects of Organic and Inorganic Amendments on Soybean Growth, Yield and Selected Soil Properties

Aqila Shaheen*, Rabia Tariq and Abdul Khaliq
Department of Soil and Environmental Sciences, The University of Poonch, Rawalakot, Azad Jammu and Kashmir

Abstract

The absolute use of inorganic fertilizers, growing of exhaustive crops, nutrient losses with runoff and leaching under mountainous sub humid conditions has declined the soil fertility and productivity. The field experiment was carried out to study the comparative and interactive effect of organic and inorganic amendments on soybean growth, yield and soil properties. The experimental design was randomized complete block design (RCBD) with three replications. The treatments combination were control (no amendments); 100 kg N ha^{-1} from urea nitrogen (UN100); 100 kg N ha^{-1} from poultry manure (PMN100); 100 kg N ha^{-1} from sawdust (SDN100); 100 kg N ha^{-1} from UN + PM (UN50 + PMN50); 100 kg N ha^{-1} from UN + SD (UN50 + SDN50); 100 kg N ha^{-1} from PM + SD (PMN50 + SDN50); 100 kg N ha^{-1} from UN + SD + PM (UN50 + SDN25 + PMN25). Results indicated higher crop growth in UN50 + SDN25 + PMN25. However, SDN100 showed lower growth but higher than control. UN100 had statistically higher grain yield (1322.7 kg ha^{-1}) and it was non-significant with UN50+PMN50 and UN50+ PMN25+SDN25. Nitrogen uptake (156.55 kg ha^{-1}) was higher in UN100, UN50+ PMN50 and UN50+ PMN25+SDN25. Post-harvest soil properties showed the minimum pH in SDN100 and higher organic matter in organic and integrated applications. The higher phosphorus contents were in UN50+PMN50. This study showed that SD and PM combined with urea have potential in soybean growth enhancement, yield increase and in improvement of soil properties.

Keywords: Poultry manure, sawdust, integrated nutrient management, soybean

*Corresponding author email:
aqeela.shaheen@gmail.com

Introduction

Soil fertility is very important for maximum crop production and sustainable agriculture. The role of organic matter (OM) is universally understood in improving and sustaining soil fertility and productivity. Smallholder farmers in the Rawalakot (Latitude 33° 51 to 33° 85 N and Longitude 73° 48 to 73° 80 E) Azad Kashmir are facing problems of declining soil fertility due to cultivation of exhaustive crops (wheat-maize), nutrient removal with crop harvest and soil erosion. The topographic features of terrain, erratic and high intensity rainfall, surface runoff and leaching of nutrients further augment the problem of soil fertility (Tiwari *et al.*, 2010) and lower the nitrogen use efficiency (Zaman *et al.*, 2009). The use of mineral N fertilizers not only diminishes soil OM but also causes soil acidification and micronutrient deficiencies (Abera *et al.*, 2012).

The soybean (*Glycine max*) is a legume and if properly inoculated, can use the nitrogen in the atmosphere (N$_2$) for plant growth. Therefore, nitrogen fertilizer is not needed for soybean production in most situations. In the previous studies conducted at Rawalakot Azad

Jammu and Kashmir, exotic Bardyrhizobium displayed a significant increase in yields of soybean compared to the non-inoculated control (Abbasi *et al.*, 2008; Abbasi *et al.*, 2010). But for higher soybean yield in N deficient soils, without inoculum higher application of N fertilizers is required. Soybean's upper limit for N fixation (considered to be about 300 lb/acre) combined with the upper limit of the soil supply (usually less than 100 lb/acre) are insufficient to meet the needs of a 100 bu/acre soybean crop (Salvagiotti et al., 2008). The higher rate of N application (100 kg ha-1) without inoculum gives higher soybean yield in sub-humid, N-poor region of Rawalakot (Khaliq et al., 2015). However, when inorganic fertilizers are applied repeatedly soil degradation due to loss of OM along with continuous cropping becomes worsen. Pandey *et al.,* (2006) reported that application of manures, irrespective of sources and rates recorded significantly higher SOC, N, P_2O_5 and K_2O compared to control.

Poultry manure is nutrient rich organic amendment compare to other organic amendments (Ano and Agwu, 2005). Organic amendments has been a rich source of nutrients and mostly added to soil on the N crop requirements (Qian *et al.*, 2004). The sawdust is another cheap and easily available organic waste and has the potential to supply nutrients to crops (Owolabi *et al.*, 2003). Sawdust immobilizes soil N (Cogger, 2005) consequently, additional N is necessary with sawdust to counterbalance the immobilization. Organic inputs are low-cost and ecofriendly but the full benefits cannot be obtained from organic amendment as its bulky hence higher transportation and management cost is required in its handling. Therefore, integrated farming system is an agricultural system conceived so as combating environmental degradation on one hand and increase productivity on other hand. The integrated application of organic and inorganic nutrients results in higher crop productivity (Adeniyan and Ojeniyi, 2005), buildup of soil organic matter and higher level of major soil nutrients especially nitrogen and phosphorus (Huang *et al.*, 2007). Earlier studies have reported higher soybean-wheat productivity with integrated application of organic manures and inorganic fertilizers (Bhattacharyya *et al.*, 2010; Shah *et al.*, 2009). Keeping in view the importance of soybean and easy availability of organic amendments i.e. poultry manure and saw dust, study was planned with the objective:

- To study the comparative effects of organic amendments (SD and PM) and urea N on soybean growth and yield.
- To study the comparative effects of organic and inorganic amendments on selective soil properties.

Materials and Methods

Experiment description and treatments
A field trial was conducted at chottagala experimental farm area of The University of Poonch Rawalakot, Azad Jammu & Kashmir-Pakistan. The experiment was randomized complete block design (RCBD) and each treatment was replicated thrice. The treatments were including application of organic and inorganic sources and the rate of sources was calculated on the N equivalent basis at the rate of 100 kg N ha^{-1} (Khaliq *et al.*, 2015). The treatment combinations were:

Control (no amendments)
UN_{100};   100 kg N ha^{-1} from urea nitrogen
PMN100;   100 kg N ha^{-1} from poultry manure
SDN100;   100 kg N ha^{-1} from sawdust
UN_{50} + PMN50;   100 kg N ha^{-1} from UN + PM
UN_{50} + SDN50;   100 kg N ha^{-1} from UN + SD
PMN50 + SDN50;   100 kg N ha^{-1} from PM + SD
UN_{50} + SDN25 + PMN25;   100 kg N ha^{-1} from UN + SD + PM

Poultry manure was collected from local poultry farms and sawdust was collected from local market. Saw dust was partially decomposed and before application it was properly mixed, air dried and was mixed in the field one month before crop sowing. Poultry manure was well decomposed and applied to field at sowing of crop. Poultry manure had 1.8% nitrogen while saw dust had 0.5% N. Soybean was sown in the month of May. The basal doses of 60 kg ha^{-1}of K_2O and 40 kg ha^{-1}of P_2O_5 were applied as sulphate of potash (SOP) and single super phosphate (SSP) respectively. Urea nitrogen was applied half at the time of sowing of crop and half at the stage of nodule formation.

Soil analysis
Soil samples were collected at 15 cm depth before sowing and after the harvest of crop for analysis. Soil mechanical analysis was done by Gee and Bauder, 1986.

The bulk density was calculated on volume basis as follows.

$$Bulk\ density = \frac{Oven\ dry\ weight\ of\ soil\ (g)}{Field\ volume\ of\ soil\ (cm3)}$$

Soil pH was determined by soil water suspension of 1:2 (Mc Lean, 1982). Soil organic matter (%) was determined by Nelson and Sommers, (1996) method. Total nitrogen in the soil was determined by Kjeldahl method of Bremner and Mulvaney, (1982). Phosphorus was determined by reducing ammonium heptamolybdate complex by ascorbic acid in the presence of antimony potassium tartrate (Murphy and Riley, 1962). The color intensity for P was measured on spectrophotometer at 880-nm. The K was measured by the method of Wright and Stuczynski, 1996 on flame photometer.

Growth characteristics

All growth parameters were recorded at the maturity of crop. The height (cm) of the selected plants was measured with a meter rod. Leaf area (cm^2) was measured by leaf area meter. Chlorophyll content (mg cm^{-2}) of selected plants from each plot was measured according to method of Lichtenthaler and wellburn (1985) with formula:

$Ca = 11.75A662 - 2.350A645$
$Cb = 18.61A645 - 3.960A662$

The shoot fresh weight (g) of selected plants from each treatment was recorded with electric balance. To record shoot dry weight weighed plants (g) were kept in an oven at 65^0C for 24 hours and dry weight was done with electric balance. The root length (cm) was determined with meter rod. The fresh root samples of the collected plants were washed thoroughly and fresh weight (g) was noted with electric balance. The roots were oven dried at 65^0C for 24 hours then root dry weight (g) was taken on electric balance. Numbers of nodules were collected from roots of three selected plants and then were averaged.

Each treatment was harvested and after threshing grain yield was recorded 1t 12% water contents as kg per hectare. Stalk yield (kg ha⁻¹) was calculated as the difference between grain yield and biological yield. Harvest index was calculated by the following formula:

$$H.I\ (\%) = \frac{Grain\ yield}{Biological\ yield} \times 100$$

Plant analysis

Plant samples were oven-dried at 65°C to a constant weight. The samples were ground and screened through 2mm sieve. These processed plant samples (stalk and grains) were analyzed for their N. Total nitrogen (%) in both stalk and grains was determined according to Bremner and Mulvaney (1982) method.

Nitrogen uptake

Nitrogen uptake (kg ha⁻¹) was calculated both in grain and stalk by multiplying concentration of nitrogen with respective yield. Nitrogen uptake in the whole plant was calculated by adding values for grains and yield.

Statistical analysis

Statistical analysis was done in Statistix 8.1 computer software. The treatments mean were compared using (LSD) at 5% level of probability (Steel et al., 1997).

Results and Discussion

Pre-sowing soil properties of study site

The pre-sowing properties of study site are given in Table 1. The properties showed that soil was loam and slightly acidic with pH of 6.53. The organic matter was low in these soils. However, available P and K were in marginal range and total N was 0.049%.

Comparative and interactive effects of urea nitrogen, poultry manure and sawdust on growth and yield of soybean

Plant Height

The comparative and interactive effects of urea nitrogen, poultry manure and sawdust on plant height (cm) of soybean are depicted in Table 2. The minimum plant height (18.76 cm) was observed in control and it was similar with SDN100 (21.73 cm). However the percent increase of UN50+SDN25+PM25 and PMN50+SDN50 was 63% and 58% respectively, over control.

The higher plant height in $UN_{50}+SDN25+PM_{25}$ may be attributed to higher nutrient use efficiency by interactive effect of organic amendments with UN. Moreover, SD rate did not show adverse effect of immobilization of nitrogen in integrated treatment. Inorganic N application alone or with organic N augmented plant height significantly (Idris et al., 2001; Singh and Agarwal, 2001) because N plays important role in cell enlargement, expansion and division. The lower plant height with SD could be due to wider C:N ratio of SD and it immobilized the nutrients (Olayinka, 2009).

Shoot fresh weight

The comparison of means (Table 2) indicated the highest shoot fresh weight (80.55g) by UN100 and it was parallel to PMN50+SDN50, UN50+PMN50 and UN50+SDN50. The uses of organic amendments like PM and SD improves the soil physico-chemical properties and ultimately increases of plant growth (Nottidge et al., 2005).

Shoot Dry Weight

The comparative and interactive effects of urea nitrogen, poultry manure and sawdust significantly affected shoot dry weight (g) of soybean (Table 2). The higher shoot dry weight 47.4 g was observed in UN_{100} and it was statistically equal with UN50+ PM25+SDN25 (47.0 g). The percent increase of UN100 and UN50+ PMN25+SDN25 was 21% over control. The SD and PM additions can show higher plant growth with slow release of (Olayinka, 2009).

Root Length

Results (Table 2) showed higher root length of 30.5 cm was UN50+ PMN25+SDN25 followed by UN50+SDN50 (29.7 cm). The percent increase of UN50+ PMN25+SDN25 and UN50+SDN50 was 62% and 58% respectively, over control. The higher root length in UN50+PMN50 and in UN50+ PMN25+SDN25 could be due to aggregated soil structure, more aeration, higher water and nutrient holding capacity, lower bulk density and higher porosity. As organic amendments improves all these soil physical properties. Studies (Moore and Edwards, 2005; Tejada et al., 2006) have shown that PM application in long term experiments improves the soil properties by increase of SOM.

Root Fresh Weight

The percent increase of UN50+PMN50 over control was 58% (Table 2). The minimum root fresh weight (11.7 g) was observed in the control and it was similar with the SDN100 (13.1 g).

The lower root fresh weight in SDN100 is might be due to less root growth because of higher rate of immobilization. The application of sawdust of high C: N ratio into the soil limits available nitrogen and cause inorganic soil N to be immobilized (Myrold, 1998).

Root Dry Weight

The data (Table 2) revealed that compare to control UN100 had 35% higher root dry weight. The minimum root dry weight (7.13g) was observed in the control and it was equivalent to the SDN100 (7.64 g).

Table – 1: Pre-sowing properties of experimental area

Soil properties	Units	Values
Bulk density	g cm^{-3}	1.42
Sand	%	39.4
Silt	%	34.6
Clay	%	26.0
Textural class	Loam	
Soil pH (1:2H$_2$O)	---	6.53
Organic matter	%	0.81
Total N	%	0.049
Available P	mg kg^{-1}	6.22
Available K	mg kg^{-1}	87.2

Table – 2: Comparative and interactive effects of urea nitrogen, poultry manure and   sawdust on the growth of soybean

Treatments	Plant Height (cm)	Shoot fresh weight (g)	Shoot dry weight (g)	Root Length (cm)	Root Fresh Weight (g)	Root Dry Weight (g)
Control (no amendments)	18.76 c	63.46 c	39.3 c	18.77 c	11.7 c	7.13 b
UN_{100}	25.40 abc	80.53 a	47.5 a	25.40 abc	16.8 ab	9.63 a
PMN_{100}	27.23 ab	69.67 bc	42.2 bc	27.23 ab	16.4 ab	9.13 a
SDN_{100}	21.73 bc	67.19 bc	41.1 bc	21.73 bc	13.1 c	7.64 b
UN_{50}+PMN50	23.66 abc	75.07 ab	44.3 ab	29.73 a	18.5 a	9.40 a
UN_{50}+SDN50	26.83 ab	74.27 ab	42.5 bc	26.83 ab	15.9ab	7.85 b
PMN50+SDN50	29.73 a	76.23 ab	39.5 c	23.67 ab	15.8 b	8.98 a
UN_{50}+ PMN25+SDN25	30.50 a	72.70 abc	47.0 a	30.50 a	16.1 ab	9.01 a
LSD	7.05	10.25	37.21	7.05	2.65	0.89

UN= Urea nitrogen; PMN= poultry manure nitrogen; SDN= Sawdust nitrogen
Means sharing same letters are statistically non-significant at $P \leq 0.05$

Leaf Area

The maximum increase in leaf area due to UN50+PMN50 was 40% (Table 3) The minimum leaf area (65.71cm) was observed in control and it was statistically equal to SDN100 (67.59 cm).

The higher leaf area with PM is attributed to its lower C: N ratio and fast nutrient release. Superior LAI under combined applications of organic and inorganic amendments also been reported by Ayoola and Makinde 2009.

Chlorophyll Contents

Results (Table 3) showed that UN50+PMN50 had statistically ($p \leq 0.05$) higher chlorophyll contents (29.2 mg cm^{-2}) and the percent increase of UN50+PMN50 was 58% over control. The minimum leaf area (18.52 mg cm^{-2}) was observed in the control and it was similar with SDN100 (19.40 mg cm^{-2}). Interaction of organic and inorganic amendments had found to increase the chlorophyll contents of plant (Yang et al., 2003).

Dry Matter Yield

Results (Table 3) showed that UN50+ PMN25+SDN25 had statistically higher dry matter yield (2863.7 kg ha^{-1}) and it was similar to UN100 (2848.3 kg ha^{-1}).

However, the percent increase of UN50+ PMN25+SDN25 and UN100 is 14% and 13% respectively, over the control. The higher dry matter yield in interactive application of PM, SD and UN showed that saw dust application rate to provide 25 kg N ha^{-1} did not show adverse effect of immobilization during growing period of soybean. The immobilization effect might have been offset by inorganic source of N. There was substantial reduction in the growth and yield when sawdust was applied without initial composting, this agreed with the former work done by Daramola et al., (2006).

Grain Yield

Results (Table 3) exhibited that UN100 had statistically higher grain yield (1322.7 kg ha^{-1}) and it was comparable with UN50+PMN50 (1242.0 kg ha^{-1}) and UN50+ PMN25+SDN25 (1222.7 kg ha^{-1}). The increase of UN100, UN50+PMN50 and UN50+ PMN25+SDN25 over control was 51%, 41% and 39% respectively. The higher yield with organic and inorganic integration has increase yield. This could be attributed to formation of favorable soil properties with addition of organic amendments in soil and ultimately higher yield (Ikpe and Powel, 2003; Ano and Agwu, 2005).

Harvest Index
The higher harvest index (31.69%) was observed in UN100 and it was statistically non-significant withUN50+ PMN50 (31.25%) (Table 3). However, the percent increase of UN100 and UN50+

PMN25+SDN25 was 22% and 21% respectively, over the control. Shahzad *et al.*, 2013 reported that the increase in harvest index is correlated with higher nitrogen uptake.

Table – 3: Comparative and interactive effects of urea nitrogen, poultry manure and sawdust on the growth and yield of soybean

Treatments	Leaf Area (cm)	Chlorophyll contents mgcm^{-2}	Dry Matter yield (kg ha^{-1})	Grain yield (kg ha^{-1})	Harvest Index (%)
Control (no amendments)	65.71 b	18.52 b	2516.3 bc	878.0 c	25.87 c
UN$_{100}$	91.52 a	29.13 a	2848.3 a	1322.7 a	31.69 a
PMN$_{100}$	87.58 a	25.90 a	2744.0 ab	1144.7 b	29.43 ab
SDN$_{100}$	67.59 b	19.40 b	2556.7 bc	981.7 c	27.86 bc
UN$_{50}$+PMN50	92.33 a	29.21 a	2726.3 abc	1242.0 ab	31.25 a
UN$_{50}$+SDN50	87.12 a	25.15 a	2655.7 abc	1158.7 b	30.44 ab
PMN50+SDN50	81.82 a	25,027a	2452.3 c	1009.7 c	29.15 ab
UN$_{50}$+PMN25+SDN25	90.88 a	27.680 a	2863.7 a	1222.7 ab	29.89 ab
LSD	11.42	5.32	281.84	132.87	3.08

UN= Urea nitrogen; PMN= poultry manure nitrogen; SDN= Sawdust nitrogen
Means sharing same letters are statistically non-significant at $P \leq 0.05$

Comparative and interactive effects of urea nitrogen, poultry manure and sawdust on nitrogen concentration and uptake

Seed Nitrogen
The UN100 had statistically ($p \leq 0.05$) higher seed nitrogen (4.60%) followed by UN50+ PMN25+SDN25 (4.41%) (Table 4). However, the percent increase of UN100 and UN50+ PMN25+SDN25 was 19% and 14% respectively, over the control. The minimum seed nitrogen (3.87%) was observed in control followed by SDN100 (4.06%).

Stalk Nitrogen
Results (Table 4) revealed that UN100 had statistically higher stalk nitrogen (95.72%) and it was equal with UN50+ PMN50 (89.54%).The increase of stalk nitrogen by UN$_{100}$ and UN$_{50}$+ PMN50 was 54% and 44% respectively, than control. The minimum stalk nitrogen (62.125%) was observed in control and it was

similar with SDN100 (64.28%) and PMN50+SDN50 (64.58%).

Grain N uptake
Statistically higher N uptake (60.84 kg ha^{-1}) was recorded in UN100 followed by PMN50+ SDN50 (55.24 kg ha^{-1}) (Table 4). The UN100 and PMN50+SDN50 had increase of 79 kg ha^{-1} and 63 kg ha^{-1}grain N uptake respectively, over the control. The minimum grain N uptake of 33.93% was observed in control and it was similar with SDN100 (39.82 kg ha^{-1}).

Stalk N uptake
Results (Table 4) indicated that UN100 had statistically higher stalk N uptake of 95.5 2 kg ha^{-1} and it was followed by UN50+ PMN50 (89.54 kg ha^{-1}). The lowest stalk nitrogen (62.13 kg ha^{-1}) was observed in control and it was similar with SDN100 (64.58 kg ha^{-1}) and PMN50+SDN50 (64.58 kg ha^{-1}). Cheatham, 2003 reported that the treatment PM gave the highest

N uptake in corn stalks and no significant difference was found in N uptake in corn stalks between PM and UN.

Total N uptake

The Results (Table 4) revealed that UN100 had statistically higher stalk N uptake (156.55 kg ha^{-1}) and it was at par with UN50+ PMN50 (144.78 kg ha^{-1}). Though, the percent increase of UN100 and UN50+ PMN50 was 63% and 51% respectively, over the control. The comparable total N uptake by integrated application of UN50+PMN50 and UN50+PMN25 +SDN25 suggests adding organic amendments at varying rates of N and dependence on chemical fertilizer can be reduced. Nitrogen and phosphorus are mainly released nutrients from organic amendments after decomposition (Paul, 2007). Moreover, lower total N uptake by SDN100 was due to the high C: N ratio of sawdust, which had limited available N and caused inorganic soil N to be immobilized (Myrold, 1998).

Table – 4: Comparative and interactive effects of urea nitrogen, poultry manure and sawdust on nitrogen concentration and uptake

Treatments	Seed Nitrogen (%)	Stalk Nitrogen (%)	Grain N uptake (kg ha^{-1})	Stalk N uptake (kg ha^{-1})	Total N uptake (kg ha^{-1})
Control (no amendments)	3.87 d	62.13 c	33.93 e	62.13 c	96.05 c
UN100	4.60 a	95.72 a	60.84 a	95.52 a	156.55 a
PMN100	4.43 ab	82.48 ab	50.61 b	82.48 ab	133.07 b
SDN100	4.06 cd	64. 28 a	39.82 de	64.58 c	104.40 c
UN50+PMN50	4.13 bcd	89.54 ab	41.88 cd	89.54 ab	144.78 ab
UN50+SDN50	4.25 abc	78.64 bc	49.19 ab	78.64 bc	127.83 b
PMN50+SDN50	4.13 bcd	64.58 c	55.24 ab	64.28 c	106.17 c
UN50+ PMN25+SDN25	4.41 abc	88.59 ab	53.81 ab	88.49 ab	142.29 ab
LSD	0.37	16.66	7.81	16.62	18.51

UN= Urea nitrogen; PMN= poultry manure nitrogen; SDN= Sawdust nitrogen
Means sharing same letters are statistically non-significant at $P \leq 0.05$

Comparative and interactive effects of urea nitrogen, poultry manure and sawdust on soil properties after soybean harvest

Soil pH

The effects of treatments on soil pH are statistically non-significant (Table 5). However the percent increase of PMN100, PMN50+SDN50 and UN50+SDN25+PMN25 over initial pH contents were 1.84%, 1.68% and 1.53% respectively. The increase in soil pH with PM and in integrated treatments of PM with organic and inorganic amendments might be due to decrease of the exchangeable Al toxicity. The increase in soil pH and decrease of soil exchangeable acidity following application of manure can be attributed to the release of organic acids (by mineralization of manure), which in turn may have inhibited Al content in the soil through chelation (Onwonga et al., 2008).

Soil Organic Matter

Statistically higher organic matter (0.84%) was found in UN50+ PMN25+SDN25, PMN50+SDN50 and UN50+SDN50 (Table 5). The minimum organic matter (0.78%) was observed in control. Compare to initial OM contents, the increase of UN50+ PMN25+SDN25, PMN50+SDN50 and UN50+SDN50 was 3.70%. In our study the higher O.M contents in integrated treatments of organic and inorganic amendments are in agreement with earlier studies. By using PM over a long period of time the soil OM would increase (Tejada *et al.*, 2006). Similarly, Bulmer (2000) indicated sawdust application showed higher organic matter level compare to untreated soil.

Soil Nitrogen

The Results (Table 5) indicated the minimum nitrogen (0.047%) in control and it was similar to SDN100 (0.048%). The percent increase of UN50+PMN50 and PMN100, UN50+ PMN25+SDN25 and UN100 was 12.24%, 8.16%, 6.12% and 4.08% respectively, over initial soil N contents. However, control, SDN100 and PMN50+SDN50 showed decrease over initial soil N contents. The higher total N contents in treatments where organic materials such as sawdust and PM were

added together with N fertilizer are in agreement with other Paustian *et al.*, 1992.The decline in total N in SD and PM+SD might be due to immobilization of inorganic N in saw dust (Myrold, 1998).

Soil Phosphorus

The Results (Table 5) showed that UN50+PMN50 had significantly higher phosphorus (6.32 mg kg^{-1}) contents and the minimum phosphorus (5.12 mg kg^{-1}) was observed in control. The increase of P contents of UN50+PMN50, PMN100, UN50+ PMN25+SDN25 and UN100 was 1.61%, 1.29%, 1.13% and 0.80% respectively, over initial P contents. The higher P contents in treatment PM and in integrated organic and inorganic amendments could be attributed to higher N and P contents of PM (Olayinka, 2009).

Table – 5: Comparative and interactive effects of urea nitrogen, poultry manure and sawdust on soil properties after soybean harvest

Treatments	pH	OM (%)	Total N (%)	Soil Phosphorus (mg kg^{-1})
Control (no amendments)	6.55 ns	0.78 c	0.047 b	5.12 d
UN100	6.56	0.83 bc	0.051 ab	6.27 a
PMN100	6.65	0.82 ab	0.053 ab	6.30 a
SDN100	6.54	0.82 ab	0.048 b	6.11 c
UN50+PMN50	6.56	0.83 ab	0.055 a	6.32 a
UN50+SDN50	6.56	0.82 a	0.050 b	6.18 b
PMN50+SDN50	6.64	0.84 a	0.048 b	6.17 bc
UN50+ PMN25+SDN25	6.62	0.84 a	0.052 ab	6.29 a

UN= Urea nitrogen; PMN= poultry manure nitrogen; SDN= Sawdust nitrogen
Means sharing same letters are statistically non-significant at $P \leq 0.05$

Conclusion

The interactive application of UN50+ PMN25+SDN25 gave statistically similar yield as UN100. The higher organic matter was found in UN50+ PMN25+SDN25. Sawdust did not decompose totally in one growing season so higher impact of interactive effect of PM and SD with UN is not pronounced in all parameters. The use of SD and PM has the potential to reduce need for inorganic fertilizers as both organic amendments are locally available.

References

Abbasi, M.K., Majeed, A., Sadiq, A., Khan, S.R., 2008. Application of Bradyrhizobium japonicum and phosphorous fertilization improved growth, yield and nodulation of soybean in the sub-humid Hilly region of Azad Jammu and Kashmir, Pakistan. Plant Prod. Sci. 11, 368-376.

Abbasi, M.K., Manzoor, M., Tahir, M.M., 2010. Efficiency of Rhizobium inoculation and P fertilization in enhancing nodulation, seed yield and phosphorous use efficiency by field grown soybean under hilly region of Rawalakot Azad Jammu and Kashmir, Pakistan. J. Plant Nutr. 33, 1080-1102.

Abera, G., E. Woldemeskel and L. Bakken. 2012. Carbon and nitrogen mineralization dynamics in different soils of the tropics amended with legume residues and contrasting soil moisture contents. Biol. Fertility Soils. 48:51-66.

Adeniyan, O. N. and S. O. Ojeniyi. 2005. Effect of poultry manure and NPK 15-15-15 and combination of their reduced levels on maize growth and soil chemical properties. Nig. J. Soil Sci., 15: 34-41.

Ano, A. O. and J. A. Agwu. 2005. Effect of animal manures on selected soil chemical properties Niger. J. Soil Sci., 15:14-19.

Ayoola, O. T. and E. A. Makinde. 2009. Maize growth, yield and soil nutrient changes with N-enriched organic fertilizers. African J. Food Agric Nut and Development. 9: 580-592.

Bhattacharyya, R., S. C. Pandey, S. Chandra, S. Kundu, S. Saha, B. L. Mina, A. K. Srivastya and H. S.Gupta. 2010. Fertilization effects on yield sustainability and soil properties under irrigated wheat–soybean rotation of an Indian Himalayan upper valley. Nutri. Cycl. Agroecosyst. 86: 255–268.

Bremner, J. M. and C. S. Mulvaney (1982). Nitrogen Total. In A. L. Page (eds.), Methods of soil analysis. Argon. No. 9, Part 2: Chemical and microbiological properties, 2nd ed., Am. Soc. Argon., Madison, WI,USA, p. 595-624.

Bulmer, C. 2000. Reclamation of forest soils with excavator tillage and organic amendments. For Ecol. Manage., 133: 157-163.

Cheatham, M. R. 2003. The impact of poultry manure on water quality using tile drained field plots and lysimeters. Master's thesis, Ames, Iowa: Iowa State University, Department of Agriculture and Biosystems Engineering.

Cogger, C. 2005. Home gardener's guide to soils and fertilizers. Pullman (WA): Washington State University.

Daramola, D. S., A. S. Adeyeye and D. Lawal. 2006. Effect of application of organic and inorganic nitrogen fertilizer on the growth and dry matter yield of amaranthus. Proc. 2nd Nat. Conf. Org. Agric. in Nigeria.27th Nov. - 1st Dec., 2006. Uni. Ibadan, Ibadan, Nigeria

Gee, G. W. and Bauder. 1986. Methods of Soil Analysis, Physical and Mineralogical Methods. Am. Soc. Agron., 9: 383-441.

Huang, B., W. Z. Sun, Y. Z. Hao, J. Hu, R.Yang, Z. Zou, F. Ding and J. Su. 2007. Temporal and spatial variability of soil organic matter and total nitrogen in an agricultural ecosystem as affected by farming practices. Geoderma. 139: 336-345.

Idris, M., S. M. Shah, M. Wisal and M. M. Iqbal. 2001. Integrated use of organic and mineral nitrogen and phosphorus on the yield, yield components, N and P uptake by wheat. Pak. J. Soil. Sci., 20: 77-80.

Ikpe, F. N. and J. M. Powel. 2003. Nutrient cycling practices and changes in soil properties in the crop livestock farming system of west Niger Republic of West Africa. Nutr.Cycl.Agroecosys. 62: 37-45.

Khaliq, A and M. K. Abbasi. 2015. Soybean Response to single or mixed soil amendments in Kashmir, Pakistan. Agronomy Journal. 107(3): 887-895.

Lichtenthaler, H. K. and A. R. Wellburn. 1985. Determination of total carotinoids and chlorophyll A and B of leaf in different solvents. Biochem. Soc. Trans., 11: 59-592.

McLean, E.O (1982). Soil pH and lime requirement. p. 199-224, In A.L. Page (ed.), Methods of Soil Analysis, part 2: Chemical and Microbiological Properties. Am. Soc. Agron, Madison, WI, USA.

Moore, P. A., Jr. and D. R. Edwards. 2005. Long-term effects of poultry litter, alum-treated litter, and ammonium nitrate on aluminum availability in soils. J. Environ. Qual., 34: 2104-2111.

Murphy, J. and J. P. Riley. 1962. A modified single solution for determination of phosphate in natural waters. Anal. Chim. Acta. 27: 35-36

Myrold, D. D. 1998. Transformations of nitrogen. In Sylvia D. M., J. J. Fuhrmann, P. G. Hartel, and D. A. Zuberer. Principles and applications of soil microbiology. 259-294pp.

Nelson, D. W. and L. E. Sommers. 1982. Total carbon, organic carbon and organic matter. In: A.L. Miller & R.H. Keeney, (eds), Methods of Soil Analysis, Part 2: Chemical and Microbiological properties. Soil Sci. Soc. Am. Madison, WI, USA. 539-579pp.

Nottidge, D. O., S. O. Ojeniyi and D. O. Asawalam. 2005. Comparative effect of plant residues and NPK fertilizer on nutrient status and yield of maize (Zea mays L.) in humid Ultisol. Niger. J. Soil Sci., 15: 1- 8.

Olayinka, A. 2009. Soil Micro organisms, Wastes and National Food Security.Inaugural Lecture Series 222 of Obafemi Awolowo University, Ile-Ife, Nigeria.

Onwonga, R. N., J. J. Lelei, B. Freyer, J. K. Friedel, S. M. Mwonga and P. Wandhawa. 2008. Low cost technologies for enhance N and P availability and maize (Zea mays L.) performance on acid soils. World J. Agric. Sci., 4: 862-873.

Owolabi, O., S. O. Ojeniyi, A. O. Amodu and K. Hassan. 2003. Response of Cowpea, Okra and Tomato to Sawdust Ash Manre. Moor J. Agric. Res., 2: 179-182.

Pandey, A. K., K. A. Gopinath, P. Bhattacharya, K. S. Hooda, S. N. Sushil, S. Kundu, G. Selvakumar and H. S. Gupta. 2006. Effect of source and rate of organic manures on yield attributes, pod yield and economics of organic garden pea (Pisum sativum) in Northwestern Himalaya. Indian. J. Agric. Sci., 76: 230-234.

Paul, E. A. 2007. Soil Microbiology, ecology and biochemistry.3rd Edition.Academic Press, Oxford, United Kingdom.

Paustian, K., W. J. Parton and J. Pesson. 1992. Modeling Soil organic matter in organic amendedand nitrogen-fertilized long-term plots. Soil Sci. Soc. Am. J., 56: 476-488.

Qian, P., J. J. Schoenau and P. Mooleki. 2004. Phosphorus Amount and Distribution in a Saskatchewan Soil after Five Years of Swine and Cattle Manure Application. Canadian .J. Soil Sci., 84: 275- 281.

Salvagiotti, F., K.G. Cassman, J.E. Specht, D.T. Walters, A. Weiss, and A. Doberman. 2008. Nitrogen uptake, fixation, and response to fertilizer N in soybeans: A review. Field Crops Res. 108:1-13.

Shah, S. A., S. M. Shah, W. Mohammad, M. Shafi and H. Nawaz. 2009. Nitrogen uptake and yield of wheat as influenced by integrated use of organic and mineral nitrogen. Int. J. Plant Prod., 3: 45−55.

Shahzad. K., A. Khan and I. Nawaz. 2013. Response of wheat varieties to different nitrogen levels under agro-climatic conditions of Mansehra. Sci., Tech. Dev., 32: 99-103.

Singh, R. and S. K. Agarwal. 2001. Growth and yield of wheat (Triticum aestivum L.) as influenced by levels of farmyard manure and nitrogen. Indian J. Agron., 46: 462-467.

Starbuck, C. 1994. Applying research in forestry. Report 6. Using sawdust as a soil amendment. Jefferson City (MO): Missouri Department of Conservation. p: 4.

Steel, R. G. D., J. H. Torrie and D. Dickey. 1997. Principles and procedures of statistics: a biometric approach. McGraw Hill Book Co, Inc, New York USA 666pp.

Tejada, M., M. T. Hernandez and C. Garcia. 2006. Application of two organic amendments on soil restoration: effects on the soil biological properties. J. Environ. Qual., 35: 1010-1017.

Tiwari, K. R., B. K. Sitaula, R. M. Bajracharya and T. Børresen. 2010. Effects of soil and crop management practices on yields, income and nutrients losses from upland farming systems in the Middle Mountains region of Nepal. Nutr. Cycl. 86: 241–253.

Wright, R. J. and T. I. Stuczynski. 1996. Atomic absorption and flame emission spectrometry. In: D.L. Sparks et al. Methods of Soil Analysis, Chemical Methods. Soil Sci. Soc. Am., 5: 65-90.

Yang, W. H., S. Peng, J. Huang, A. L. Sanico, R. J. Buresh and C. Witt. 2003. Using leaf color charts to estimate leaf nitrogen status of rice. Agron. J., 95: 212-217.

Zaman, M., S. Saggar, J. D. Blennerhassett and J. Singh. 2009. Effect of urease and nitrification inhibitors on N gaseous emissions of ammonia and nitrous oxide, pasture yield and N uptake in grazed pasture system. Soil. Biol. Biochem., 41: 1270–1280.

17

Effect of soil tillage and mycorrhiza application on growth and yields of upland rice in drought condition

Laila Nazirah[1, 2*], Edison Purba[3], Chairani Hanum[3], Abdul Rauf[3]
[1]Doctoral Program of Agricultural Sciences, Faculty of Agriculture, Universitas Sumatera Utara, Padang Bulan, Medan 20155, Indonesia.
[2]Lecture Faculty of Agriculture, Malikussaleh University, Indonesia.
[3]Lecture Program Study of Agriculture, Universitas Sumatera Utara, Padang Bulan, Medan 20155 Indonesia.

Abstract

Dry land management technology for food crop agriculture with soil conservation, organic matter management, and water management. This study aims to determine the effect of soil treatment and mycorrhiza on growth and yield of upland rice in drought stress conditions. This experiment uses the Split Split Plot design consists of 3 factors: The first factor as the main plot of mycorrhizas consisting of no mycorrhizal and mycorrhizal administration. The second factor as a plot is the soil cultivation consisting of no soil preparation) and treatment. The third factor as Multiplication Children is Varieties consisting of three groups of varieties namely Toleran group (Ciapus Varieties, Inpago Varieties 4 and Varieties inpago 8) moderate varieties group include (Inpago Varieties 5, Varietas situ bagendit, Inpago Varieties 7 and Varietas towuti) and the susceptible varieties are (In jari 6 varieties, Inpari 33 varieties and synthetic varieties). Treatment without tillage and without mycorrhiza decreased leaf area, root canopy ratio, leaf proline content, degree of root infection and dry grain production. Soil sampling and mycorrhizal fertilization of Inpago 4 tolerant varieties showed a mechanism of avoidance against drought stress by increasing leaf area, root canopy ratio, leaf proline content and root infection. The highest dry grain production was found in the tolerant (Inpago 4) varieties group of 7.5 tons per ha and can be planted in drought stress conditions at rainfall ± 3.2 mm / day.

*Corresponding author email:
laila_nazirah@yahoo.co.id

Keywords: Upland, Rice, Land Preparation, Mycorrhiza, Soil treatment

Introduction

Potential of upland rice production in Indonesia has reached up to 2.69 t/ha (BPTP, 2012) or about 5.2% of the total national rice production. Although the proportion is low, yet it has a high value. This is due to the harvest time of upland rice generally falls during the dearth rice periods. Upland rice is generally harvested time comes earlier than rainfed lowland rice and limited irrigated rice fields (Toha, 2010).

Farming on dry land comes with a certain situation such as lack of water, high acidity, but low nutrients (Gunawan, 1993). One effort which had been taken to overcome the problems is the utilization of arbuscular mycorrhizal fungi. Mycorrhiza is a fungus that lives in symbiosis with the root system of plants (Grant et al., 2011). Utilization of mycorrhiza not only increase the absorption zone and the availability of nutrients but also water stress resistance and pest resistance (Setiadi, 1999).

Mycorrhiza has a potential for alternative technology to increase plant growth and productivity, especially on the less fertile soil. Inoculation of mycorrhizal fungi in food crops can increase water uptake and provide sufficient water requirements for the plant's physical needs, especially in dry conditions (Thangadurai et al., 2010).

Crop treatment by always covering the land surface by vegetation and/or plant or litter remains also plays a certain role in soil conservation. Intensive soil cultivation is the cause of production decrease of dry land. The results show that excessive soil tillage can damage soil structure (Larson and Osborne 1982) and lead to soil organic matter crudeness (Rachman et al., 2004). Land conservation practice (LCP) is an alternative to land preparation to heighten land productivity (Brown et al., 1991) LCP is characterized by reduced discharge or reversal of the soil, intensifying the use of crop residues or other ingredients as mulch, sometimes (but not recommended) with the use of herbicides to suppress the growth of weeds or other nuisance plants. This study aims to determine the effect of soil tillage and mycorrhiza application on the growth and yield of upland rice in drought stress conditions.

Material and Methods

The study was conducted in Alue Mudem Lhoksukon village, North Aceh District, Nanggoe Aceh Darussalam Province, Indonesia from January 2015 to May 2015. The experiment in the field using split split plot design consists of three factors: The first factor as the main plot is mycorrhizas and no mycorrhizas. The second factor as a subplot is land cultivation consisting of no tillage and tillage (one round cultivation). The third factor as sub-subplot is rice varieties consisting of three groups of varieties, namely tolerant varieties (Ciapus, Inpago 4 and Inpago 8), moderate varieties (Inpago-5, Situbagendit, Inpago-7 and towuti) and sensitive (Inpari-6 jete , inpari-33 and Sintanur).

On soil tillage treatment, the soil was cultivated one week before planting whereas for no soil tillage the land was conducted by scrapping soil surface using small hoe. There were 120 plots with the size of 1 m x 2.5 m. Manure at 10 ton ha^{-1}was applied two weeks before rice planted. Fertilizer of Phonska at the dose of 300 kg.ha^{-1}and urea 200 kg.ha^{-1}were applied at the day of rice planted. Arbuscular mycorrhizal fungi, 10 g. planting hole^{-1}wasinoculated before the rice seeds

were planted. Prevention of pests and diseases is done intensively from seed treatment to harvest. Fungicides in use are Oksiklorida 50 % and insecticides used are BPMC 480, MIPC Carbaryl 85% and Diazinon. Measurement of leaf area was carried out using leaf area meter at 6, 8, 10 WAP weeks after planting (WAP). Shoot-root ratio was determined by measuring the ratio of dry weight of shoot compared to the dry weight of roots. Leaf proline content was analyzed at 10 and 13 WAP.

Degree of root infection caused by micorrhizas was assessed at 10WAP by staining according to Kormanik and Graw (1982). Grain production was determined at a moisture content of 14%.

Data Analysis

Data were analyzed using Windows SAS statistical program (Version 9) for analysis of variance at α test = 0.05 and continued with Duncan's Multiple Range Test (DMRT) test for significant difference.

Results and Discussion

Results

Groundwater level of the soil during the study was ranging from 33.55% to 36% of field capacity. The development of leaf area increased from age 6 and 10 WAP and at harvest. Leaf area of Inpago 4 (tolerant) variety is the highest while the lowest is of Inpari 33 (susceptible) variety (Figure 1).

Leaf area of upland rice treated with mycorrhizal fertilizer at 10 WAP showed more leaf area with no mycorrhizal fertilizer (Table 1).While the effect of soil tillage on leaf area is presented in Tabel 2. Leaf area of rice is much higher if soil is cultivated before planting (Table 2).

Root-Shoot Ratio

Root bearing ratios were significantly influenced by the interaction of upland rice and mycorrhizal varieties at age 6 WAP and 8 WAP Inpari 33 varieties (susceptible varieties) with no tillage and no mycorrhizal fertilizer showed the highest ratios while the inpago 4 varieties (tolerant varieties groups) on no soil treatment with no mycorrhizal fertilizers showed the lowest root canopy ratios as listed in Table 3.

The interaction of soil preparation and mycorrhizal fertilizer gave a significant effect on the root-shoot ratio. The highest root-shoot ratio was observed at no mycorrhiza and no soil tillage (Table 4).

Proline content

Inpago 4 varieties (tolerant varieties group) have the highest proline content available without tillage with no mycorrhizal biochemical fertilizer whereas the lowest proline content ratio was found in inpari 33 varieties (sensitized) without soil treatment without mycorrhizal fertilizers, as shown in Table 5.

Figure 2 shows the mechanism difference of 10 varieties in accumulating proline. Increased proline is a crop mechanism for dealing with drought stress conditions. In varieties of inpago 4 (tolerant varieties group) shows the highest prolina accumulation in the band with other varietal groups.

Degree of root infection

The degree of root infestation by mycorrhiza was measured based on the proportion of the mycorrhizal infected field. Infected category was based on Rajapakse and Miller (1992) in Prafithriasari (2010) as follows: <5% infected is categorized very low (Class

The content of proline is significantly influenced by the interaction of soybean and mycorrhizal biofertilizer varieties at age 10 MST and 12 MST. 1), 6 - 25% is categorized low (Class 2), 26 - 50% is categorized moderate (Class 3) 51 - 75%is categorized high (Class 4), and> 75% is categorized very high (Class 5). Roots of Inpago 4 variety (tolerant group) were 46.33 percent infected whereas the number of uninfected roots was found in untreated with mycorrhiza.

Rice grain production

The production of dried un hulled rice harvested from different land preparation and different mycorrhiza treatment is shown in Table 7. The production is significantly influenced by the interaction of soil tillage and mycorrhizal treatment. Inpago 4 variety (tolerant variey group) gave the highest dry grain production in soil tillage plus mycorrhiza treatment whereas the lowest grain production was found in sintanur varieties (sensitive variety group) grown on no tillage and mycorrhiza (Table 7).

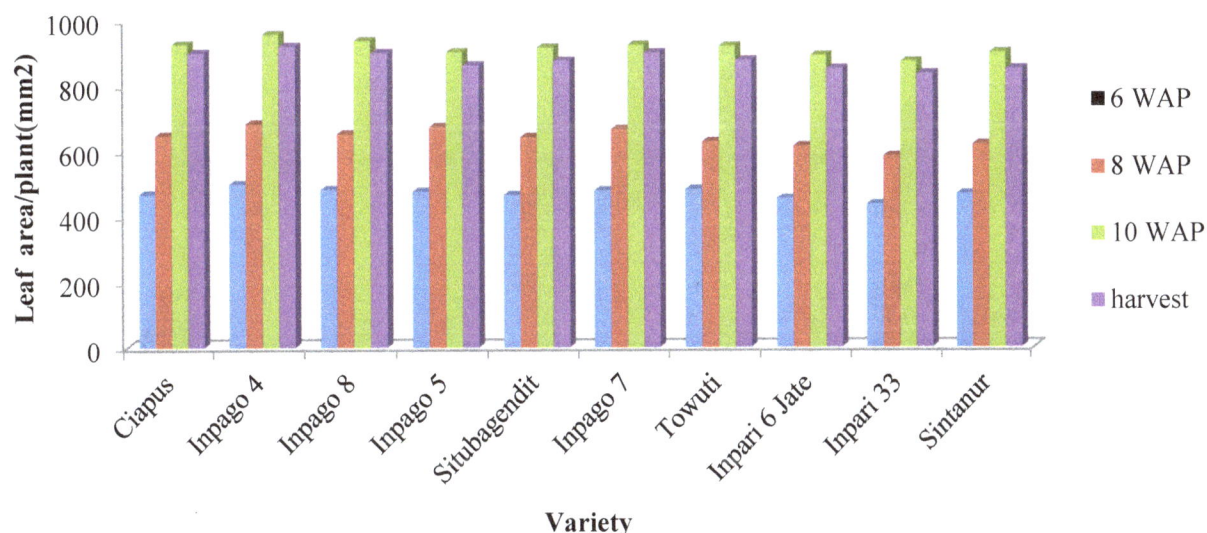

Figure 1. Leaf area of 10 upland rice varieties at 6, 8, 10 WAP and harvest time

Table 1. The average of leaf area of rice plant treated with mycorrhiza 10 WAP

Leaf Area[1]	
Treatment	10 WAP (mm^2)
Control (no mycorrhiza)	995.35 a
Mycorrhiza	833.03 b

1) Numbers followed by the same letters in the same column are not significantly different based on DMRT atα 0.05

Table 2. Leaf area of upland rice grown on soil tillage and no tillage 6 WAP

Treatment	Leaf area[1]
	---------------------- mm^2----------------------
No tillage	851.53 b
Soil tillage (plowed)	976.85 a

1) Numbers followed by the same letters in the same column are not significantly different based on DMRT at α 0.05

Table 3. Root-shoot ratio of 10 upland rice varieties on different land preparation and mycorrhizal fertilizer at the age of 6 and 8 WAP

Land Preparation	Upland rice variety	Root-shoot ratio[†]			
		6 WAP		8 WAP	
		No mycorrhiza	With mycorrhiza	No mycorrhiza	With mycorrhiza
No tillage	Ciapus	2.77 b-d	1.87 e-j	1.82 e-j	2.18 c-i
	Inpago 4	1.09 n-p	1.36 j	1.53 g-j	1.48 h-j
	Inpago 8	1.88 g-k	1.68 f-j	1.98 e-j	1.32 j
	Inpago 5	1.73 h-l	1.90 e-j	1.80 e-j	1.99 e-j
	Situbagendit	2.07 f-i	1.77 e-j	2.07 c-j	2.77 a-d
	Inpago 7	1.70 h-l	1.92 e-j	2.28 c-g	2.26 d-h
	Towuti	1.26 l-o	1.90 e-j	2.06 c-i	2.01 d-j
	Inpari 6 Jate	2.35 d-g	2.03 d-j	1.46 ij	1.55 g-j
	Inpari 33	2.26 d-h	2.15 c-i	2.82 a-c	1.58 g-j
	Sintanur	2.24 d-h	1.96 f-j	2.15 c-i	1.68 f-j
Tillage	Ciapus	1.06 n-p	2.77 b-d	1.87 e-j	1.82 e-j
	Inpago 4	0.58 p	1.09 n-p	1.36 j	1.53 g-j
	Inpago 8	0.59 p	1.88 g-k	1.68 f-j	1.98 e-j
	Inpago 5	1.47 j-n	1.73 h-l	1.90 e-j	1.80 e-j
	Situbagendit	1.37 k-o	2.07 f-i	1.77 e-j	2.07 c-j
	Inpago 7	1.42 k-n	1.70 h-l	1.92 e-j	2.28 c-g
	Towuti	1.30 k-o	1.26 l-o	1.90 e-j	2.06 c-i
	Inpari 6 Jate	2.11 f-i	2.35 d-g	2.03 d-j	1.46 ij
	Inpari 33	1.25 l-o	2.26 d-h	2.15 c-i	2.82 a-c
	Sintanur	1.58 i-n	2.24 d-h	1.96 f-j	2.15 c-i

[†]Numbers followed by similar letter in the same column are not significantly different based on DMRT test at α 0.05

Table 4. The impact of the interaction of land preparation (soil tillage) and mycorrhizal fertilizer on root-canopy ratio 10 WAP and at harvest time

Land preparation (tillage)	Root canopy ratio[†]			
	10 WAP		Harvest	
	No mycorrhiza	Mycorrhiza	No mycorrhiza	Mycorrhiza
No tillage	2.96 a	2.24 b	2.81 a	2.19 b
Soil tillage	2.33 b	2.24 b	2.42 b	2.30 b

[†]Numbers followed by similar letter on the age with no significant effect based on DMRT test at α 0.05

Table 5. Proline content in 10 varieties on different soil tillage and mycorrhiza

| Land Preparation | Upland rice variety | Mycorrhiza | | | |
| | | 10 WAP | | 13WAP | |
		No mycorrhiza	With mycorrhiza	No mycorrhiza	With mycorrhiza
No tillage	Ciapus	38.84 hi	40.34 gh	16.68 j-l	24.65 g-i
	Inpago 4	62.23 a	79.01 a	44.55 b	51.33 a
	Inpago 8	57.27 c	50.82 d	37.88 c	43.21 b
	Inpago 5	28.45 l	25.19 m	17.43 jk	15.99 kl
	Situbagendit	24.16 mn	38.59 h-i	18.70 j	23.18 i
	Inpago 7	27.46 l	22.11 op	13.29 m-o	13.58 mn
	Towuti	45.14 e	42.81 f	26.10 f-h	24.29 h-i
	Inpari 6 Jate	16.23 q	24.20 mn	9.43 rs	17.84 jk
	Inpari 33	11.87 r	17.16 q	7.62 s	9.58 q-s
	Sintanur	22.99 no	23.46 m-o	12.07 n-q	18.32 jk
Tillage	Ciapus	30.39 k	37.77 i	16.68 j-l	22.66 i
	Inpago 4	62.00 b	51.43 d	31.33 d	36.66 c
	Inpago 8	46.66 e	42.88 f	30.35 de	28.14 ef
	Inpago 5	28.15 l	27.36 l	12.80 m-p	11.81 n-q
	Situbagendit	27.57 l	24.35 nm	11.62 n-r	14.63 lm
	Inpago 7	22.24 op	33.17 j	10.28 p-r	27.84 f
	Towuti	42.55 f	41.50 fg	26.85 fg	28.48 ef
	Inpari 6 Jate	34.01 j	33.93 j	18.21 jk	23.67 i
	Inpari 33	24.21 mn	20.75 p	10.02 rq	10.94 o-r
	Sintanur	20.39 p	28.87 kl	10.65 p-r	18.32 jk

†Numbers followed by similar letter on the age with no significant effect based on DMRT test at α 0.05

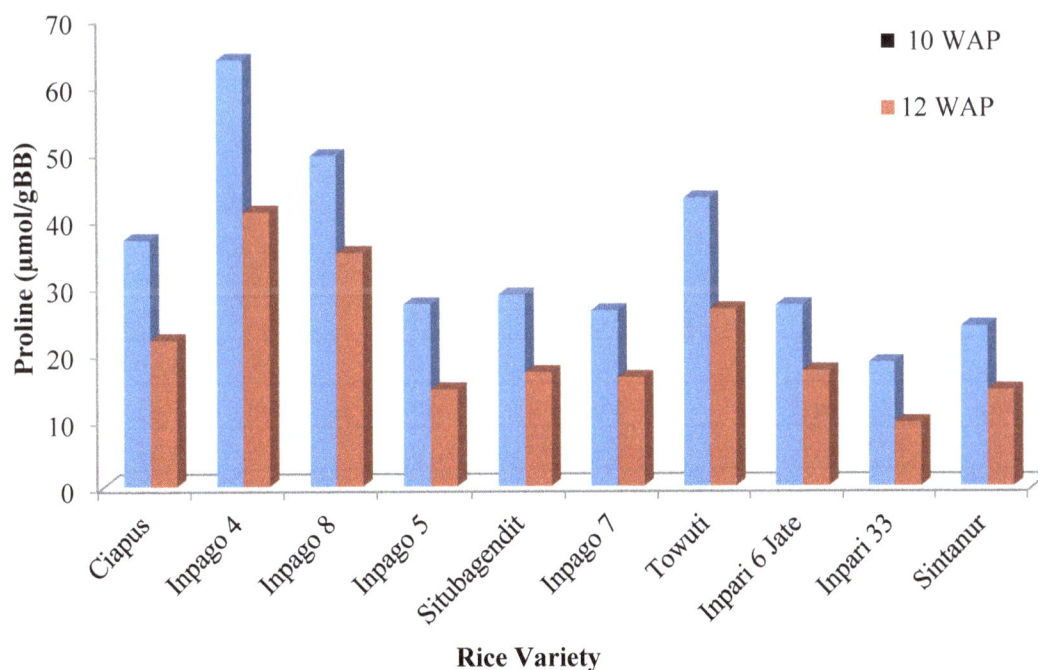

Figure 2. Proline content found in each rice variety 10 WAP and 12 WAP

Table 6. Root infection by mycorrhiza in 10 upland rice varieties on mycorrhizal treated soil

Rice Variety	Land preparation	Root infected (%) in soil treated with mycorrhiza	Category
Ciapus	No tillage	22.25 j	level 2(low)
Inpago 4		36.00 a-d	level 3 (low)
Inpago 8		33.33 b-c	level 3 (medium)
Inpago 5		22.75 e- j	level 2 (low)
Situbagendit		21.83 f-j	level 2 (low)
Inpago 7		28.66 c-g	level 3 (medium)
Towuti		30.66 c-d	level 3 (medium)
Inpari 6 Jate		19.50 g- j	level 2 (low)
Inpari 33		13.00 j	level 2 (low)
Sintanur		27.83 c-d	level 3 (medium)
Ciapus	Soil tillage	41.83ab	level 3 (medium)
Inpago 4		46.33 a	level 3(medium)
Inpago 8		35.33 b-c	level 3(medium)
Inpago 5		36.16 a-d	level 3(medium)
Situbagendit		37.33 a-c	level 3 (medium)
Inpago 7		25.33 d-i	level 2 (low)
Towuti		26.16 d-h	level 2 (low)
Inpari 6 Jate		16.50 h-j	level 2 (low)
Inpari 33		14.83 ij	level 2 (low)
Sintanur		13.66 j	level 2 (low)

†Numbers followed by similar letter on the age with no significant effect based on DMRT test atα 0.05

Table 7. Rice grain production of 10 upland rice varieties growing on soil tillage and treated with mycorrhiza

Rice variety	No tillage and no mycorrhiza (Control)	No tillage-mycorrhiza	Soil tillage (ploughed) and no mycorrhiza	Soil tillage (ploughed) and mycorrhiza
ciapus	3.41 o-r	4.18 ef	3.78 j-m	4.31 de
Inpago 4	3.80 i-m	5.37 d	4.53 fg	7.55 a
Inpago 8	3.32 o-p	4.21 d	3.79 g-l	4.73 d
Inpago 5	2.91 op	3.68 k-m	3.86 h-m	6.84 h-m
Situ Bangendit	3.55 l-m	3.92 fg	3.61 fg	4.63 k-m
Inpago 7	2.45 p-q	3.83 h-m	4.37 f-h	6.32 f-h
Towuti	3.52 lm	4.21 de	3.84 g	4.60 bc
Inpari 6 Jate	2.23 rq	4.33 f-i	3.44 mn	3.65 lm
Inpari 33	2.91 s	3.51 no	3.19 r-q	3.65 mn
Sintanur	2.07 rs	3.69 k-m	2.75 o-q	4.34 f-i

Discussion

The results showed that the difference among the 10 varieties tested showed that Inpago 4 variety seemed to be more adaptive to unfavorable environments, where the water shortage did not interfere the growth (Table 3, 5 7). Inpago 4 variety representing new superior varieties (VUB) is tolerant to environmental stress, including to drought. Soil treatment arrangements and mycorrhizal applications of varieties may be reflected in leaf area characteristics, root-shoot ratios, degree of root infection, leaf proline content and grain production that can be used in determining drought tolerant varieties. The tolerant variety (Inpago 4 variety) showed that the character of leaf area, root-shoot ratio, degree of root infection, leaf proline content and highest production in soil tillage and mycorrhizal treatment. This indicated that inpago 4 upland rice variety is more adaptable to these conditions and did not indicate any significant inhibition of long root growth. This variety could grow better than other varieties due to its ability to absorb water from deeper depth in the soil. Root–shoot ratio is an important indicator for plant ability to take water from deeper soil (Abdallah et al., 2016). Fukai Lilley (1995) reported that the depth of root achievement and root diameter is an important indicator in determining tolerance to drought stress in upland rice

Root establishment in the tolerant variety group is better than that of moderate varieties and susceptible varieties which therefore increase the ability to reach water level. Serraj et al. (2004) reported that root response to drought stress will increase the depth of root achievement and wider root development

Asch (2005) stated that relatively tolerant varieties increase root growth greater in drought stress conditions. The mechanisms of tolerance in plants as a response to drought stress include (i) the ability of plants to continue to grow in the condition of water shortage is to decrease the leaf area and shorten the growth cycle (ii) the ability of the root to absorb water in the deepest layer of soil. (iii) the ability to protect the root meristem from drought by increasing the accumulation of certain compounds such as glycine, betaine, alcohol sugar and proline for adjustment and (iv) optimizing the role of stomata to prevent leaf water loss (Nguyen et al., 1997) in the presence of such osmotic adjustment allowing growth to continue and stomata remains open.

Inpago 4 tolerant varieties showed higher physiological agronomic characteristics, root-shoot ratios, compared to other varieties in soil tillage and mycorrhizalization (Tables 3 and 4). Growth ratio of root-shoot in drought conditions will result in different responses for each variety. Root length will increase when faced with drought associated with mechanism of resistance of plant genotype to drought not by influence of moisture content. Sitompul and Guritno (1995) stated that, plants that grow in a state of water shortage will form longer and more root quantities with lower yields than plants grow in water adequacy.

Root shoot character is important to keep the potential of leaf water root remains high and to maintain evapotranspiration in water deprivation (Peng and Ismail 2004).

Drought stress caused increased accumulation of proline in ten varieties of upland rice and mycorrhizal biofertilizer treatment showed different response of varieties to drought stress conditions (Table 2). Drought stress associated with drought tolerance is an increase in prolina accumulation. According to Yue et al (2006) the mechanism of tolerance through osmotic adjustment is increased proline accumulation but under normal circumstances, proline would be reoxidized into glutamic acid (Widyasari and Sugiyarta, 1997). In this study proline content was higher in …variety at 10 to 13 WAP (Fig. 2).

The occurrence of mycorrhizal infection with rooting is also the beginning of symbiosis between the mycorrhiza arbuscular fungus (CMA) and the roots of upland rice plants. In this study, a higher rate of root infections was found in the treatment of mycorrhiza and soil application on inpago 4 tolerant varieties of about 46.3%, while inpari 33 varieties (sensitive varieties) in mycorrhizal and no till fertilizer applications showed the lowest root infection 13% (table 6). Mycorrhiza is a symbiosis between fungi and plant roots. Mycorrhizas are beneficial for plants that increase nutrient uptake especially of phosphorus and increase plant resistance to drought stress. Upland rice is a plant that has a positive response to the development of mycorrhiza and has the ability as a mass of mycorrhizal mass propagation. This is in line with Sastrahidayat (2010) which stated that the ability of spores adapting to the environment greatly determine the effectiveness of inoculation in host plants. The absence of good root infections in no mycorrhizal condition

Soil cultivation and mycorrhizal application of Inpago 4 showed the ability to adapt drought conditions by avoiding drought stress which is shown in the ratio of root-shoot, leaf proline content, root infection and increased grain production of 7.5 tons per hectare.

References

Abdoellah S, 1997. The threat of water stress in long dry season on coffee and cocoa plants. News Coffee and Cocoa Research Center. 13 (2):77-82

Asch F, Dingkuhn M, Sowc A and Audebert A, 2005. Drought-induced changes in rooting patterns and assimilate partitioning between root and shoot in upland rice. Field Crops Res. 93: 223-236

BPTP,   2012. http://pangan.litbang.pertanian.go.id/berita-706- waste- immersion- salinity-and-resistance-pest- disease-main.html. (9-02-2016).

BPPTP,   2015. http://pangan.litbang.pertanian.go.id/berita-706- padi-toleran drought-salinity-and-resistance-pest- disease-main.html. (9-02-2016)

Brown RE, Havlin JL, Lyons DJ, Fenster CR and Peterson GA, 1991. Long term tillage and nitrogen effects on wheat production in a wheat fallow rotation. In Agronomy Abstracts. Annual Meetings ASA, CSSA, and SSSA, Denver Colorado, 27 October–1 November 1991. 326 pp.

Fukai S and Lilley JM, 1994. Effects of timing and severity of water deficit on four diverse rice cultivars. Field Crop Res. 37: 225-234.

Gowdaa VRV, Henrya BA, Yamauchie A, Ahashidharb HE and Serraj RA, 2011. Root biology and genetic improvement for drought avoidance in rice. Field Crops Res. 122:1-13.

Grant C, Shabtai M, Marcia P, Christian and Christian M, 2011. Soil and Fertilizer Phosphorus: Effects on Plant P Supply and Mycorrhizal Development. Canadian J. Plant Sci. 85(1): 3-14, https://doi.org/10.4141/P03-182

Gunawan AW, 1993. Arbuscular Mycorrhiza. PAU IPB Life Sciences. Bogor. Larson WE and Osborne GJ 1982. Tillage accomplishments and potential. In Predicting Tillage Effects on Soil Physical Properties and Processes. ASA Special Publ. No. 44.

Larson WE and Osborne GJ, 1982. Tillage accomplishments and potential. In Predicting Tillage Effects on Soil Physical Properties and Processes. ASA Special Publ. No. 44.

Mangoendidjojo W, 2005. Fundamentals of Plant Breeding. Publisher Kanisius, Yogyakarta.

Nguyen HT, Babu RC and Blum A, 1997. Breeding for Drought Resistance in Rice Physiology and Molecular Genetic Considerations. Crop Sci. 37: 1426-1434.

Peng S and Ismail AM, 2004. Physiological basis of yield and environmental adaptation in rice. In Nguyen HT and Blum A (eds). Physiology and biotechnology integration for plant breeding. Marcel dekker, Inc. New York

Rachman A, Dariah A and Husen E, 2004. Land conservation. In soil conservation technology on dried up land. Central Research and Development of Land and Agroclimate, Bogor. p. 189-210.

Sastrahidayat IR, 2011. Engineering of Mycorrhizal Fertilizer in Increasing Agricultural Production. Universitas Brawijaya Press, Malang, Indonesia

Setiadi Y, 1999. Development of CMA as a biological fertilizer in the field of forestry, AMI mycorrhizal work paper. Bogor

Sarukhan N, Terzi R and Kadioglu K, 2006. The effects of exogenous polyamines on some biochemical changes during drought Ctenanthe setosa. Acta. Biol. Hungarica 57:221-229

Sitompul M and Guritno B, 1995. Plant Growth Analysis. Gadjah Mada University Press. Yogyakarta, Indonesia.

Serraj R, 2009. Improvement of drought resistance in rice. Advanc. Agron. 103:41-99.

Setiadi Y, 1999. Development of CMA as biological fertilizer in the field forestry, AMI mycorrhizal work paper. Bogor.

Thangadurai D, Carlos AB, dan Mohamed H, 2010. Mycorrhizal Biotechnology. Science Publishers. Enfield, USA

Toha HM, 2010. Development of Gogo Rice Overcome Food Insecurity Region Marginal. Rice Research Institute Sukamandi, Indonesia

Yang J, Zhang J, Wang Z, Zhu Q and Liu L, 2006. Involvement of abscisic acid and cytokinins in the senescence and remobilization of carbon reserves in wheat subjected to water stress during grain filling. Plant Cell Environ. 26: 1621-1631

Yue B, Xue W, Xiong L, Yu Q, Luo L, Cui K, Jin D, Xing Y and Zhang Q, 2006. Genetic basis of drought resistance at reproductive stage in rice: Separation of drought tolerance from drought avoidance. Genetics. 172:1213-1228

Widyasari WB and Sugiyarta E, 1997. The accumulation of proline in leaf tissue of dried cane-resistant varieties. Sugar Res. Mag. 33: 1-10.

18

Influence of copper and lead on germination of three Mimosoideae plant species

Ahmed M. Abbas[1], Sabah Hammad[1], Wagdi Saber Soliman[2]
[1]Department of Botany, Faculty of Science, South Valley University, Qena 83523, Egypt
[2]Horticulture Department, Faculty of Agriculture and Natural Resources, Aswan University, Aswan 81528, Egypt

Corresponding author email:
wagdi79@aswu.edu.eg

Abstract

Contamination with heavy metals is a critical problem facing large areas of agricultural soils. Seed, a developmental stage, is considered highly protective against environmental stresses. This study aimed to examine the influence of two heavy metals; copper and lead, on the germination of *Acacia tortilis*, *A. raddiana*, and *Prosopis juliflora*.

Seeds were exposed to different concentrations of copper and lead including control, low (1000 ppm) and high (2000 ppm) copper or lead, low mix (500 ppm of copper and lead), or high mix (1000 ppm of copper and lead).

High copper and high mix had highly negative effects on germination of *A. tortilis*. Germination of *A. raddiana* was slightly affected under stress. While all stress treatment showed significantly negative effects on germination rate of *Prosopis juliflora*. Both Acacia species were not significantly affected after recovery, while the germination of stressed seeds of *Prosopis juliflora* has been induced after recovery. The 1st germination day was greatly affected with treatments, especially for *Prosopis juliflora*, where 1st germination day delayed about eight days under high mix treatment compared to control.

The results suggested that heavy metals had negative impacts on germination rate. *Prosopis juliflora* was more sensitive to heavy metal stress compared to Acacia species.

Keywords: Copper; Lead; Germination; *Acacia tortilis*; *Prosopis juliflora*

Introduction

Nowadays, large areas of agricultural soils are facing a critical problem of contamination with heavy metals. The environmental contamination is due to different sources such as wastewater, sewage sludge, metallurgy and manufacturing. This contamination has deleterious impacts on the ecosystems and human beings themselves (Lombardi and Sebastiani, 2005; Yuswir et al., 2015). Some plant species can spread where the vegetation is limited by environmental factors such as heavy metal contamination (Yang et al., 2007; Mateos-Naranjo et al., 2011). Over the last

10 years, increasing attention was received to the efficiency of using trees as a suitable vegetation cover in heavy metal-contaminated soils (EPA, 2000). Tree species have been suggested as a sustainable, eco-friendly, and low-cost solution to remediate heavy metal-contaminated soils (Dickinson, 2000).

Some heavy metals are essential micronutrients for plants, such as copper, zinc and nickle, however these metals have toxic effects on organisms at high levels (Munzuroglu and Geckil, 2002). On the other hand, some other toxic heavy metals are non-essential including Hg^{+2}, Cd^{+2}, and Pb^{+2}. These metals are naturally accumulated in water and soil or as contaminations due to human activities. Among toxic heavy metals, Cu and Pb are pollutants presented mostly in free ions form (Cu^{+2} and Pb^{+2}). Copper (Cu) is essential nutrition for plant with important role in plant metabolism, enzyme activity, and detoxification (Taiz and Zeiger, 2002). Copper is transported through the xylem in bounding with amino acids (Reichman, 2002). The excess accumulation of Cu can cause toxic effects such as suppression of photosynthesis and metabolism as well as chromosome injury (Påhlsson, 1989). In contrast, lead (Pb) is not essential to plant function. Extensive studies focused on soil contamination with Pb in agricultural, rural, and urban regions (Markus and McBratney, 2001). Lead contamination occurs as a result of air pollution from motor traffic in addition to industrial materials (Bogdanov, 2005), so certain species that grow near to roadsides accumulate such metal (Price et al., 1974). On the List of Comprehensive Environmental Response, Compensation, and Liability Act Priorities, lead is listed the second hazardous substance (US EPA, 2011). Lead cause a wide range of toxic effects on plants, whether morphological, physiological, and/or biochemical effects. Lead has negative impacts on plant growth, root development, germination, transpiration, photosynthetic tissues, cell division, and protein content (Krzesłowska et al., 2009; Maestri et al., 2010; Pourrut et al., 2011). The toxic effects depend on lead content, period of exposure, and phase of plant development. Fortunately, plants develop various mechanism to cope the toxic metal effects including internal detoxification, excretion, complication by specific glands, and compartmentalization (Gupta et al., 2009; Jiang and Liu, 2010; Maestri et al., 2010). Recently, several studies investigated the molecular mechanism underlying heavy metal uptake and transport (Song et al., 2003). Further investigation is necessary to explore the effect of essential and non-essential metals on plant during different growth stages.

Acacia tortilis is geographically distributed in the Northern Africa including the Sahel, Eastern and Southern Africa, and Arabia regions (Brenan, 1983; Fagg and Barnes, 1990; Barnes et al., 1996). Brenan (1983) distinguishes four subspecies; *raddiana*, *spirocarpa*, *heteracantha*, and *tortilis*. *Raddiana* and *tortilis* are found in Northern Africa and parts of the Arabic peninsula, whereas *spirocarpa* and *heteracantha* are found in Eastern Africa and Southern Africa, respectively. *Acacia tortilis* and *A. raddiana* are recommended in semiarid areas that face severe shortage of forage, particularly where soil erosion, desertification, and/or sand stabilization are problems. *Prosopis juliflora* was considered among the top 100 least wanted species worldwide in the rating of Invasive Species Specialist Group of (IUCN, 2004). It characterizes with rapid growth, great rooting system, effective uptake of nutrients, high evapotranspiration rate, great biomass production, and pronounced clone specific capacity for heavy metal uptake (Nasr et al., 2012), which makes it a very effective tree for phytoremediation purposes (Pegado et al., 2006). *P. juliflora* was introduced to Egypt intentionally into Gabel Elba National Park during the late 80s by local people of Old-Hala'ib village for agro-forestry purposes. After eight years, local people alike showed a problematic and noxious of *P. juliflora* tree due to its strongly aggressive invasion ability (Abbas et al., 2016). In Egypt, *P. juliflora* spread into new areas of invasion, including populations in different dry (ephemeral) riverbeds such as wades in South Egypt where Acacias widely spread, and where many factories causing environment pollution with heavy metals, specially copper and lead.

This study was carried out to assess the effect of copper, lead, and/or their combination on germination of the invasive tree *P. juliflora* and two native *Acacia tortilis* and *Acacia tortilis* subspecies *raddiana* as these are the key events for the establishment of plants under any prevailing environment. We hypothesized that a high proportion of *Prosopis* and *Acacia* seeds would not survive when treating with the heavy metals, at the same time that their germination would be depressed.

Materials and Methods

Fruit collection

Fruits of three species belonging to the subfamily Mimosoideae, family Fabaceae; *Acacia tortilis*, *Acacia tortilis* subspecies *raddiana* (*A. raddiana*, herafter), and *Prosopis juliflora*, were collected randomly from multiple mature individuals in May 2016 from Wadi Merikwan, Wadi Rahaba, Wadi Khoda, and Gebel Elba National Park located in the southeastern part of Egypt ($22°14'2''N - 36°36'30''E$. Collected fruits were stored in dry and dark conditions in the laboratory at 25 °C until using.

Seed germination

In the end of August 2016, seeds were subjected to copper (Cu) and lead (Pb) in seven groups of concentrations; 1) control (distilled water); 2) low Cu (1000 ppm); 3) high Cu (2000 ppm); 4) low Pb (1000 ppm); 5) high Pb (2000 ppm); 6) low mix (500 ppm Cu + 500 ppm Pb); and 7) high mix (1000 ppm Cu + 1000 ppm Pb). Seeds were disinfected before treatments by soaking in a solution of 1% sodium hypochlorite for two minutes and completely washed with sterile water before the germination treatment (Mancilla-Leytón et al., 2011).

In petri dishes, seeds were sowed on a 9 cm filter paper (10 seeds per dish and 4 replicated dishes per treatment). Dishes were placed in a germinator after wrapping with parafilm (Lab-Line Biotronette, Instruments, INC, Meleose, Park, ILL, USA) for 25 days under 12 h of light (at 25 °C, 35 μmol m^{-2}s^{-1} of 400–700 nm wavelength) and 12 h of darkness (at 20 °C). This is the normal condition at the end of autumn in the South of Egypt, where these species are presented (El-Keblawy and Al-Rawai, 2005). The germinated seeds were counted daily and the seed was considered as germinated when its radicle emerged (Abbas et al., 2012). Final germination rate, first germination time, and mean time to germination (MTG) were calculated. MTG was calculating using the equation:

$$MTG = \frac{\Sigma ini \times di}{N}$$

where n is the number of germinated seeds at day i, d the incubation period in days and N the total number of germinated seeds in the treatment (Brenchley and Probert, 1998).

Recovery experiment

Recovery test was conducted to determine if high heavy metals concentration inhibited or caused damage to the seeds. After continuous exposure to different treatments, non-germinated seeds were washed using distilled water, and submerged in new Petri dishes with 5 ml of sterile water under same conditions as previous for twenty days. The three germination parameters were recorded; final rate of germination, first germination time, and MTG.

Viability test

The viability of the embryo was determined after the germination experiment using the tetrazolium test to non-germinated seeds (MacKay, 1972; Cui et al.,

2007). Seeds were maintained in water for 16 h at a 25° C, and then in darkness for one day in 1% aqueous solution of 2,3,5-triphenyl- tetrazolium chloride, pH 7. The seeds were dissected and through a magnifying glass, the embryo was analyzed (Bradbeer, 1998).

Statistical analysis

Statistical analysis was carried out using JMP software (version 4.0; SAS institute, Cary, NC, USA). The statistical differences were analyzed by ANOVA test. Differences in germination rate parameters among treatments were detected using Tukey's Honest Significant Difference (HSD) test. A factorial experiment was set up base on complete randomized design incorporating four replications.

Results

Analysis of variance (ANOVA) showed that copper (Cu) and lead (Pb) treatments significantly affected on germination rate of all species (Table 1). For *Acacia tortilis*, the increments of germination rate differed among treatments. High Cu and high mix (Cu+Pb) treatments had the least germination rate compared to other treatments. The differences were obviously appeared after eight days of treatments. For *A. raddiana*, germination rate slightly increased with time. High Cu and Pb concentrations showed slightly lower germination rate compared to control and low Cu and Pb concentrations. On the other hand, germination rate of *Prosopis juliflora* decreased significantly under Cu and Pb treatments compared to control, and this was significantly evident on the tenth day of treatments and beyond (Figure 1).

After 20 days of treatments, non-geminated seeds were successively washed and treated with distilled water for recovery. Germination rate significantly differed for *Prosopis juliflora* after recovery compared to before recovery under Cu and Pb treatments, but not under control conditions. The differences among treatments were significant at 22 day, and the differences have decreased over time to become insignificant starting from the 28th day. The other species, *A. raddiana* and *A. tortilis*, showed no significant changes in germination rate after recovery compared to before recovery (Table 1 and Figure 1).

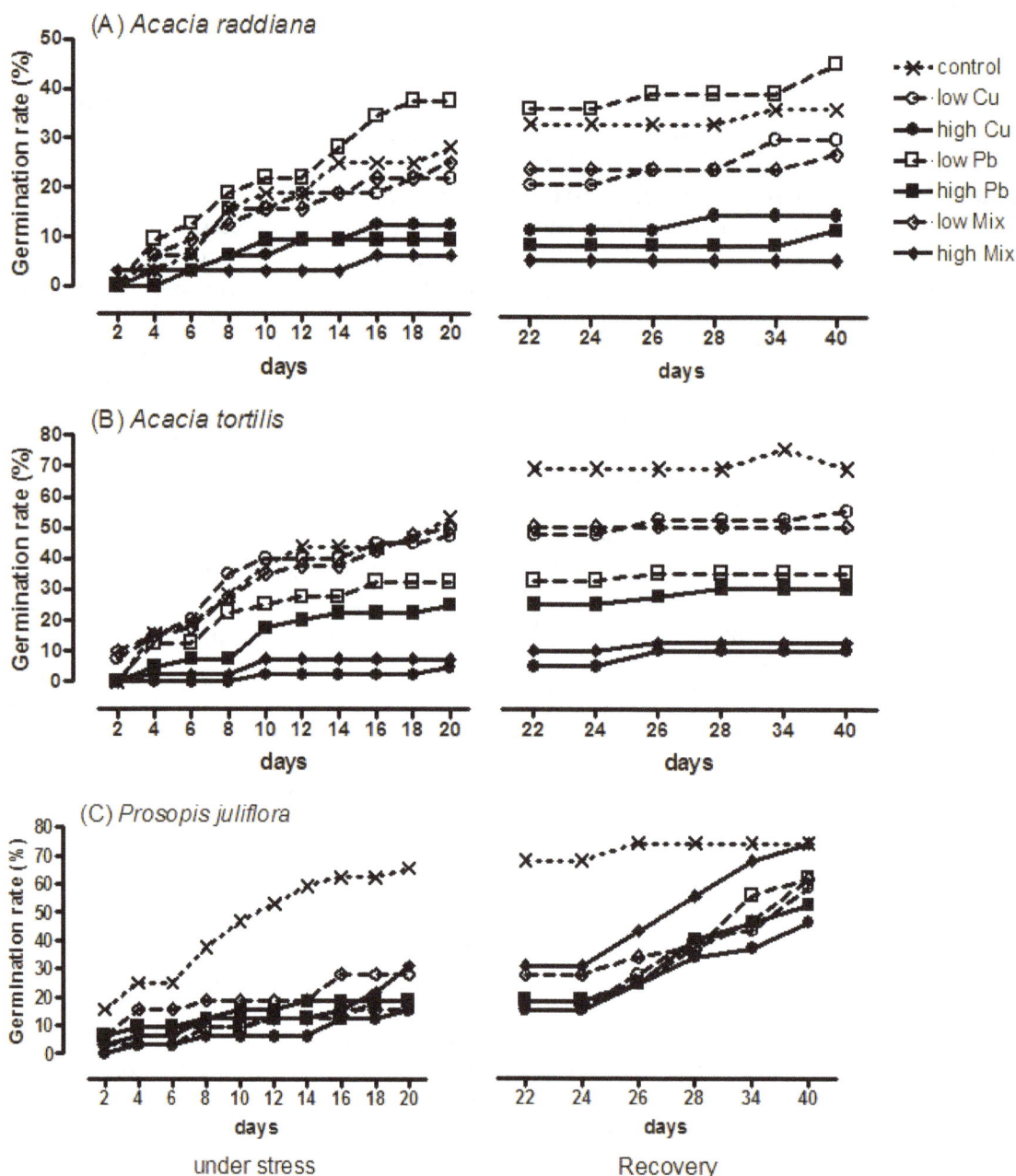

Fig. – 1: Germination rate of *Acacia tortilis*, *A. raddiana*, and *Prosopis juliflora* under copper (Cu) and lead (Pb) treatments; low Cu or Pb (1000 ppm), high Cu or Pb (2000 ppm), low mix (500 ppm of both Cu and Pb), and high mix (1000 ppm of both cu and Pb) as well as after recovery (treatment with distilled water).

The 1st germination day differed greatly among species and among treatments. The 1st germination day was during the second day for *Prosopis juliflora*, during the fourth day for *A. tortilis*, and during the sixth day for *A. raddiana* under control conditions (Table 2). For *Prosopis juliflora*, the 1st germination day ranged between the second day under control and the 14th day under high mix treatment. For *A. tortils*

and *A. raddiana*, the earliest germination was under low Cu treatment and the latest germination was under high mix (10th, and 9th day, respectively). Mean time of germination (MTG) showed slight differences among treatments for *A. tortilis* and *A. raddiana*, and the differences were obviously clear for *Prosopis juliflora* (Table 2). For *Prosopis juliflora*, the highest MTG was under high mix treatments (1.84).

The Viability test showed that heavy metal treatment did not show clear effects on the viability of seeds. Few numbers of seeds were dead which were not related to specific treatment (Table 2).

Table - 1: Analysis of Variance (F value) of germination rate under different copper and Lead treatments as well as under recovery of the three species; _Acacia tortilis, A. raddiana,_ and _Prosopis juliflora_

	A. tortilis	A. raddiana	P. juliflora
Under stress			
Treatments	30.2***	13.33***	34.9***
days	11.2***	9.99***	7.49***
Recovery			
Treatments	63.5***	19.0***	23.8***
days	0.70	0.40	19.1***

*** Significant at P = 0.001.

Table - 2: Number of days to first germination, mean time to germination (MTG) and viability under control, low Cu, high Cu, low Pb, high Pb, low mix and high mix for the invasive _Prosopis juliflora_ and two native; _A. tortilis, A. raddiana_ from Southeast Egypt.

Treatment	1st germination	MTG (d)	Viability (%)
P. juliflora			
Control	1.75±0.48	0.97±0.11	96.9±3.1
Low Cu	2.50±0.50	0.94±0.15	100±0.0
High Cu	8.67±3.53	1.38±0.52	100±0.0
Low Pb	8.00±3.67	1.26±0.48	100±0.0
High Pb	2.00±1.00	0.75±0.22	100±0.0
Low mix	4.75±3.09	1.00±0.28	100±0.0
High mix	13.75±2.9	1.84±0.42	100±0.0
A. tortilis			
Control	4.00±2.03	1.19±0.13	90.6±6.0[ab]
Low Cu	1.00±1.25	3.25±0.15	97.5±2.5[a]
High Cu	9.00±0.00	1.50±0.38	100±0.0[ab]
Low Pb	4.38±1.30	0.70±0.04	100±0.0[ab]
High Pb	7.63±2.75	1.05±0.23	97.5±2.5[ab]
Low mix	4.75±2.14	0.95±0.22	100±0.0[ab]
High mix	10.00±3.8	1.10±0.35	95.0±5.0[b]
A. raddiana			
Control	6.00±2.08	1.25±0.18	93.8±6.0
Low Cu	5.50±2.02	1.04±0.26	96.9±3.0
High Cu	7.33±2.03	1.17±0.31	100±0.0
Low Pb	6.00±1.58	1.19±0.25	100±0.0
High Pb	7.67±1.45	0.96±0.27	100±0.0
Low mix	5.33±2.33	1.25±0.31	100±0.0
High mix	9.00±7.00	1.13±0.48	100±0.0

Discussion

Seed, a developmental stage, is considered highly protective against external environmental stresses. In contrast, vegetative development stage is generally sensitive to stress (Li et al., 2005). The toxicity of metals on _Arabidopsis_ seeds depended on the physiological state of seeds and the metal ions selective permeation through tissues surrounding the embryo (Li et al., 2005). Among common pollutants, plants frequently encountered the toxicity of copper and lead (Påhlsson, 1989; Cecchi et al., 2008; Shahid et al. 2011). Exposure of plants to lead has negative effects on germination and growth even at micro-molar levels (Islam et al., 2007).

In this study, heavy metal treatments showed significant decreases in germination rate (Table 1). The germination rate under high stress treatments was lower than that under control and low stress treatments (Figure 1). These results showed the negative impacts of both Cu and Pb on germination rate especially under high concentrations. This is consistent with previous findings in general. High concentration of copper (Cu) has negatively affected on germination rate of wheat and rice (Mahmood et al., 2007), alfalfa (Aydinalp and Marinova, 2009), _Vigna mungi_ (Solanki et al., 2011), and Crambe (Hu et al., 2015). Also, there are several reports about lead inhibition of species seed germination including _Brassica penkinensis, Elsholtzia argyi, Hordeum vulgare, Oryza sativa, Pinus halepensis, Spartina alterniflora,_ and _Zea mays_ (Islam et al., 2007; Sengar et al., 2009). In this study, Germination rate ranged between 6-40% for _A. raddiana_ at 20 day of stress treatments, to be between 6-47% after recovery. For _A. tortilis_, the germination rate ranged between 7-55% at 20 day of stress, and between 10-70% after recovery. On the other hand, germination rate of _Prosopis juliflora_ ranged between 15-66% at 20 day of stress, and 45-75% after recovery. These results indicated the negative impact of heavy metals, Cu and Pb, on germination especially for _Prosopis juliflora,_ which showed significant increases in germination rate after recovery for stressed seeds. These results suggested that _Prosopis juliflora_ was more sensitive to both Cu and Pb stress. On the other hand, _Acacia tortilis_ was more sensitive to Cu rather than Pb. _A. raddiana_ was the least affected to Cu and Pb stress compared to others species.

In this study, the 1st germination day affected significantly with heavy metal treatments (Table 2). _Prosopis juliflora_ was the most affected species compared to others. The earliest germination was during the second day under control, and the latest germination was during the 14th day under high mix (1000 ppm Cu+1000 ppm Pb) treatment. The 1st germination day for *A. tortilis* and *A. raddiana* were during the 4th day and 6th day, respectively, under control. The latest germination was under high mix treatment for both *A. tortilis* and *A. raddiana* (10th, and 9th day, respectively). The mean time of germination and viability test showed no clear effects on all species. These results suggested the role of heavy metal absorption during the germination stage on slowdown of germination, but not on viability of seeds.

From the results of this study, it is concluded that heavy metals had negative impacts on germination especially for *Prosopis juliflora* species which showed sensitivity to heavy metal stress compared to Acacia species. The mean time of germination, MTG, and seed viability were not significantly affected by heavy metals which reflect the role of heavy metal absorption on slowdown of germination, but not on viability of seeds.

References

Abbas AM, Rubi-oCasal AE, De Cires A, Figueroa ME, Lambert AM and Castillo JM, 2012. Effects of flooding on germination and establishment of the invasive cordgrass *Spartina densiflora*. Weed Res. 52: 269-276.

Abbas AM, Soliman WS, El Taher AM, Hassan IN, Mahmoud M, Youssif MF, Mansour MH and Abdelkareem M, 2016. Predicting the spatial spread of invasive *Prosopis juliflora* (SW.) D.C. along environmental gradients in Gabel Elba National Park, Egypt. Int. J. Scientific Engin. Res. 7(2): 596-599.

Aydinalp C and Marinova S, 2009. The effects of heavy metals on seed germination and plant growth on Alfalfa Plant (*Medicago sativa*). Bulg. J. Agric. Sci. 15: 347-350.

Barnes RD, Filer DL and Milton SJ, 1996. *Acacia karroo*. Tropical Forestry Papers 32. Oxford Forestry Institute, Oxford University.

Bogdanov S, 2005. Contaminants of bee products. Apidologie. 37: 1-18.

Bradbeer JW, 1998. Seed Dormancy and Germination. New York, NY: Chapman and Hall.

Brenan JPM, 1983. Manual on Taxonomy of Acacia species. Food and Agriculture Organization of the United Nations, Rome, p. 47.

Brenchley J and Probert R, 1998. Seed germination responses to some environmental factors in the seagrass *Zostera capricorni* from eastern Australia. Aquat. Bot. 62: 177-188.

Cecchi M, Dumat C, Alric A, Felix-Faure B, Pradere P and Guiresse M, 2008. Multi-metal contamination of a calcic cambisol by fallout from a lead-recycling plant. Geoderma. 144(1–2): 287-298.

Cui J, Li Y, Zhao H, Su Y and Drake S, 2007. Comparison of Seed Germination of *Agriophyllum squarrosum* (L.) Moq. and *Artemisia halodendron* Turcz. Ex Bess, Two Dominant Species of Horqin Desert, China. Arid Land Res. Manag. 21: 165-179.

Dickinson NM, 2000. Strategies for sustainable woodland on contaminated soils. Chemosphere. 41: 259-263.

El-Keblawy A and Al-Rawai A, 2005. Effects of salinity, temperature and light on germination of invasive *Prosopis juliflora*. J. Arid Environ. 61: 555-565.

EPA, 2000. Introduction to Phytoremediation. Washington: U.S. Environmental Protection Agency; EPA/600/R-99/107.

Fagg CW and Barnes RD, 1990. African Acacias: Study and Acquisition of the Genetic Resources. Final report, ODA Research Scheme R.4348, Oxford Forestry Institute, UK. p. 170.

Gupta D, Nicoloso F, Schetinger M, Rossato L, Pereira L, Castro G, Srivastava S and Tripathi R, 2009. Antioxidant defense mechanism in hydroponically grown *Zea mays* seedlings under moderate lead stress. J. Hazard Mater. 172(1): 479-484.

Hu J, Deng Z, Wang B, Zhi Y, Pei B, Zhang G, Luo M, Huang B, Wu W and Huang B, 2015. Influence of heavy metals on seed germination and early seedling growth in *Crambe abyssinica*, a potential industrial oil crop for phytoremediation. Am. J. Plant Sci. 6: 150-156.

Islam E, Yang X, Li T, Liu D, Jin X and Meng F, 2007. Effect of Pb toxicity on root morphology, physiology and ultrastructure in the two ecotypes of *Elsholtzia argyi*. J. Hazard Mater. 147(3): 806-816.

Jiang W and Liu D, 2010. Pb-induced cellular defense system in the root meristematic cells of *Allium sativum* L. BMC Plant Biol. 10: 40-40.

Krzeslowska M, Lenartowska M, Mellerowicz EJ, Samardakiewicz S and Wozny A, 2009. Pectinous cell wall thickenings formation–a response of moss protonemata cells to lead. Environ. Exp. Bot. 65(1): 119-131.

Krzesłowska M, Lenartowska M, Samardakiewicz S, Bilski H and Wo′zny A, 2010. Lead deposited in the cell wall of *Funaria hygrometrica* protonemata is not stable–a remobilization can occur. Environ. Pollut. 158(1): 325-338.

Li W, Khan MA, Yamaguchi S and Kamiya Y, 2005. Effects of heavy metals on seed germination and early seedling growth of *Arabidopsis thaliana*. Plant Growth Regul. 46: 45-50.

Lombardi L and Sebastiani L, 2005. Copper toxicity in *Prunus cerasifera*: growth and antioxidant enzymes responses of *in vitro* grown plants. Plant Sci. 168: 797-802.

MacKay DB, 1972. The Measurement of Viability. Pages 172–208 in E.H. Roberts, ed. Viability of Seeds. Springer Netherlands.

Maestri E, Marmiroli M, Visioli G and Marmiroli N, 2010. Metal tolerance and hyperaccumulation: costs and trade-offs between traits and environment. Environ. Exp. Bot. 68(1): 1-13.

Mahmood T, Islam KR and Muhammad S, 2007. Toxic effects of heavy metals on early growth and tolerance of cereal crops. Pak. J. Bot. 39: 451-462.

Mancilla-Leytón JM, Fernández-Alés R and Vicente AM, 2011. Plant-ungulate interaction: Goat gut passage effect on survival and germination of Mediterranean shrub seeds. J. Veg. Sci. 22: 1031-1037.

Markus J and McBratney AB, 2001. A review of the contamination of soil with lead II. Spatial distribution and risk assessment of soil lead. Environ. Int. 27: 399-411.

Mateos-Naranjo E, Andrades-Moreno L and Redondo-Gomez S, 2011. Comparison of germination, growth, photosynthetic responses and metal uptake between three populations of *Spartina densiflora* under different soil pollution conditions. Ecotoxicol. Environ. Safe. 74: 2040-2049.

Munzuroglu O and Geekil H, 2002. Effects of metals on seed germination, root elongation, and coleoptile and hypocotyls growth in *Triticum aestivum* and *Cucumis sativus*. Arch. Environ. Contamin. Toxicol. 43: 203-213.

Nasr SMH, Parsakhoo A, Naghavi H and Koohi SKS, 2012. Effect of salt stress on germination and seedling growth of *Prosopis juliflora* (Sw.). New Forests. 43: 45-55.

Påhlsson AB, 1989. Toxicity of heavy metals (Zn, Cu, Cd, Pb) to vascular plants. Water Air Soil Pollut. 47: 287-319.

Pegado CMA, Andrade LA, Felix LP and Pereira LM, 2006. Efeitos da invasão biológica de algaroba— *Prosopis juliflora* (Sw.) DC. sobre a composição e a estrutura do estrato arbustivo-arbóreo da caatinga no Município de Monteiro, PB, Brasil. Acta Botanica Brasilica. 20: 887-898.

Pourrut B, Shahid M, Dumat C, Winterton P and Pinelli E, 2011. Lead uptake, toxicity, and detoxification in plants. Rev. Environ. Contam. T. 213: 113-136.

Price PW, Rathcke BJ and Gentry DA, 1974. Lead in terrestrial arthropods: evidence for biological concentration. Environ. Entomol. 3: 370-372.

Reichman SM, 2002. The Responses of Plants to Metal Toxicity: A Review Focusing on Cu, Manganese and Zinc. The Australian Minerals and Energy Environment Foundation, Melbourne.

Sengar RS, Gautam M, Sengar RS, Garg SK, Sengar K and Chaudhary R, 2009. Lead stress effects on physiobiochemical activities of higher plants. Rev. Environ. Contam. T. 196: 1-2.

Shahid M, Pinelli E, Pourrut B, Silvestre J and Dumat C, 2011. Lead-induced genotoxicity to *Vicia faba* L. roots in relation with metal cell uptake and initial speciation. Ecotoxicol. Environ. Safe. 74(1): 78-84.

Solanki R, Arjun, Poonam and Dhankhar R, 2011. Zinc and copper induced changes in physiological characteristics of *Vigna mungo* L. J. Environ. Biol. 32: 747-751.

Song WY, Sohn EJ, Martinoia E, Lee YJ, Yang YY, Jasinski M, Forestier C, Hwang I and Lee Y, 2003. Engineering tolerance and accumulation of lead and cadmium in transgenic plants. Nat. Biotechnol. 21: 914-919.

Taiz L and Zeiger E, 2002. Plant Physiology. 3rd edn. Sinauer, Sunderland.

US EPA, 2011. CERCLA Priorities List of Hazardous Substances. http:// www.atsdr.cdc.gov/spl/ Accessed on 11 February 2017.

Yang RY, Tang JJ, Yang YS and Chen X, 2007. Invasive and noninvasive plants differ in response

19

Quality response of maize fodder cultivars to harvest time

Abdul Rehman[1], Aurangzeb[1], Rafi Qamar[1*], Atique-ur-Rehman[2], Muhammad Shoaib[3], Jamshaid Qamar[1] and Farwa Hassan[1]

[1]Department of Agronomy, University College of Agriculture, University of Sargodha, Pakistan
[2]Department of Agronomy, Bahauddin Zakariya University, Multan, Pakistan
[3]Maize and Millets Research Institute, Yousufwala, Sahiwal, Pakistan

Corresponding author email:
drrafi1573@gmail.com

Abstract

The study was conducted to investigate the fodder quality of four maize cultivars; DK919, 30R50, 31R88 and 6621 as influenced by harvest time at Agronomic Research Area of University of Sargodha. Maize cultivars were harvested at three different times viz. 80, 90 and 100 days after sowing (DAS). Significant differences were recorded among the cultivars for plant height, acid and neutral detergent fiber contents, lignin and crude protein. Maximum acid detergent fiber content, neutral detergent fiber content and lignin were observed at 100 DAS while crude protein was maximum at 80 DAS. However, plant height was remained unaffected with respect to harvest times. Moreover, maize cultivars had distinct differences in plant height and fodder quality parameters. Maximum plant height and crude protein were recorded in cultivar 31R88. The cultivar DK919 showed maximum values of acid detergent fiber content and neutral detergent fiber while lignin content was higher in V6621. Fodder quality parameters of cultivars 31R88, DK919 and V6621 were superior than 31R88 under the present climatic conditions of Sargodha.

Keywords: Harvest Time, Cultivars, Fodder Yield, Maize, Fodder Quality

Introduction

Maize (*Zea mays* L.) is a dual-purpose crop universally grown for grain and forage. It produces ample quantity of green herbage with high nutritional and appetizing value (Akdeniz et al., 2004; Erdal et al., 2009). It has a distinct position in the national economic system of Pakistan and have 6.4% share in the total grain production. Moreover, it is a quality source of food, feed and fodder (Abdullah et al., 2007). In Pakistan, at least two crops of maize can be harvested in a year i.e., in spring and autumn seasons. According to the Economic Survey of Pakistan, the maize cultivation during the year (2014-15) was 1.13 million hectares with grain production of 4.69 million tons with average production was 4155 kg ha^{-1} (GOP, 2014-15). Maize production in Pakistan is still low compared to other countries in spite of favourable environmental conditions and high yielding varieties. There are various limiting factors like water shortage, unpredictable rainfall, unavailability and high cost of fertilizers, less significance of fodder production and human population pressure adversely affecting fodder production (Rashid et al., 2007).

Quality of grain and fodder is considered as most important in maize production. Environment, planting time, stage at harvest, type of hybrids, agronomic management, hygienic quality, digestibility and consumption by animal are the most important grain, fodder and silage quality determining and limiting factors (Bal et al., 2000; Widdicombe and Thelen, 2002; Geren, 2000; Yilmaz et al., 2003). Maize fodder and feed products provides all forms of elementary nutrients and source of energetic nutrients with relatively low content of crude protein (Mlynár et al., 2004). Fodder harvesting at appropriate time is a main aspect for a successful forage production. Fodder cutting at maturity resulted in higher lignin content while lower concentration of fodder quality traits like plant protein, neutral detergent fiber (NDF), acid detergent fiber (ADF) and leaf proportion (Atis et al., 2012). Forage quality is directly influenced by the stage of maturity, which decreases as plant advances towards maturity and results in lower forage digestibility and consumption by animals (Ball et al., 2001). The neutral detergent fiber and acid detergent fiber are the most important fodder quality constituent, which are used as standard forage testing techniques. Moreover, these quality parameters are used to calculate fodder digestibility and intake potential (Ball et al., 2001).

Fodder quality characters may vary among different maize cultivars. Similarly, the time of fodder harvesting is affecting the fodder quality of each cultivar. The present study, therefore, was undertaken to evaluate the quality of four maize cultivars harvested at three different harvest time of autumn sown maize under semi-arid condition of Pakistan.

Materials and Methods

Study site
The current study referred to know the fodder quality response of four autumn maize cultivars to three harvest times was conducted at the research area of University College of Agriculture, University of Sargodha, Sargodha during the year 2012. Sargodha lies at 32.08° N and 72.67° E. General elevation of land from sea level is 193 m.

Soil collection and analysis
The soil on which experiment was conducted was clay loam. Before starting the experiment, the samples of soil were collected to a depth of 30 cm and were analyzed for various physical and chemical properties (Jackson, 1962; Moodie et al., 1959; Watanable and Olsen, 1965). Soil characteristic recorded are presented in Table-1.

Experimental design and crop husbandry
Experiment was conducted applying randomized complete block design in a split plot arrangement having three replications with net plot size assigned to single treatment of 4 m x 6 m. Three harvest times 80, 90 and 100 days were allocated to main plots while four cultivars (DK919, 30R50, 31R88 and 6621) were assigned to subplots.

Table – 1: Pre-sowing analysis of experimental soil during 2012

Characteristic	Unit	Values	
		0-15 cm	15-30 cm
Chemical analysis			
Saturation	%	37	36
pH	----	7.8	7.9
EC	dS m^{-1}	1.64	1.65
Organic matter	%	0.96	0.84
Total nitrogen	%	0.047	0.043
Available phosphorous	Ppm	12.5	8.5
Mechanical analysis			
Sand	%	29	25
Silt	%	47	46
Clay	%	22	33
Soil texture	----	Silt loam	Loam

Field was prepared by using tractor mounted cultivator, cultivating thrice which resulted in well pulverized. There were 36 plots prepared and each plot had 5 ridges in East-West direction. Sowing was done manually on July 26, 2012. Maize hybrids seed was used at the rate of 10 kg acre^{-1}. Nitrogen fertilizer was applied in three splits in the form of Urea i.e. 1/3rd at the time of sowing, 1/3rd when crop was at knee height and 1/3rd at tasseling stage. Whole of the phosphorus and potash fertilizers, in the form of Di-ammonium phosphate and murate of potash were applied as basal dose. On the basis of need of the crop, seven irrigations were applied in addition to rainfall during the whole growth period. First irrigation was applied after 05 days of sowing and later irrigations were applied as when needed. To maintain inter plant spacing thinning was done when crop reached at the height of 15 cm. Plant protection measures like application of weedicide and insecticide etc. were kept normal for all treatments. Harvesting of different samples was done 80, 90 and 100 days after sowing respectively.

Fodder quality parameters of maize
Ten plants were selected at random from each plot to record individual plant observation like plant height (cm) by using standard procedure. On each harvest date one plot of each genotype was harvested. Dried samples were ground using a hammer mill to pass a 1 mm screen. Whole plants samples were analyzed for NDF, ADF and CP content. A 0.5 g sample was used

for sequential detergent analysis to determine NDF and ADF contents (Soest et al., 1991). Total N was determined by the Kjeldahl procedure and CP content was calculated by multiplying total N by 6.25. All compositional data were calculated on a dry matter basis. The concentration of lignin was measured by using the simplified method adopted by Goering and Van Soest, (1970).

Statistical analysis

Data were analyzed statistically using SAS (SAS Institute, 2008). The effects of harvest time and cultivars and their interaction were evaluated by the Duncan's Multiple Range test (DMRT) at $p \leq 0.05$ unless otherwise mentioned. The computer package MS-Excel was used to prepare the graphs.

Results

Different maize cultivars were affected by harvest time, cultivars and their interactive effect were significantly different ($p > 0.05$) on plant height (cm), acid and neutral detergent fiber (%), lignin (%) and crude protein (%). Plant height is an important yield-contributing as well as fodder production factor. It is apparent from Figure-1 that there is significant difference in plant height of maize cultivars when harvested at various times. Statistically taller (212 cm) plant height was noted at 100 DAS. Cultivars 31R88

cultivars. The interaction between harvest time and cultivars showed that cultivar 31R88 recorded (262.4 cm and 259.5 cm) taller plants ($p > 0.05$) at harvest time of 100 and 80 DAS, respectively. However, shorter (174.2 cm) plant height was recorded in V6621 at 100 DAS compared than other treatments. An overview of the acid detergent fiber data (Figure-2) revealed that maximum (32.7 and 32.3 %) ADF was attained in V6621 and DK919 at 100 DAS respectively. Neutral detergent fiber is the percent of cell wall material of the forage cellulose hemicelluloses and unavailable protein. It is the evident from the results (Figure 3) that harvest time and varieties significantly affected the neutral detergent fiber. Moreover, significantly maximum (50.7, 48.3 and 48.3 %) neutral detergent fiber was recorded in DK919, 31R88 and V6621 at 100 DAS respectively. Lignin is an indigestible plant structural constituent which binds the other plant molecules and makes them indigestible. Statistically maximum lignin contents (3.6 and 3.4 %) were attained in variety V6621 and DK919 at 100 DAS while statistically minimum lignin (2.4 %) was observed in 31R88 at 80 DAS (Figure 4). Crude protein affects the nutritional value and palatability of the forage crop. It is apparent from Figure 5 that harvest time and cultivars had a significant effect on crude protein of maize. The interaction between harvest time and cultivars showed maximum crude protein (8.8, 8.7, 8.3 and 8.1 %) in 31R88, V6621, 30R50 and DK919 at 80 DAS

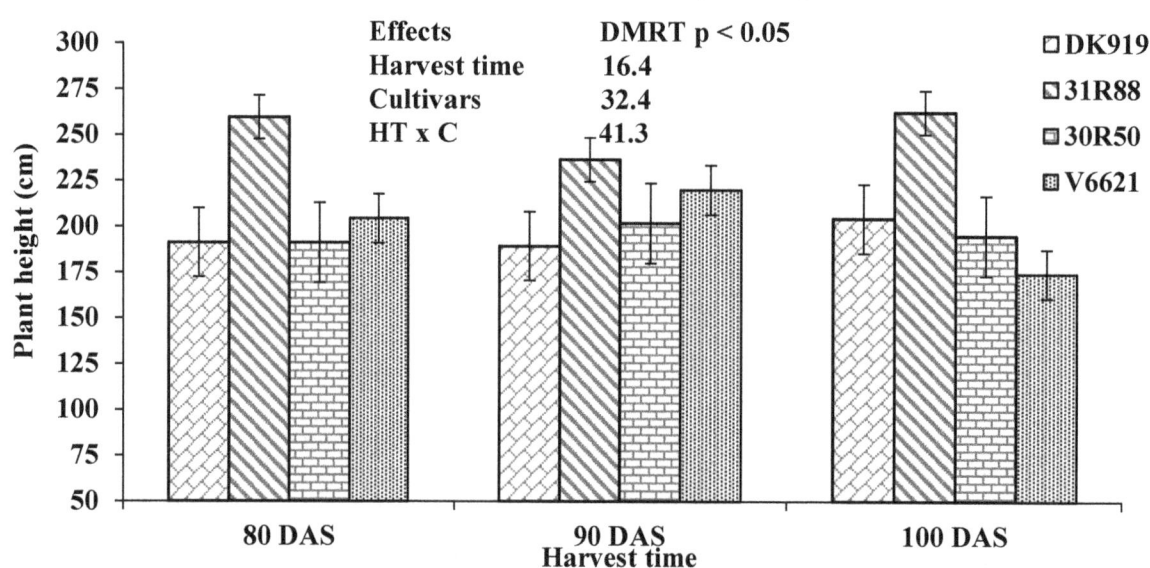

showed taller (252.8 cm) plant height among other respectively.

Figure – 1: Effect of harvest time and cultivars on plant height (cm) of autumn maize.

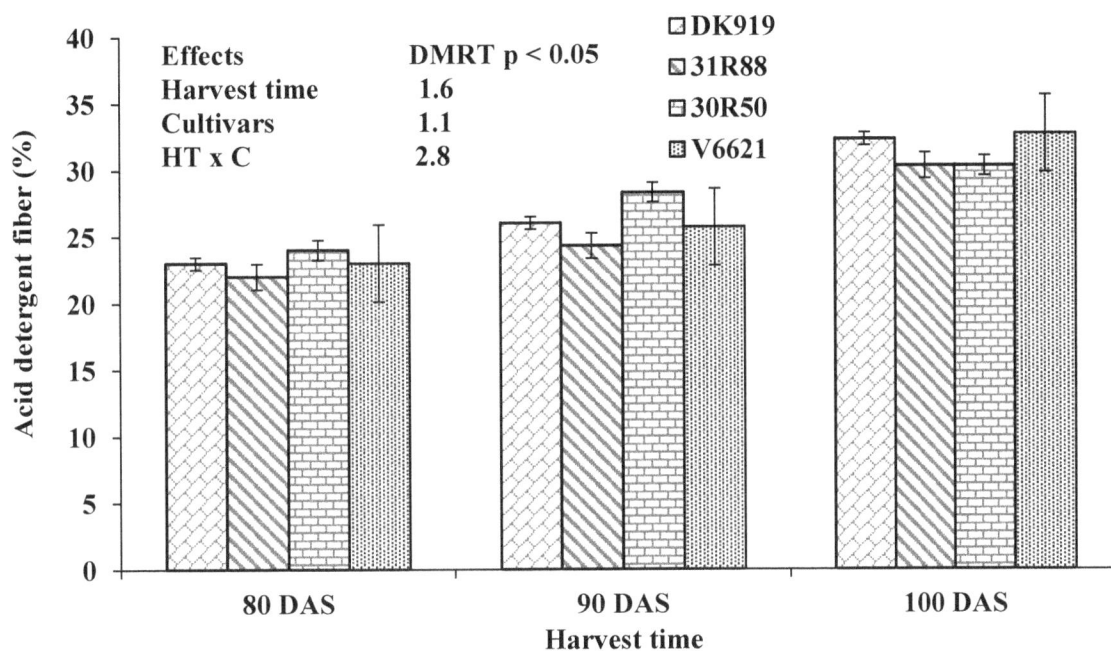

Figure – 2: Effect of harvest time and cultivars on acid detergent fiber (%) of autumn maize.

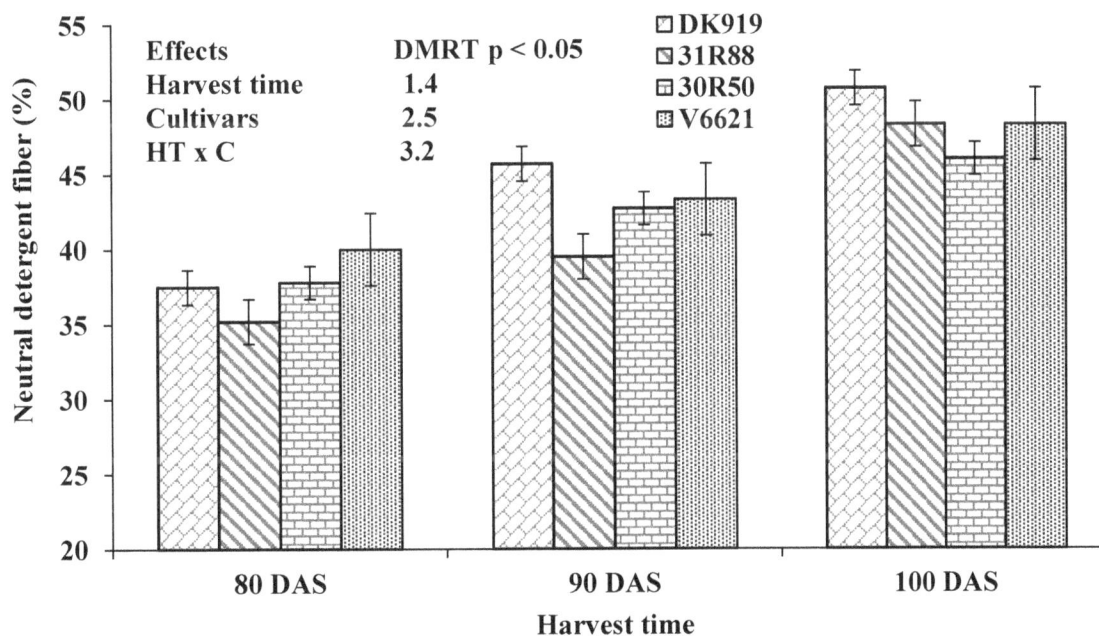

Figure – 3: Effect of harvest time and varieties on neutral detergent fiber (%) of autumn maize.

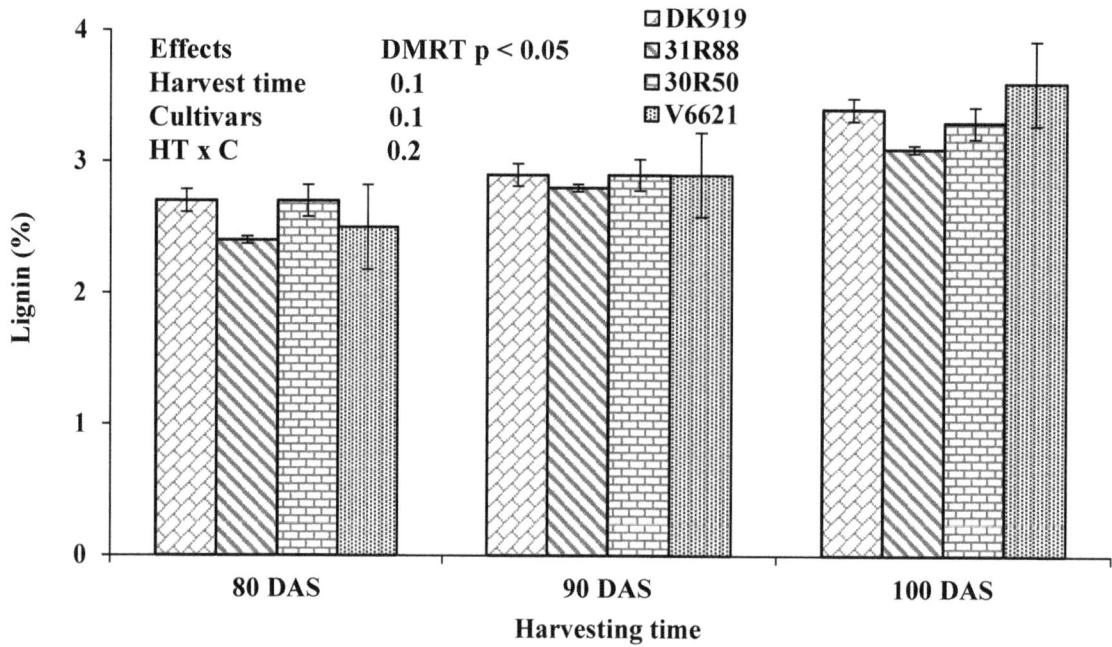

Figure – 4: Effect of harvest time and cultivars on lignin (%) of autumn maize.

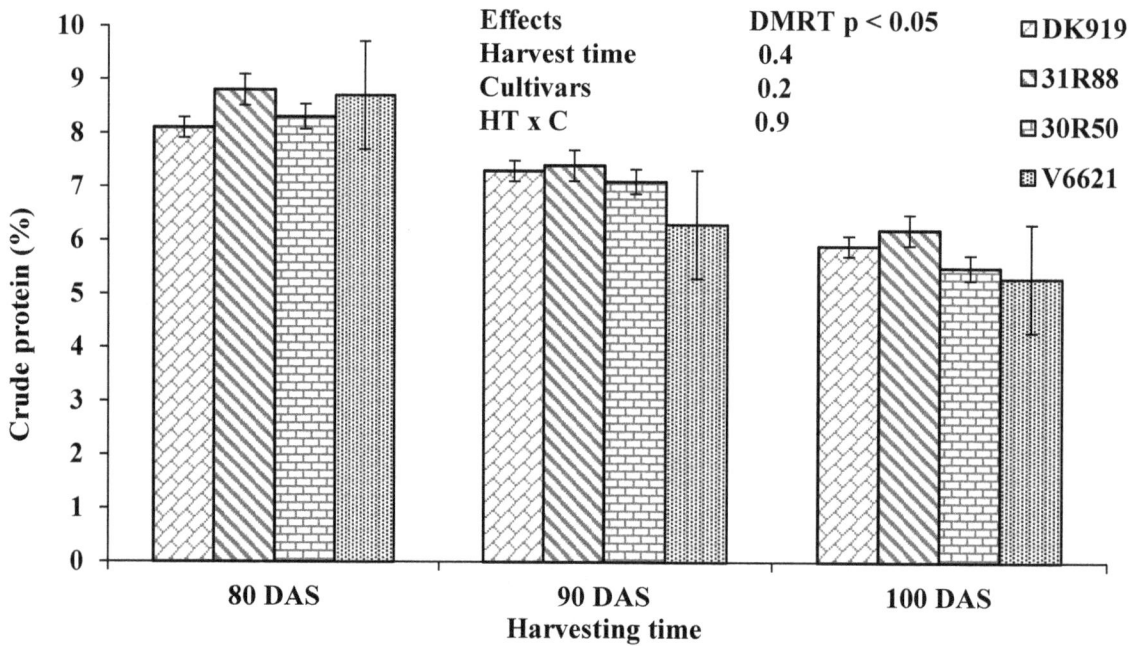

Figure – 5: Effect of harvest time and cultivars on crude protein (%) of autumn maize.

Discussion

Maize forage yield and quality were influenced by harvest time and cultivars. During plant maturity the trend in plant height was increased from early to late harvest due to prolonged exist in field (Ayub et al., 2002; Xie et al., 2012). On the contrary, Carmi et al. (2006) testified that plant height at early harvest was not changed from late harvest. The inconsistent findings might be due to genetic variation in plants traits while in the current study shorter plant height was recorded. The variation among cultivars might be due to environment, soil fertility, harvesting stage and cultivars genome. Moreover, under similar agronomic and climatic conditions genetic character dominated and showed significant variation in cultivars tallness among various maize cultivars (Hussain et al., 2010; Awan et al., 2001). Forage quality and yield must be optimized to determine the best time for harvest. If forage is harvested too early, excessive loss of nutrients, from soil run off, occurs due to poor starch development in the kernel and low energy concentration. Fodder harvested too late above 100 DAS has decreased nutritive value due to poor starch and fiber digestion of silages (Neylon and Kung, 2003). The effect of harvest time and cultivars were significant for quality parameters. Fraction of leaves was constantly reduced as improvement in maturity (Carmi et al., 2005 and 2006). Harvest time significantly influenced the fodder quality parameters like acid detergent fiber, neutral detergent fiber and lignin which were maximum at late harvest stage (Butler and Muir, 2003; Carmi et al., 2005). Lignin accumulation and synthesis occur at the stage of secondary cell wall development (Carmi et al., 2006). The reason might be that at early harvest, the moisture contents in the plant were high and the concentration of dry matter was less as the plant matures dry matter accumulation increased in the plant, which resulted in maximum content of ADF, NDF and lignin (Carmi et al., 2006). However, the crude protein content was minimum observed in our study due to late harvest stage. Our results supported the findings of Huang et al. (2012) they reported that delayed harvest stage produced lower crude protein concentration which might be due to the higher dry matter yield per land area. Differences in crude protein content among genotypes were also reported by Carmi et al. (2005); Miron et al. (2005); Miron et al. (2006); Yosef et al. (2009). They reported that crude protein concentration was maximum at the first harvest stage and declined with maturity of plant due to increase in concentration of acid and neutral detergent fiber and lignin. They have a reverse trend and reached at maturity. Additionally, cultivars may affect rates of nutrient translocation (nutrients moving from the stalk and leaves to the ear) and rates of maturation (Lewis et al., 2004; Owens, 2005).

Conclusion

The results of the research showed that investigated fodder quality parameters of cultivars were influenced by harvest time. Fodder growers who are interested in increased concentration of ADF, NDF and lignin content than maize fodder was harvested late. Keeping in view the yield and silage quality the cultivars 30R50 and 31R88 should be preferred respectively over other cultivars under the present climatic conditions of Sargodha.

References

Abdullah G, Hassan I Khan A and Munir M, 2007. Effect of planting methods and herbicides on yield and yield components of maize. Pak. J. Weed Sci. Res. 13: 39-48.

Akdeniz H, Yilmaz I Andic N and Zorer S, 2004. A Study on Yield and Forage Values of Some Corn Cultivars. Univ. of Yuzuncuyil. J. Agric. Sci. 14: 47-51.

Atis I, Konuskan O Duru M Gozubenli H and Yilmaz S, 2012. Effect of harvesting time on yield, composition and forage quality of some forage sorghum cultivars. Int. J. Agric. Biol. 14: 879-886.

Awan TH, Mahmood MT Maqsood M Usman M and Hussain MI, 2001. Studies on hybrid and synthetic cultivars of maize for forage yield and quality. Pak. J. Agri. Sci. 38: 50-52.

Ayub M, Nadeem MA Tanveer A and Husnain A, 2002. Effect of different levels of nitrogen and harvesting times on the growth, yield and quality of sorghum fodder. Asian J. Plant Sci. 1: 304-307.

Bal MA, Shaver RD Shinners KJ Coors JG Lauer JG Straub RJ and Koegel RG, 2000. Stage of maturity, processing and hybrid effects on ruminal in situ disappearance of whole-plant corn silage. Anim. Feed Sci. Technol. 86: 83-94.

Ball DM, Collins M Lacefield GD Martin NP Mertens DA Olson KE Putnam DH Undersander DJ and Wolf MW, 2001. Understanding Forage Quality.

American Farm Bureau Federation Publication 1-01, Park Ridge, Illinois, USA.

Butler TJ and Muir JP, 2003. Row Spacing and Maturity of Forage Sorghum Silage in North Central Texas. Forage Research in Texas, http://forageresearch.tamu.edu/2003/Forage Sorghum.

Carmi A, Umiel N Hagiladi A Yosef E Ben-Ghedalia D and Miron J, 2005. Field performance and nutritive value of a new forage sorghum variety 'Pnina' recently developed in Israel. J. Sci. Food Agric. 85: 2567–2573.

Carmi A, Aharoni Y Edelstein M Umiel N Hagiladi A Yosef E Nikbachat M Zenou A and Miron J, 2006. Effects of irrigation and plant density on yield, composition and in vitro digestibility of a new forage sorghum variety, Tal, at two maturity stages. Anim. Feed. Sci. Tech. 131: 120-132.

Erdal S, Pamukcu M Ekiz H Soysal M Savur O and Toros A, 2009. The determination of yield and quality traits of some candidate silage maize hybrids. Univ. of Akdeniz. J. Agric. Sci. 22: 75-81.

Geren H, 2000. Investigations on the effect of sowing dates on the forage yields and agronomical characteristics related to silage of different maize (Zea mays L.) cultivars grown as main and second crops. Univ. of Ege, Graduate School of Natural and Applied Sciences, (Unpublished Ph.D. Thesis), Izmir Turkey, pp: 36-43.

Goering HK and Van Soest PJ, 1970. Forage fiber analysis. In: Agricultural Handbook No 379. Agricultural Research Service, United States Department of Agriculture, Washington, D.C. pp. I-20.

Govt. of Pakistan, (2014-15). Economic Survey of Pakistan. Finance Division Economic Advisor's Wing, Islamabad.

Huang H, Faulkner DB, Singh V Danao MC and Eckhoff SR, 2012. Effect of harvest date on yield, composition and nutritive value of corn stover and DDGS. Transactions of the ASABE. 55: 1859-1864.

Hussain N, Zaman Q Nadeem MA and Aziz A, 2010. Response of maize varieties under Agro-ecological conditions of Dera Ismail Khan. J. Agri. Res. 48: 59-63.

Jackson ML, 1962. Soil Chemical Analysis. Printice Hall. Inc., Englewood Cliffs, New Jerrsy. USA.

Lewis AL and Cherney WJH, 2004. Hybrid, maturity and cutting height interactions on corn forage yield and quality. Agron. J. 96: 267-274.

Miron J, Zuckerman E Sadeh D Adin G Nikbachat M Yosef E Ben-Ghedalia D Carmi A Kipnis T and Solomon R, 2005. Yield, composition and in vitro digestibility of new forage sorghum varieties and their ensilage characteristics. Anim. Feed Sci. Tech. 120: 17-32.

Miron J, Solomon R Adin G Nir U Nikbachat M Yosef E Carmi A Weinberg ZG Kipnis T Zuckerman E and Ben-Ghedalia D, 2006. Effects of harvest stage and re-growth on yield, composition, ensilage and in vitro digestibility of new forage sorghum varieties. J. Sci. Food Agric. 86: 140-147.

Mlynár R, Rajčáková Ľ and Gallo M, 2004. The effect of Lactobacillus buchneri on fermentation process and aerobic stability of maize silage. Risk factors of food chain, Nitra, SPU. 170-172.

Moodie CD, Smith HW and McCreery RA, 1959. Laboratory Manual for Soil Fertility. Washington State College, Mimeograph, USA.

Neylon JM and Jr. Kung L, 2003. Effects of Cutting Height and Maturity on the Nutritive Value of Corn silage for Lactating Cows. J. Dairy Sci. 86: 2163-2169.

Owens F, 2005. Corn genetics and animal feeding value. 66th Minnesota Nutrition Conference. St. Paul, MN. 20-21.

Rashid M, Ranjha AM and Rehim A, 2007. Model based P fertilization to improve yield and quality of sorghum (Sorghum bicolor L.) fodder on an ustochrept soil. Pak. J. Agri. Sci. 31: 27-29.

SAS Institute, 2008. SAS online doc 9.13. SAS Institute, Inc., Cary, NC.

Soest Van PJ, Robertson JB and Lewis BA, 1991. Methods for dietary fiber, neutral detergent fiber and non-starch polysaccharides in relation to animal nutrition. J. Dairy Sci. 74: 3583–3597.

Watanabe FS and Olsen SR, 1965. Test of ascorbic acid method for determining phosphorous in water and $NaHCO_3$ extract from soils. Soil Sci. Soc. Am. Pro. 29: 677-678.

Widdicombe WD and Thelen KD, 2002. Row width and plant density effect on corn forage hybrids. Agron. J. 94: 326-330.

Xie T, Su P Shan L and Ma J, 2012. Yield, quality and irrigation water use efficiency of sweet sorghum [Sorghum bicolor (Linn.) Moench] under different land types in arid regions. Austra. J. Crop Sci. 6: 10-16.

Yilmaz S, Gozubenli H Can E and Atis I, 2003. Adaptation and silage yield of some maize (*Zea mays* L.) lines in a milk plain condition. Turkey 5th Field Crops Congress, 13-17 October, Diyarbakr-Turkey. Volume: I, ISBN: 975-7635-19-7. pp: 341-345.

Yosef E, Carmi A Nikbachat M Zenou A Umiel N and Miron J, 2009. Characteristics of tall *versus* short-type varieties of forage sorghum grown under two irrigation levels, for summer and subsequent fall harvests, and digestibility by sheep of their silages. Anim. Feed Sci. Tech. 152: 1-11.

20

Economics of direct seeding methods of upland rice production in the Northern Guinea Savanna

Nwokwu Gilbert Nwogboduhu
Department of Crop Production and Landscape Management, Faculty of Agriculture and Natural Resources Management, Ebonyi State University, Abakaliki, Nigeria.

Abstract

This study assessed the economic impacts of direct seeding of rice as an alternative crop establishment method for farmers in Samaru, Zaria in the Northern Guinea Savanna of Nigeria. Specifically, it examined the changes in farmers' inputs (labour and inputs) and level of productivity and incomes among direct-seeded methods such as broadcasting, drilling and dibbling and measured the economic returns on investment in direct seeding. Analyses included cost and return, and economic surplus framework. The economic analysis of upland rice production at both locations indicated that production of NERICA 8 and JAMILA by either broadcasting or drilling method at the seed rate of 80 kg ha^{-1} gave the highest gross margin as well as return on investment. The result revealed that NERICA 8 and JAMILA sown by broadcasting method at 80 kg ha^{-1} was the most profitable with gross margin of ₦246, 166.50 with return on investment of ₦6.72. This was followed by broadcasting JAMILA at 120 kg ha^{-1} seed rate which gave a gross margin of ₦194, 583.50 and return on investment of ₦4.32. However, the least gross margin of ₦61, 249.85 was observed when NERICA 4 was dibbled at 120 kg ha^{-1} which brought a loss of ₦16, 716.50 and ₦ 0.62k was lost per every naira invested.

Keywords: Economics, Direct Seeding, Upland Rice, Production

Corresponding author email:
g.nwokwu@yahoo.com

Introduction

Rice is the world's most important crop and is a staple food for more than half of the world's population. Worldwide, rice is grown on 161 million hectares, with an annual production of about 678.7 million tons of paddy (FAO, 2009). About 90% of the world's rice is grown and produced (143 million ha of area with a production of 612 million tons of paddy) in Asia (FAO, 2009). Rice provides 30–75% of the total calories to more than 3 billion people (Khush, 1997; Von Braun and Bos, 2004). To meet the global rice demand, it is estimated that about 114 million tons of additional milled rice need to be produced by 2035, which is equivalent to an overall increase of 26% in the next 25 years (Yamano et al., 2016). The possibility of expanding the area under rice in the near future is limited. Therefore, this extra rice production needed has to come from a productivity gain. The major challenge to achieve this gain is less water, labour and chemicals (Hira, 2009)

A major reason for farmers' interest in direct seeding is the rising cost of cultivation and decreasing profits with conventional practice. Farmers likely prefer a technology that gives higher profit despite similar or slightly lower yield. Studies shows that various methods of direct seeding reduced the cost of production by US$9125 ha^{-1} compared with conventional practice of transplanting (Inayat-Ali et al., 2012; Santhi et al., 1998). The largest reductions in cost occurred in practices in which zero tillage was combined with direct seeding. These cost reductions were largely due to either reduced labour cost or tillage cost or both under direct seeding method. In areas where wages are high the labour cost savings in rice establishment can reach US$50 ha^{-1} (Kumar et al., 2009).

The conventional method of rice growing is not only water-guzzling but also cumbersome and laborious. Rice transplanting requires 200-250 man-hour ha^{-1}, which is 25% of the total requirement for the rice crop production (Anoop et al., 2007). The problem has further been intensified with the timely unavailability of labour. Delay in transplanting beyond optimum time due to labour scarcity is creating a reduction in rice yield. Further, reduced labour availability is increasing the cost of transplanting and squeezing the farmer's profit as the cost of transplanting is increasing continuously. Paddy transplanting by labour also results in low and non-uniform plant population due to which crop yields are reduced (Mahajan et al., 2009). The productivity and sustainability of rice-based systems are threatened because of the inefficient use of inputs; increasing scarcity of resources, especially water and labour; changing climate; the emerging energy crisis and rising fuel prices; the rising cost of cultivation and emerging socio-economic changes such as urbanization, migration of labour, preferences of non-agricultural work, concerns about farm-related pollution (Kumar and Ladha, 2011). Conventional tillage and crop establishment by transplanting is the most input intensive process in crop production and, therefore, more efficient alternatives are urgently needed. Potential solutions is a shift from manual transplanting to direct seeded rice. Direct seeded rice with zero tillage system performed as well as the conventional practice but with significant savings in water and labour use (Bhusan et al., 2007). Direct-seeding is cost-effective, can save water through earlier rice crop establishment (Ladha et al., 2003a; Singh et al.,, 2003). With alternate wetting and drying cycles in direct seeding, the crop is subjected to greater weed competition than transplanted rice because weeds emerge before or at the same time as the rice (Chuhan, 2012). Therefore, heavy weed infestation is a major problem in direct seeded rice and its success lies in effective weed control measures (Singh et al.,, 2003; Rao et al.,, 2007), as failure to eliminate weeds may result in low or no yield (Estorninos and Moody, 1988). Therefore, the study aimed at evaluating the economics of direct seeding methods of upland rice varieties with a goal for finding most suitable ones

with a potential to cover large area with similar agro ecological conditions.

Materials and Methods

The field trials were conducted on the experimental farm of the Institute for Agricultural Research, Ahmadu Bello University Samaru, Zaria and on the Research Farm of the Kaduna State Agricultural Development Programme, Maigana in 2011 and 2012 cropping seasons. Samaru is on the Latitude 11°11^{1} N and Longitude 7 °38^{1} E and is 686 m above sea level while Maigana is located on Latitude 11°11.06^{1} N and Longitude 7.54° 7.58$^{1 E}$ both in the Northern Guinea Savannah Agro ecological zone of Nigeria. Random samples of soils were taken at depth of 0-30cm from the experimental sites using an auger of 10 cm diameter before land preparation and were analysed for physical and chemical properties.

The treatments consisted of three seeding methods (broadcasting, drilling and dibbling), three seed rates (40, 80 and 120 kg ha^{-1}) and three upland rice varieties (NERICA 4, 8 and JAMILA). The treatments were laid out in a split plot design with the combination of sowing methods and seed rates in the main plots and three rice varieties in the sub plots measuring 3 m x 3 m with a net plot of 2 m x 2 m and were replicated three times. The main plots were separated by a distance of 1 m and the sub plots by 0.5 m. Pre-planting herbicides, glyphosate (round up) was applied to the experimental sites at the rate 2 kg active ingredients ha^{-1} two weeks before land preparation in each year of the study in order to control the prevalent weeds on the field. Thereafter, the field was harrowed twice to ensure fine tilth of the soil and the soil levelled manually.

Seeds of each variety were treated with Apron star as seed dressing chemical at the rate of 1.0 g of metalaxy to 3.0 kg seed to prevent pest attack and the three rice varieties were planted on 13th July 2011 and 9th June 2012 at Samaru while at Maigana, the varieties were planted on 30th July 2011 and 12th June 2012 when the rains were fully stabilized using direct seeding such as broadcasting, drilling and dibbling methods. Also hand pulling methods of weed control was used to control the weeds that later emerged at four and eight weeks after sowing. Fertilizers were applied at the recommended rate of 100 kg N ha^{-1}, 50 kg P$_2$O$_5$ ha^{-1} and 50 kg K$_2$O ha^{-1}. The nitrogen fertilizer were applied as split, half of the nitrogen fertilizer together with 50 kg P$_2$O$_5$ and 50 kg K$_2$O ha^{-1} were applied once at two weeks after sowing using NPK (15:15:15) while the second half of the nitrogen were applied at panicle initiation stage using urea (46% N). Economics and energy analysis was done by taking pooled data of both the years. The economic analysis was measured using the partial budget procedure to determine the economic return. The data obtained was subjected to economic analysis where the returns and variable costs were calculated and gross margin (GM) and return on investment (ROI) was also determined. This depended on the prevailing market prices of inputs, labour and outputs. This was computed as:

GM=TR-TVC (Olukosi and Erhabor, 1988).

Where;
TR= Total revenue per hectare.
TVC=Total variable cost (sum of labour and material input cost) per hectare.

ROI =Net return /TVC X 100/1

Where; Net return = Total revenue (yield x output price) – TVC

Results and Discussion

The physical and chemical properties of the soil used during the experimental periods at Samaru and Maigana are summarized in Table1. In 2011 at Samaru, the physico-chemical properties of the field on which the experiment was conducted showed that the soil was clay loam, slightly acidic with moderate total nitrogen and available phosphorus and low potassium and CEC and high organic carbon. In 2012 cropping season the soil was silt loam, slightly acidic with high total nitrogen; available phosphorus and potassium were moderate while organic carbon and CEC were low. At Maigana in 2011, the soil was silt clay loam, slightly acidic with high total nitrogen and organic carbon while available phosphorus, potassium and CEC were low. In 2012, the soil was sandy loam, slightly acidic with high total nitrogen while available P, K was moderate and organic carbon was high

The cost and return analysis per hectare on investment of growing upland rice varieties using different sowing methods and seed rates in 2011, 2012 wet seasons and the combined at Samaru and Maigana are presented in Tables 3 to 8. At Samaru in 2011, the result revealed that NERICA 8 sown by broadcasting method at 80 kg ha^{-1} was the most profitable with gross margin

of N129, 793.25 and Return on Investment of N3.02 which implied that for every N1.00 invested N2.02 was realized. Similarly, it was followed by NERICA 8 sown by broadcasting method at 120 kg ha^{-1} seed rate which gave a gross margin of N95, 787.65 and Return on Investment of N2.23 which implied that for every N1.00 invested N1.23 was realized. However, the sowing of JAMILA with dibbling method at 120 kg ha^{-1} brought a loss of N17, 932.50 and N 0.34k was lost per naira invested.

In 2012, the use of broadcasting method to sow JAMILA at 80 kg ha^{-1} seed rate was the most profitable with gross margin of N198, 750.00 and return on investment of of N4.42 indicating that for N1.00 invested N3.42 was realized. This was followed by broadcasting JAMILA at 120 kg ha^{-1} seed rate which gave a gross margin of N194, 583.50 and return on investment of N4.32. There was no loss incurred during this year's production. However, the least gross margin of N61, 249.85 and return on investment of N1.11 was observed when NERICA 4 was dibbled at 80 kg ha^{-1} seed rate.

The combined economic analysis indicated that broadcasting NERICA 8 at 80 kg ha^{-1} seed rate was the most profitable with gross margin of N 146, 215.00 and return on investment of N4.32 which indicated that for every N1.00 invested, N 3.32 was realized. This was closely followed by either drilled NERICA 8 or JAMILA at 80 kg ha^{-1} seed rate which gave a gross margin of N141, 194.30 and return on investment of N3.88.The least gross margin of N35, 467.65 and return on investment of N1.66 was observed when NERICA 4 was dibbled at 120 kg ha^{-1} seed rate. Net returns in broadcasting was at par with drilling method and were higher than dibbling method. Kumar (2011) also observed similar findings and found higher B:C ratio in direct seeding as compared to manual puddled transplanted rice .

At Maigana in 2011, the result indicated that the highest gross margin of N119, 581.50 and return on investment of N3.85 was achieved by broadcasting NERICA 8 at 80 kg ha^{-1} seed rate. This was followed by drilled NERICA 8 at 80 kg ha^{-1} which gave a gross margin of N30, 832.50 and return on investment of N1.73. However, dibbled JAMILA at 120 kg ha^{-1} brought a loss of N16, 716.50 and 0.62k was lost per every naira invested.

In 2012 and the combined, the result indicated that the highest gross margin was by drilled JAMILA at 80 kg ha^{-1} seed rate. The gross margin of 2012 and the combined were N246, 166.50 with return on investment of ₦ 6.72 and ₦182, 874.00 with return on investment of ₦5.30 respectively. This was followed by drilled JAMILA at 120 kg ha⁻¹ seed rate with gross margin of ₦198, 666.50 and return on investment of ₦ 5.62, gross margin of ₦98, 547.50 and return on investment of ₦3.32 for 2012 and the combined data. However, the least gross margin was observed when NERICA 8 was dibbled at 40 kg ha⁻¹ seed rate in 2012 and the combined were ₦24, 375.15 with return on investment of ₦1.54 and ₦11, 187.50 with return on investment of ₦1.25. In paddy, a labour saving of 95-99 percent in broadcasting or drilling methods were observed as compared to bibbling method during both years of the study and total labour use mainly depends on the weed management. In present study we made two hand weeding to cope up with weeds as the herbicide used were not so effective, which ultimately resulted in more labour use and higher cost of production. In overall, broadcasting and drilling methods had more percent labour saving as compared to dibbling method in both years of the study. Sehrawat et al. (2010) also observed 13-16% labour saving in direct seeding as compared to manual puddled transplanted rice.

Economic analysis is the ultimate yard stick to recommend a production technology. The loss incurred at both locations in 2011 was due to low rainfall recorded in that year.

Generally, throughout the study, increase in total revenue resulted an increase in gross margin in all the three sowing methods used in the study. In 2011 at Samaru, the maximum gross margin obtained from broadcasting method was due to the fact that crops established with these methods matured early than drilling and dibbling method. This helped the crops to escape the stress as a result of late planting as well early cessation of rainfall in that year and thereby resulted in highest yield. Despite the higher cost of production in 2012 wet season and the combined at Samaru, the highest gross margin obtained from broadcasting and drilling method were due to higher grain yield as a result of better rain fall recorded in that year at the experimental sites. Generally, dibbling method consistently produced the least gross margin throughout the study at both locations. This was due to the low yield recorded when rice was sown using this method throughout the study period. In 2011 at both locations, the lowest gross margin was recorded as well as the loss incurred was due to low yield as a result of late planting. However, the highest gross margin in that year was obtained from the

broadcasting method at 80 kg of NERICA 8. The differences in gross margin in the two years at both locations could be attributed to the differences in cost of production as well as the market price. In this country it was discovered that prices of materials at different locations increase as the years passed by without necessarily increase the price of the farm produce. In this study, the cost of production at Maigana was far cheaper than Samaru especially the cost of labour and as well as some inputs such as fertilizers.

In both locations, the cost of labour such as weed control in broadcasting and drilling methods were lower than that of dibbling method. The highest difference in labour cost was in crop establishment. This reduced need for labour not only saves time and money of farmers but also allows greater flexibility so that farmers can attend other activities either in farm or at home. Higher expenditures on seeds were expected in direct seeding methods because higher seeding rates are required for direct seeding relative to transplanting. Farmers who practiced direct seeding were more reliant on herbicides simply because they cannot rely on flooding to suppress weeds during the crucial initial period of crop establishment (Johnson 2006). On the other hand, expenditures on fertilizers, and rent in land preparation were lower for direct seeding than transplanting. This was because farmers were inclined to use more fertilizers as a treatment or preventive measure against transplanting shock.

Previous studies had found that direct seeding may obtain a lower yield due to the unstable establishment of rice seedlings and slow growth during the early growing stage (Kimio et al., 1999). Yield in direct seeding can be also reduced by weed problems. Yield losses (due to weeds) largely depend on season, weed species, weed density, rice cultivar, and growth rate and density of weeds and rice (Azmi et al., 2005). Another factor affecting yield in direct seeding is seed rate. A seed rate higher than the recommended rate can result in lower yield of direct seeding since it may lead to nitrogen deficiency, thus reducing tillering and increasing the proportion of ineffective tillers, to attacks of brown plant hoppers, and to crop lodging. However, yield of direct seeding is not always lower than that of transplanting. Rice yields of wet- or dry-seeded crops have been higher than those of transplanted crops, provided weeds are adequately controlled (Johnson et al., 2003) .Higher rice yield resulting from direct seeding is due to the shorter time it takes for direct seeding rice to reach maturity. This allows for on-time planting, thus saving farmers from a 1 percent (or more) reduction in yield per day (Hobbs 2001).

Rainfall at Samaru during the experimental period was 945.2 and 1333.3mm in 2011 and 2012 respectively (Table 2). Rain started in 2011 on 16th April and ended on 11th October with a total of 73 rainy days. In 2012, rain commenced on10th April and terminated on 24tth October with a total of 76 rainy days. Rainfall at Maigana during the experimental period was 636.6 and 762.9 in 2011 and 2012 respectively. Rain started in 2011 on 25th April and ended on 12th October with a total of 67 rainy days. In 2012, rain commenced on 17th April and ended on 22nd October with a total of 70 rainy days. Generally, rainfall was not evenly distributed during the conduct of this research. Rain was low at both locations in 2011 with Maigana recording lower rainfall in the two years of this study when compared with Samaru.

Table – 1: Physical and Chemical Properties of soil (0-30cm) collected from the Experimental Sites

Soil Properties	Samaru		Maigana	
	2011	2012	2011	2012
Physical Properties				
Sand (g/kg)	222.00	240.00	192.00	520.00
Silt (g/kg)	474.00	540.00	504.00	380.00
Clay (g/kg)	304.00	220.00	304.00	100.00
Textural Class	Clay loam	Silt loam	Silty Clay loam	Sandy loam
Chemical Properties				
pH in 0.01M $CaCL_2$	4.28	5.98	4.35	4.55
Organic Carbon (g/kg)	13.6	2.10	16.00	20.40
Total Nitrogen (g/kg)	1.70	9.00	2.20	10.10
Available P mg/Kg	3.05	1.42	4.48	2.84
Exchangeable Cation (cmol/kg)				
K	0.15	0.22	0.17	0.22
Mg	0.59	2.00	0.63	1.76
Ca	0.59	3.00	0.61	5.00
Na	7.32	1.60	9.10	1.60
CEC	4.28	5.20	4.35	2.90

Table – 2: Rainfall Distribution at Samaru and Maigana during 2011 and 2012 Cropping Seasons

MONTHS	Samaru Amount of Rainfall	2011 Rainy days	2012 Amount of Rainfall	Rainy days	Maigana Amount of Rainfall	2011 Rainy days	2012 Amount of Rainfall	Rainy days
January	0	0	0	0	0	0	0	0
February	0	0	0	0	0	0	0	0
March	0	0	0	0	0	0	0	0
April	27.8	4	7.3	3	36.4	1	21.1	4
May	123.3	11	263.4	15	63.4	8	109.6	14
June	162.5	12	120.7	8	102.5	15	145.7	13
July	223.9	14	165.3	12	113.2	12	150.4	15
August	239.9	16	426.7	16	165.2	16	130.5	19
September	113.8	12	270.3	17	89.1	19	191	18
October	54	4	79.6	5	66.8	5	14.6	3
November	0	0	0	0	0	0	0	0
December	0	0	0	0	0	0	0	0
TOTAL	945.2	73	1333.3	76	636.6	76	762.9	86

Table – 3: Economic analysis of cost and return on investment of growing upland rice varieties using seed rate and sowing method at Samaru in 2011 wet season.

Variety	Sowing method	Seed rate (kg ha⁻¹)	Total yield (Kg ha⁻¹)	No. of bags (100kg)	Average Price/bag 100kgha¹ ₦	Total revenue (TR) (₦ ha⁻¹)	Total variable cost (TVC) (₦ ha⁻¹)	Gross margin (TR- TVC) (₦ ha⁻¹)	Return on investment (ROI) ₦
NERICA 4	Broadcasting	40	1088.67	10.89	4500	48990.15	43000	5990.15	0.14
NERICA 4	Broadcasting	80	2413.08	24.13	4500	108588.60	43000	65588.60	1.53
NERICA 4	Broadcasting	120	1395.77	13.96	4500	62809.65	43000	19809.65	0.46
NERICA 4	Drilling	40	1517.92	15.18	4500	68306.40	48000	20306.40	0.42
NERICA 4	Drilling	80	1710.75	17.11	4500	76983.75	48000	28983.75	0.60
NERICA 4	Drilling	120	1895.43	18.95	4500	85294.35	48000	37294.35	0.78
NERICA 4	Dibbling	40	1565.17	15.65	4500	70432.65	53000	17432.65	0.33
NERICA 4	Dibbling	80	1699.48	16.99	4500	76476.60	53000	23476.60	0.44
NERICA 4	Dibbling	120	1101.33	11.01	4500	49559.85	53000	-3440.15	-0.06
NERICA 8	Broadcasting	40	1666.58	16.67	4500	74996.10	43000	31996.10	0.74
NERICA 8	Broadcasting	80	3839.85	38.40	4500	172793.25	43000	129793.25	3.02
NERICA 8	Broadcasting	120	3084.17	30.84	4500	138787.65	43000	95787.65	2.23
NERICA 8	Drilling	40	2289.02	22.89	4500	103005.90	48000	55005.90	1.15
NERICA 8	Drilling	80	2994.75	29.95	4500	134763.75	48000	86763.75	1.81
NERICA 8	Drilling	120	2757.83	27.58	4500	124102.35	48000	76102.35	1.59
NERICA 8	Dibbling	40	1876.5	18.77	4500	84442.50	53000	31442.50	0.59
NERICA 8	Dibbling	80	2537.92	25.38	4500	114206.40	53000	61206.40	1.15
NERICA 8	Dibbling	120	1279.67	12.80	4500	57585.15	53000	4585.15	0.09
JAMILA	Broadcasting	40	917.33	9.17	5000	45866.50	43000	2866.50	0.07
JAMILA	Broadcasting	80	1342.16	13.42	5000	67108.00	43000	24108.00	0.56
JAMILA	Broadcasting	120	969.58	9.70	5000	48479.00	43000	5479.00	0.13
JAMILA	Drilling	40	868.01	8.68	5000	43400.50	48000	-4599.50	-0.10
JAMILA	Drilling	80	1080.83	10.81	5000	54041.50	48000	6041.50	0.13
JAMILA	Drilling	120	1094.85	10.95	5000	54742.50	48000	6742.50	0.14
JAMILA	Dibbling	40	967.85	9.68	5000	48392.50	53000	-4607.50	-0.09
JAMILA	Dibbling	80	1014.55	10.15	5000	50727.50	53000	-2272.50	-0.04
JAMILA	Dibbling	120	701.35	7.01	5000	35067.50	53000	-17932.50	-0.34

Calculation of total revenue was based on market prevailing prices of ₦4,500.00 and ₦5000.00 per 100 kg bag of NERICAs and JAMILA respectively at Samaru and environs.

Table – 4: Economic analysis of cost and return on investment of growing upland rice varieties using seed rate and sowing method at Samaru in 2012 wet season.

Variety	Sowing method	Seed rate kg ha^{-1}	Total yield Kg ha^{-1}	No. of bags (100kg)	Average Price/bag ₦	Total revenue (TR) ₦ ha^{-1}	Total variable cost (TVC) ₦ ha^{-1}	Gross margin (TR- TVC) ₦ ha^{-1}	Return on investment (ROI) ₦
NERICA 4	Broadcasting	40	3125.00	31.25	4500	140625.00	45000	95625.00	2.13
NERICA 4	Broadcasting	80	3500.00	35.00	4500	157500.00	45000	112500.00	2.50
NERICA 4	Broadcasting	120	2833.33	28.33	4500	127499.85	45000	82499.85	1.83
NERICA 4	Drilling	40	3541.67	35.42	4500	159375.15	50000	109375.15	2.19
NERICA 4	Drilling	80	4208.33	42.08	4500	189374.85	50000	139374.85	2.79
NERICA 4	Drilling	120	4208.33	42.08	4500	189374.85	50000	139374.85	2.79
NERICA 4	Dibbling	40	2708.33	27.08	4500	121874.85	55000	66874.85	1.22
NERICA 4	Dibbling	80	2583.33	25.83	4500	116249.85	55000	61249.85	1.11
NERICA 4	Dibbling	120	2875.00	28.75	4500	129375.00	55000	74375.00	1.35
NERICA 8	Broadcasting	40	3083.33	30.83	4500	138749.85	45000	93749.85	2.08
NERICA 8	Broadcasting	80	4666.67	46.67	4500	210000.15	45000	165000.15	3.67
NERICA 8	Broadcasting	120	3750.00	37.50	4500	168750.00	45000	123750.00	2.75
NERICA 8	Drilling	40	3583.33	35.83	4500	161249.85	50000	111249.85	2.22
NERICA 8	Drilling	80	5458.33	54.58	4500	245624.85	50000	195624.85	3.91
NERICA 8	Drilling	120	3958.33	39.58	4500	178124.85	50000	128124.85	2.56
NERICA 8	Dibbling	40	4125.00	41.25	4500	185625.00	55000	130625.00	2.38
NERICA 8	Dibbling	80	4125.00	41.25	4500	185625.00	55000	130625.00	2.38
NERICA 8	Dibbling	120	3000.00	30.00	4500	135000.00	55000	80000.00	1.45
JAMILA	Broadcasting	40	3208.00	32.08	5000	160400.00	45000	115400.00	2.56
JAMILA	Broadcasting	80	4875.00	48.75	5000	243750.00	45000	198750.00	4.42
JAMILA	Broadcasting	120	4791.67	47.92	5000	239583.50	45000	194583.50	4.32
JAMILA	Drilling	40	4875.00	48.75	5000	243750.00	50000	193750.00	3.88
JAMILA	Drilling	80	4208.33	42.08	5000	210416.50	50000	160416.50	3.21
JAMILA	Drilling	120	5150.00	51.50	5000	257500.00	50000	207500.00	4.15
JAMILA	Dibbling	40	3458.33	34.58	5000	172916.50	55000	117916.50	2.14
JAMILA	Dibbling	80	4833.33	48.33	5000	241666.50	55000	186666.50	3.39
JAMILA	Dibbling	120	3625.00	36.25	5000	181250.00	55000	126250.00	2.30

Calculation of total revenue was based on market prevailing prices of ₦4,500.00 and ₦5000.00 per 100 kg bag of NERICAs and JAMILA respectively at Samaru and environs.

Table – 5: Combined economic analysis of cost and return on investment of growing upland rice varieties using seed rate and sowing method at Samaru in 2011 and 2012 wet seasons

Variety	Sowing method	Seed rate (kg ha^{-1})	Total yield (Kg ha^{-1})	No. of bags (100kg)	Average Price/ bag ₦	Total revenue (TR) (₦ ha^{-1})	Total variable cost (TVC) (₦ ha^{-1})	Gross margin (TR- TVC) (₦ ha^{-1})	Return on investment (ROI) ₦
NERICA 4	Broadcasting	40	2106.83	21.07	4500	94807.35	44000	50807.35	2.15
NERICA 4	Broadcasting	80	2956.54	29.57	4500	133044.30	44000	89044.30	3.02
NERICA 4	Broadcasting	120	2114.55	21.15	4500	95154.75	44000	51154.75	2.16
NERICA 4	Drilling	40	2529.79	25.30	4500	113840.55	49000	64840.55	2.32
NERICA 4	Drilling	80	2959.54	29.60	4500	133179.30	49000	84179.30	2.72
NERICA 4	Drilling	120	3051.88	30.52	4500	137334.60	49000	88334.60	2.80
NERICA 4	Dibbling	40	2136.75	21.37	4500	96153.75	54000	42153.75	1.78
NERICA 4	Dibbling	80	2141.41	21.41	4500	96363.45	54000	42363.45	1.78
NERICA 4	Dibbling	120	1988.17	19.88	4500	89467.65	54000	35467.65	1.66
NERICA 8	Broadcasting	40	2374.96	23.75	4500	106873.20	44000	62873.20	2.43
NERICA 8	Broadcasting	80	4226.54	42.27	4500	190215.00	44000	146215.00	4.32
NERICA 8	Broadcasting	120	3694.92	36.95	4500	166275.00	44000	122275.00	3.78
NERICA 8	Drilling	40	2936.18	29.36	4500	132128.10	49000	83128.10	2.70
NERICA 8	Drilling	80	4226.54	42.27	4500	190194.30	49000	141194.30	3.88
NERICA 8	Drilling	120	3358.08	33.58	4500	151113.60	49000	102113.60	3.08
NERICA 8	Dibbling	40	3000.75	30.01	4500	135033.75	54000	81033.75	2.50
NERICA 8	Dibbling	80	3331.46	33.31	4500	149915.70	54000	95915.70	2.78
NERICA 8	Dibbling	120	2139.83	21.40	4500	96292.35	54000	42292.35	1.78
JAMILA	Broadcasting	40	2062.83	20.63	5000	103141.50	44000	59141.50	2.34
JAMILA	Broadcasting	80	3066.92	30.67	5000	153346.00	44000	109346.00	3.49
JAMILA	Broadcasting	120	2922.29	29.22	5000	146114.50	44000	102114.50	3.32
JAMILA	Drilling	40	2871.51	28.72	5000	143575.50	49000	94575.50	2.93
JAMILA	Drilling	80	3804.58	38.05	5000	190229.00	49000	141229.00	3.88
JAMILA	Drilling	120	3122.43	31.22	5000	156121.50	49000	107121.50	3.19
JAMILA	Dibbling	40	2213.09	22.13	5000	110654.50	54000	56654.50	2.05
JAMILA	Dibbling	80	2923.94	29.24	5000	146197.00	54000	92197.00	2.71
JAMILA	Dibbling	120	2163.18	21.63	5000	108159.00	54000	54159.00	2.002

Calculation of total revenue was based on market prevailing prices of ₦4,500.00 and ₦5000.00 per 100 kg bag of NERICAs and JAMILA respectively at Samaru and environs.

Table – 6: Economic analysis of cost and return on investment of growing upland rice varieties using seed rate and sowing methods at Maigana in 2011 wet season.

Variety	Sowing method	Seed rate (kg ha⁻¹)	Total yield (Kg ha⁻¹)	No. of bags (100kg)	Average Price/ bag ₦	Total revenue (TR) (₦ ha⁻¹)	Total variable cost (TVC) (₦ ha⁻¹)	Gross margin (TR- TVC) (₦ ha⁻¹)	Return on investment (ROI) ₦
NERICA 4	Broadcasting	40	718.67	7.19	4500	32340.15	40000	-7659.85	-0.81
NERICA 4	Broadcasting	80	1451.20	14.51	4500	65304.00	40000	25304.00	1.63
NERICA 4	Broadcasting	120	946.10	9.46	4500	42574.50	40000	2574.50	1.06
NERICA 4	Drilling	40	1199.73	11.10	4500	53987.85	42000	11987.85	1.29
NERICA 4	Drilling	80	1372.73	13.73	4500	61772.85	42000	19772.85	1.47
NERICA 4	Drilling	120	732.80	7.33	4500	32976.00	42000	-9024.00	-0.79
NERICA 4	Dibbling	40	818.72	8.19	4500	36842.40	44000	-7157.60	-0.84
NERICA 4	Dibbling	80	1621.83	16.22	4500	72982.35	44000	28982.35	1.66
NERICA 4	Dibbling	120	1119.62	11.20	4500	50382.90	44000	6382.90	1.15
NERICA 8	Broadcasting	40	1143.77	11.44	4500	51469.65	40000	11469.65	1.29
NERICA 8	Broadcasting	80	3231.63	32.32	5000	161581.50	42000	119581.50	3.85
NERICA 8	Broadcasting	120	637.02	6.37	4500	28665.90	40000	-11334.10	-0.72
NERICA 8	Drilling	40	1246.91	12.47	4500	56110.95	42000	14110.95	1.34
NERICA 8	Drilling	80	1618.5	16.19	4500	72832.50	42000	30832.50	1.73
NERICA 8	Drilling	120	634.83	6.35	4500	28567.35	42000	-13432.65	-0.68
NERICA 8	Dibbling	40	933.33	9.33	4500	41999.85	44000	-2000.15	-0.95
NERICA 8	Dibbling	80	1253.13	12.53	4500	56390.85	44000	12390.85	1.28
NERICA 8	Dibbling	120	810.13	8.10	4500	36455.85	44000	-7544.15	-0.83
JAMILA	Broadcasting	40	691.83	6.92	5000	34591.50	40000	-5408.50	-0.86
JAMILA	Broadcasting	80	873.33	8.73	5000	43666.50	40000	3666.50	1.09
JAMILA	Broadcasting	120	1000.00	10.00	5000	50000.00	40000	10000.00	1.25
JAMILA	Drilling	40	674.17	6.74	5000	33708.50	42000	-8291.50	-0.80
JAMILA	Drilling	80	1380.97	13.81	5000	69050.00	42000	27050.00	1.55
JAMILA	Drilling	120	808.57	8.09	5000	40428.50	42000	-1571.50	-0.96
JAMILA	Dibbling	40	873.00	8.73	5000	43650.00	44000	-350.00	-0.99
JAMILA	Dibbling	80	978.80	9.79	5000	48940.00	44000	4940.00	1.11
JAMILA	Dibbling	120	545.67	5.46	5000	27283.50	44000	-16716.50	-0.62

Calculation of total revenue was based on market prevailing prices of ₦4,500.00 and ₦5000.00 per 100 kg bag of NERICAs and JAMILA respectively at Samaru and environs.

Table – 7: Economic analysis of cost and return on investment of growing upland rice varieties using seed rate and sowing method at Maigana in 2012 wet season.

Variety	Sowing method	Seed rate (kg ha⁻¹)	Total yield (Kg ha⁻¹)	No .of bags (100kg)	Average Price/ bag ₦	Total revenue (TR) (₦ ha⁻¹)	Total variable cost (TVC) (₦ ha⁻¹)	Gross margin (TR- TVC) (₦ ha⁻¹)	Return on investment (ROI) ₦
NERICA 4	Broadcasting	40	2500.00	25.00	4500	112500.00	41000	71500 00	2.74
NERICA 4	Broadcasting	80	2916.67	29.17	4500	131250.15	41000	90250.15	3.20
NERICA 4	Broadcasting	120	3916.67	39.17	4500	176250.15	41000	135250.15	4.30
NERICA 4	Drilling	40	1666.67	16.67	4500	75000.15	43000	32000.15	1.74
NERICA 4	Drilling	80	2625.00	26.25	4500	118125.00	43000	75125.00	2.75
NERICA 4	Drilling	120	3425.00	34.25	4500	154125.00	43000	111125.00	3.58
NERICA 4	Dibbling	40	2075.00	20.75	4500	93375.00	45000	48375.00	2.08
NERICA 4	Dibbling	80	1750.00	17.50	4500	78750.00	45000	33750.00	1.75
NERICA 4	Dibbling	120	1875.00	18.75	4500	84375.00	45000	39375.00	1.88
NERICA 8	Broadcasting	40	2000.00	20.00	4500	90000.00	41000	49000.00	2.20
NERICA 8	Broadcasting	80	4166.67	41.67	4500	187500.15	41000	146500.15	4.57
NERICA 8	Broadcasting	120	3375.00	33.75	4500	151875.00	41000	110875.00	3.70
NERICA 8	Drilling	40	1625.00	16.25	4500	73125.00	43000	30125.00	1.71
NERICA 8	Drilling	80	3000.00	30.00	4500	135000.00	43000	92000.00	3.14
NERICA 8	Drilling	120	3458.33	34.58	4500	155624.85	43000	112624.85	3.62
NERICA 8	Dibbling	40	1541.67	15.42	4500	69375.15	45000	24375.15	1.54
NERICA 8	Dibbling	80	2416.67	24.17	4500	108750.15	45000	63750.15	2.42
NERICA 8	Dibbling	120	2375.00	23.75	4500	106875.00	45000	61875.00	2.38
JAMILA	Broadcasting	40	2583.33	25.83	5000	129166.50	41000	88166.50	3.15
JAMILA	Broadcasting	80	4333.33	43.33	5000	216666.50	41000	175666.50	5.28
JAMILA	Broadcasting	120	3541.67	35.42	5000	177083.50	41000	136083.50	4.32
JAMILA	Drilling	40	2291.67	22.92	5000	114583.50	43000	71583.50	2.66
JAMILA	Drilling	80	5783.33	57.83	5000	289166.50	43000	246166.50	6.72
JAMILA	Drilling	120	4833.33	48.33	5000	241666.50	43000	198666.50	5.62
JAMILA	Dibbling	40	2916.67	29.17	5000	145833.50	45000	100833.50	3.24
JAMILA	Dibbling	80	3041.67	30.42	5000	152083.50	45000	107083.50	3.38
JAMILA	Dibbling	120	4375.00	43.75	5000	218750.00	45000	173750.00	4.86

Calculation of total revenue was based on market prevailing prices of ₦4,500.00 and ₦5000.00 per 100 kg bag of NERICAs and JAMILA respectively at Samaru and environs.

Table – 8: Combined economic analysis of cost and return on investment of growing upland rice varieties using seed rate and sowing method at Maigana in 2011 and 2012 wet seasons

Variety	Sowing method	Seed rate (kg ha⁻¹)	Total yield (Kg ha⁻¹)	No. of bags (100kg)	Average Price/bag ₦	Total revenue (TR) (₦ ha⁻¹)	Total variable cost (TVC) (₦ ha⁻¹)	Gross margin (TR- TVC) (₦ ha⁻¹)	Return on investment (ROI) ₦
NERICA 4	Broadcasting	40	1609.33	16.09	4500	72419.85	40500	31919.85	1.79
NERICA 4	Broadcasting	80	2233.93	22.34	4500	100526.85	40500	60026.85	2.48
NERICA 4	Broadcasting	120	2431.38	24.31	4500	109412.10	40500	68912.10	2.70
NERICA 4	Drilling	40	1433.20	14.33	4500	64494.10	42500	21994.00	1.52
NERICA 4	Drilling	80	1998.87	19.99	4500	89949.15	42500	47449.15	2.12
NERICA 4	Drilling	120	2078.90	20.79	4500	93550.50	42500	51050.50	2.20
NERICA 4	Dibbling	40	1446.86	14.47	4500	65108.70	44500	20608.70	1.46
NERICA 4	Dibbling	80	1685.92	16.86	4500	75866.40	44500	31366.40	1.70
NERICA 4	Dibbling	120	1497.31	14.97	4500	67378.95	44500	22878.95	1.51
NERICA 8	Broadcasting	40	1571.88	15.72	4500	70734.60	40500	30234.60	1.75
NERICA 8	Broadcasting	80	2773.82	27.74	4500	124821.90	40500	84321.90	3.08
NERICA 8	Broadcasting	120	2006.01	20.06	4500	90270.45	40500	49770.45	2.23
NERICA 8	Drilling	40	1435.95	14.36	4500	64617.75	42500	22117.75	1.52
NERICA 8	Drilling	80	2309.25	23.09	4500	103916.25	42500	61416.25	2.45
NERICA 8	Drilling	120	2046.58	20.47	4500	92096.10	42500	49596.10	2.17
NERICA 8	Dibbling	40	1237.50	12.38	4500	55687.50	44500	11187.50	1.25
NERICA 8	Dibbling	80	1834.90	18.35	4500	82570.50	44500	38070.50	1.86
NERICA 8	Dibbling	120	1592.57	15.93	4500	71665.65	44500	27165.65	1.61
JAMILA	Broadcasting	40	1637.58	16.38	5000	81879.00	40500	41379.00	2.02
JAMILA	Broadcasting	80	2207.50	22.08	5000	110375.00	40500	69875.00	2.73
JAMILA	Broadcasting	120	2666.67	26.67	5000	133333.50	40500	92833.50	3.29
JAMILA	Drilling	40	1482.92	14.83	5000	74146.00	42500	31646.00	1.74
JAMILA	Drilling	80	4507.48	45.07	5000	225374.00	42500	182874.00	5.30
JAMILA	Drilling	120	2820.95	28.21	5000	141047.50	42500	98547.50	3.32
JAMILA	Dibbling	40	1894.83	18.95	5000	94741.50	44500	50241.50	2.13
JAMILA	Dibbling	80	2010.23	20.10	5000	100511.50	44500	56011.50	2.26
JAMILA	Dibbling	120	2460.33	24.60	5000	123016.50	44500	78516.50	2.76

Calculation of total revenue was based on market prevailing prices of ₦4,500.00 and ₦5000.00 per 100 kg bag of NERICAs and JAMILA respectively at Samaru and environs.

Conclusion

In this study, direct seeding of rice practice was evaluated and it is clear that direct seeded rice practices may not perform similarly in all agro ecological conditions because of rainfall distribution. At both location, the result indicated that the highest gross margin was by drilled JAMILA at 80 kg ha⁻¹ seed rate. The economic analysis of upland rice production at both locations indicated that production of NERICA 8 and JAMILA by either broadcasting or drilling method at the seed rate of 80 kg ha⁻¹ gave the highest gross margin as well as return on investment. The result revealed that NERICA 8 and JAMILA sown by broadcasting method at 80 kg ha⁻¹ was the most profitable with Gross margin of ₦246, 166.50

with return on investment of ₦6.72. This was followed by broadcasting JAMILA at 120 kg ha^{-1} seed rate which gave a gross margin of ₦194, 583.50 and return on investment of ₦4.32. However, the least gross margin of ₦61, 249.85 was observed when NERICA 4 was dibbled at 120 kg ha^{-1} which brought a loss of ₦16, 716.50 and ₦ 0.62k was lost per every naira invested. The data presented in the study shows that broadcasting and drilling method can also be a viable solution under scarcity of labour and water. but, there is need to develop proper weed management practices and requires further study to access the long term effects of herbicides on soil, water and development of weed flora.

References

Anoop D, Khurana R, Jastarn S and Gususahib SD, 2007. Farm Power and Machinery, Punjab Agricultural University, Ludhiana 141001. Comparative performance of Different Paddy Transplanter Development in India, A review. Agric. Rev. 28(4): 262-269.

Azmi M, Chin DV, Vongsaroj P and Johnson DE, 2005. Emerging Issues in Weed Management of Direct-Seeded Rice in Malaysia, Vietnam, and Thailand. In Rice is Life: Scientific Perspectives for the 21st Century. K. Toriyama, K.L. Heong (eds), Tokyo and Tsukuba, Japan. Proceedings of the World Rice Research Conference pp. 196-198.

Bhusan L, Ladha JK, Gupta RK, Singh S Tirole-Padre A, Sehrawat YS, Gathala M and Pathak H, 2007. Saving of water and labour in rice-wheat system with no tillage and direct seeding technologies. Agron. J. 99: 1288-1296.

Chuhan BS, 2012. Weed ecology and weed management strategy for dry seeded rice in Asia. Weed Technol. 26: 1-13.

Estorninos Jr. LE and Moody K, 1988. Evaluation of herbicides for weed control in dry-seeded wetland rice (*Oryza sativa* L.). Philipp. J. Weed Sci. 15: 50–58.

FAO, 2009. FAOSTAT Database FAO, Rome www.faostat.fao.org (accessed in June 2010).

Hira GS, 2009. Water Management in Northern and food Security of Indian. J. Crop Improve. 23: 36-47.

Inyat Ali R, Nadeem I, Usman MS and Akhtar M, 2012. Effect of different planting methods on economic yield and grain quality of rice. Int. J. Agri. Sci. 4(1): 28-34.

Kumar V, Ladha JK and Gathala MK, 2009. Direct drill-seeded rice: A need of the day. In Annual Meeting of Agronomy Society of America, Pittsburgh, November 1–5, 2009, http://a-c-s.confex.com/crops/2009am/webprogram/Paper53386.html

Kumar V and Ladha JK, 2011 Direct seeding of rice: Recent developments and future research needs. Advances in Agron. 111: 297-413.

Kumar R. 2011. Comparative performance evaluation of mechanical transplanting and direct seeding of rice under puddled and unpuddled condition. MTech Thesis, Deptt. of Farm Machinery and Power Engineering, CCS HAU, Hisar, Haryana, India.

Mahajan G, Bharaj TS and Timsina J, 2009. Yield and water productivity of rice as affected by time of transplanting in Punjab, India. Agri. Water Manage. 96: 525-532.

Kimio I, Ko N and Hiromitsu K, 1999. An Automatic Irrigation System for Direct Seeding. Rice Cultivation. J. Agri. Meteorol. 55(2): 127–135.

Johnson D, 2006. The Direct Approach. Rice Today 5(2). International Rice Research Institute Rogers, E.M. 1995. Diffusion of Innovations. 4th ed. New York: The Free Press.

Rao AN, Johnson DE, Shivaprasad B, Ladha JK and Mortimer AM, 2007. Weed management in direct-seeded rice. Adv. Agron. 93: 153-255.

Reeves T, 2009. The impacts of climate change on wheat production in India – adaptation, mitigation and future directions. Food and Agriculture Organization, Rome. Hobbs, P Present Practices and Future Options. J. Crop Prod. 4(1): 1–22.

Saharawat YS, Singh B, Malik RK, Ladha JK, Gathala M, Jat ML and Kumar V, 2010. Evaluation of alternative tillage and crop establishment methods in a rice–wheat rotation in North Western IGP. Field Crops Res. 116: 260–267.

Khush GS, 1997. Origin, dispersal, cultivation and variation of rice. Plant Mol. Biol. 35: 25-34.

Santhi PK, Pannuswamy and Kempuchetty N, 1998. A labour saving techniques in direct sown and transplanted rice. Int. Rice Res. Notes. 23: 35-36.

Singh Y, Singh B, Nayyar VK and Singh J, 2003. Nutrient management for sustainable rice-wheat cropping system. NATP, ICAR, New Delhi and PAU, Ludhiana, Punjab, India.

Olukosi JO and Erhabor PO, 1988. Introduction to Farm Management Economics. Ajitab Publishers, Zaria, pp. 35-36 and 71-80.

Von Braun J and Bos MS, 2004. The changing economics and politics of rice: Implications for food security, globalization, and environmental sustainability. In "Rice Is Life: Scientific Perspectives for the 21st Century" (K. Toriyama, K. L. Heong, and B. Hardy, Eds.), pp. 7–20. International Rice Research Institute, Los Ban ˜os,

Philippines and Japan International Research Center for Agricultural Sciences, Tsukuba, Japan.

Yamano T, Aroma A, Labarta RA, Huelgas ZM and Mohanty S, 2016. Adoption and impacts of International rice research technologies. Global Food Sec. 8: 1-8.

21

Rohi Sarsoon: A new high yielding rapeseed variety released for general cultivation in south Punjab (Pakistan)

Muhammad Aslam Nadeem, Hafiz Muhammad Zia Ullah*, Abdul Majid Khan, Fida Hussain, Muhammad Jamil, Idrees Ahmad
Oilseeds Research Station, Khanpur, District Rahim Yar Khan, Pakistan

Corresponding author email:
ziaghazali@gmail.com

Abstract

This paper reports the release of new variety "Rohi Sarsoon" which is high yielding with erect growth habits. This variety was released in the year 2016 for arid and irrigated areas of agricultural lands. Rohi Sarsoon has been evolved from an elite line selected from progeny of a cross between KN-120 and KN-131 at Oilseeds Research Station, Khanpur during 2001-2002. Progenies F_2-F_5 were advanced by pedigree selection method. It is tolerant to Alternaria Blight, Powdery mildew, Downy mildew and white rust with yield potential 3927 Kg/hectare. The sowing time of this variety is 1-15 October with seed rate 0.60-0.80 Kg/acre. Plant color is light green with height of 160-173 cm and growth habits is determinate type. Its 1000 seed weight has been observed about 4-4.5 g. Leaf color is light green and seed contains 44-47% oil contents in it. Its meal contains 30-35% protein. This variety takes 155-160 days to mature and due to good performance it is recommended for the Southern Punjab and Cholistan areas in meeting of Federal Seed Certification and Registration Department held on 12 August 2016. Moreover, Rohi Sarsoon has got resistance against lodging, tolerant to aphid and best suited for Wheat, Mung bean, Rapeseed, Bt Cotton, Wheat, Fodder, Rapeseed, Sugarcane crop rotation.

Keywords: *Brassica napus*, Disease resistant, Rohi Sarsoon, Variety, Yield

Introduction

Rapeseed and mustard is traditional oilseed crop in Pakistan and considered as main source of edible oil. The contribution of these crops in Pakistan towards edible oilseed production is about 16-20 % (Pakistan Economic Survey, 2015-16). The production of rapeseed is 21 thousand tons from an area of 4 thousand hectares. Rapeseed is popular in Punjab where its production is 120.0 thousand tons from an area of 135.6 thousand hectares. The consumption of domestic edible oil is higher than its production, 2.821 million tons, in the country out of which about 0.684 million tons are met through local resources and rest of it (2.1 million tons) is met by import (Pakistan Economic Survey, 2015-16). The bill on imported oil of Pakistan is considered to be the second largest after petroleum (Ahmad et al., 2013). Less than 2 percent erucic acid in oil and less than 20 µmol/g glucosinolates in oil free meal is described as the recommended amount of these compounds. Seed meal of Brassica contains 40 % protein with balanced amino acid but the quantity of protein percentage is lower than its demand. A huge amount of foreign exchange is being spent over the import of rapeseed oil with forcible compromise on its quality for human consumption (Pakistan Economic Survey, 2015-16). There is a dire need to reduce the gap between domestic edible oil production and its import to maximum extent. It is the need of hour to develop new varieties with less erucic acid and glucosinolates to meet the demands of increasing population for local edible oil production.

Rohi Sarsoon is a variety suitable for arid and irrigated areas of Punjab. The Commercial release of this new rapeseed variety "Rohi Sarsoon" will certainly contribute much for the growers as well as for the country to bridge up the gap between production and demand. Rohi Sarsoon possesses good genetic potential and agronomic traits as it has performed better in yield trials.

Material and Method

Rohi Sarsoon was evolved as a result of cross KN-120 x KN-131 at Oilseeds Research Station, Khanpur. Hybridization was started during 2001-2002 for the development of this variety through pedigree method. The evaluation of elite line was done against particular traits and it was advanced to F_5 generation. During 2006-2007, best performed progeny was selected against yield and disease resistance character in F_5. This progeny/line was given the name KN-259 and evaluated against different parameters in yield trials at various locations. Finally the selected line was evaluated in National Uniform Yield Trials during 2012- 14. The average data of replication of each location was converted to kg/ha for comparison (Mustafa et al., 2008). Yield data were subjected to ANOVA and means were compared by using LSD (Steel et al., 1997). Various steps involved in the development of Rohi Sarsoon are given in Table 1.

Hybridization

KN-120, a locally developed high yielding strain, and local line KN-131 (disease resistant) were selected for hybridization. KN-120 was kept as a female parent and KN-131 as a male parent. Both parents were sown alternately, in October in crossing block at Oilseeds Research Station during crop season 2001-02. At flowering, emasculation was done in the morning, and emasculated flowers were covered with butter paper bag to avoid any foreign pollen contamination. On the next day the emasculated flower was pollinated with pollen of male parent by removing the butter paper bag. Crossed seed harvested from female parent was stored for raising generations.

Filial Generation Development

Crossed seed was sown in the field of Oilseeds Research Station Khanpur by drill method to get F_1 seed at the end of February. F_1 plants were sown during 2002-2003 and were self-pollinated at flowering to get seed for F_2 generation. During 2003-16 filial generations were developed on the basis of superior attributes selection. From F_2 generation, 62 single plants with better performance were harvested separately. Out of these single plants, 59 plants were used to grow F_3 progeny rows. Forty three plants were harvested separately from the selected rows on the basis of yield and disease resistant parameters, and were sown as progenies rows of F_4. Thirty desirable selections were made in F_4 for raising F_5 generation, through which 19 plants were selected from progeny rows to grow F_6. Twelve superior rows were selected for further study on the basis of better performance. The lines with high yield and disease resistance were evaluated in replicated yield trials for two years at different locations. Erucic acid and glucosinolates concentration of these rows were determined by National Agricultural Research Centre (N.A.R.C), Islamabad and National Institute of Food and Agriculture (N.I.F.A), Peshawar following the procedure used by Sadat et al., 2010. The best performing line (KN-259) was evaluated in Zonal Varietal Trial, Micro Yield Trial and National Uniform Rapeseed Yield Trial for two years.

Planting of Trial

Randomized Complete Block Design (RCBD) was used in all the experiments with four replications. Each plot was maintained a size of 5 m x 1.8 m. Seed was sown with the help of drill maintaining an inter row and inter plant distances of 45 cm and 10 cm respectively. All standards of agronomic and cultural practices were applied uniformly for all treatments from sowing to harvesting in each experiment. Data regarding days to flowering, days to maturity, plant height, seed yield and yield related traits were recorded and analyzed statistically on the basis of mean values (Singh and Chaudhry, 2004).

Table – 1: Various steps involved in the development of rapeseed variety Rohi Sarsoon

Year	Generations/Trials	Remarks
2001-02	Original Cross attempted (KN-120 x KN-131)	Hybrid seed was harvested for plantation.
2002-03	F_1 was raised	F_1 hybrid seed was harvested for F_2 plantation.
2003-04	F_2 generation was grown	F_2 hybrid seed was harvested for F_3 plantation.
2004-05	F_3 generation was grown	F_3 hybrid seed was harvested for F_4 plantation
2005-06	F_4 generation was grown	F_4 hybrid seed was harvested for F_5 plantation
2006-07	F_5 generation was grown	Superior line for yield and disease resistance were selected and given the number KN-259 and were forwarded to yield trials
2010-11	Zonal Varietal Yield Trials	These trials were conducted at five different locations.
2011-12	Micro Yield Trial	These trials were grown at seven different locations in Punjab under coded numbers handled by Director, Oilseed Research Institute Faisalabad.
2012-14	National Uniform Rapeseed Yield Trial (N.U.R.Y.T)	These trials were conducted by National Coordinator Rapeseed, NARC Islamabad throughout Pakistan.
2012-14	Agronomic Trials	These trials were conducted at Oilseeds Research Station Khanpur.
2011-13	Entomological Trials	These trials were conducted at Oilseeds Research Station Khanpur and resistance against different diseases were observed.
2012-13 & 2015-16	DUS Studies	Data of particular line was recorded by Federal Seed Certification and Registration Department.
2016	Spot Examination	Candidate line was evaluated by Committee and was recommended for varietal approval
2016		On the basis of better performance, it was released by Punjab Seed Council, Lahore for general cultivation with the Particular commercial identity.

Results and Discussions

KN-259 (Rohi Sarsoon) is an advanced line of cross, KN-120 x KN-131, selected from segregating population on the basis of plant height, plant and pod shape, seed size and seed color. It showed best performance in local environmental conditions.

Station Trials

Rohi Sarsoon was tested in replicated trials for two years, Preliminary Yield Trial (2008-09) and Advanced Yield Trial (2009-10) were conducted at Oilseeds Research Station, Khanpur. Mean data of grain yield of particular variety was compared with Punjab Sarsoon (check) (Table 2). It was observed that average yield of Rohi Sarsoon was 2470 kg/ha compared with Check which yielded 2080 kg/ha showed 19% higher yield over check variety.

Zonal Varietal Trials

KN-259 was evaluated at five different locations i.e. Faisalabad, Sahiwal, Khanpur, Piplan and Bahawalpur of Punjab in Zonal Varietal Trial during 2010-11 (Table 3). Grain yield was observed at these different location and found average yield of Rohi Sarsoon 2176 kg/ha as compared to yield (1934 kg/ha) of check variety. The results depicted that particular variety gave 12% more yield that Punjab Sarsoon.

Micro Yield Trials

The performance of promising line was further evaluated in Micro Yield Trial at seven different location of Punjab during 2011-12. The average grain yield of Rohi Sarsoon was 1450 kg/ha and Punjab Sarsoon gave 1288 kg/ha. Results of new variety revealed 13% higher yield as compared to check (Table 4).

Yield performance of KN-259 at National Uniform Rapeseed Yield Trial (NURYT)

Rohi Sarsoon was evaluated in national testing system through Nation Uniform Rapeseed Yield Trial (NURYT) consecutively for two years during 2012-13 and 2013-14 across the country. The location wise comparison of yield performance of promising line with standard variety is given in Table 5 and 6. Results of Rohi Sarsoon at eight different locations during 2012-13 showed that it gave yield 1895 kg/ha and check variety (Punjab Sarsoon) yielded 1438 kg/ha which revealed its better performance over check. During 2013-14 performance of new variety was compared with check named as Hyola-401. The mean performance depicted it was high yielder that Hyola-401. The two years performance based evaluation of KN-259 confirmed the results found on-station studies that it was high yielder as compared to check varieties. The overall good performance of new variety across the country proved that it is well adapted to various climatic conditions of Punjab and Pakistan than earlier released varieties.

Table – 2: Yield Performance of KN-259 in Station Yield Trials at Oilseeds Research Station, Khanpur

Variety/Line	Seed yield in Kg/ha			% increase over check
	P.Y.T (2008-09)	A.Y.T (2009-10)	Avg.	
KN-259	2430	2510	2470	19%
Punjab Sarsoon	2120	2040	2080	-
L.S.D 5%	110	107		

Table – 3: Yield Performance of KN-259 in Zonal Varietal Yield Trial during 2010-11

Variety/Line	Seed yield in Kg/ha						% increase
	FSD	SWL	KPR	PIPLAN	BWP	Avg.	
KN-259	2170	2020	2510	1700	2480	2176	12%
Punjab Sarsoon	1910	1850	2220	1440	2250	1934	-
L.S.D 5%	102	150	172	140	130		

Table – 4: Yield Performance of KN-259 in Micro Yield Trial during 2011-12

Variety/Line	Seed yield in kg/ha								% increase
	FSD	BWP	KPR	Bhakar	F/Jg	CHK	Karor	Avg.	
KJ-259	1605	2426	1026	2093	696	785	1520	1450	13%
Punjab Sarsoon	1530	2050	860	1940	625	624	1391	1288	-
L.S.D 5%	120	200	175	115	133	125	180		

Table – 5: Yield Performance of KN-259 in N. U.R.Y.T. (8 locations) during 2012-13

Variety /Line	Seed yield in kg/ha								Mean
	NARC Isd	BARI ChK	ORI Fsd	NIA T-Jam	Pioneer Swl	ORS K-Pur	RARI B-pur	NIFA P-War	
KN-259	2045	825	1844	1101	2177	2552	2052	2565	1895
Punjab Sarsoon	1212	540	1607	576	1835	2506	2016	1212	1438
L.S.D 5%	210	357	375	284	530	515	476	129	

Table – 6: Yield Performance of KN-259 in N. U.R.Y.T. (8 locations) during 2013-14

Variety /Line	Seed yield in kg/ha								Mean
	ORI Fsd	ORS K-Pur	Pioneer Swl	RARI B-pur	NIFA P-War	Tarnab P-war	BARS Kohat	ARI T-jam	
KN-259	1975	1625	2434	3208	3861	3927	2500	1113	2580
Hyola-401	1439	1396	1426	1983	3417	2594	1833	988	1885
L.S.D 5%	169	194	283	173	379	245	253	105	

Agronomic Performance

Agronomic trials of elite line were conducted against nine different level of fertilizer application during 2013-14. Maximum seed yield (2670 kg/ha) was obtained by treatment No. 5 in which N: P was applied @ 75:75 followed by treatment No.6 with yield of 2640 kg/ha (Table 7). The average yield performance of KN-259 in sowing date trail was assessed from 20th September to 30th October with 10 days intervals. The data (Table 8) showed that highest mean yield 2159 kg/ha was recorded when KN-259 was sown on 10th October

Table – 7: Response of KN-259 to different levels of NP at Oilseeds Research Station Khanpur during 2013-14.

Treatment	Nitrogen (Kg/ha)	Phosphorus (Kg/ha)	Seed Yield (Kg/ha)
1	60	60	2030
2	75	60	2140
3	90	60	2180
4	60	75	2560
5	**75**	**75**	**2670**
6	90	75	2640
7	60	90	2510
8	75	90	2540
9	90	90	2470
LSD5%		185	

Table – 8: Response of KN-259 to different sowing dates at Oilseeds Research Station Khanpur during 2012-13 & 2013-14.

Sowing Date	Yield (Kg/ha) 2012-13	Yield (Kg/ha) 2013-14	Av.(Kg/ha)
20th September	--	1850	-
30th September	1927	2292	2110
10th October	**2002**	**2315**	**2159**
20th October	1538	1366	1452
30th October	1493	824	1159
LSD 5%	136	128	

Screening against Insects and Diseases

The response of variety Rohi Sarsoon to various diseases and insects was studied at Oilseeds Research Station, Khanpur during two consecutive years 2011-12 and 2012-13. It was tested against Alternaria blight, Powdery mildew and White rust. Results of two years (Table 9) revealed that particular advance line was highly resistant against diseases. KN-259 escaped from aphid attack when it was sown in optimum time in i.e. 1st fortnight of October. Sarwar et al., (2004) has

also observed that crop sown in first week of October can escape aphid attack.

Table – 9: Disease response of KN-259 at Oilseeds Research Station Khanpur during 2011-12 & 2012-13

Variety/ Line	Year	Alternaria blight	Powdery mildew	White rust
KJ-259	2011-12	0	0	0
	2012-13	0	0	0

***0= resistant, 5= susceptible**

Botanical Description of Rohi Sarsoon

Rohi Sarsoon is erect with plant height 160-173 cm. Its plant color is dark green having determinate growth habit. The color of leaf is light green with absence of hairs. Its leaf size is medium. Rohi Sarsoon takes 75-79 days to flower. Its petal color is yellow. Pod has long length conical beak shape. Its seed color is dark black having bold size.

Quality characteristics

Rohi Sarsoon contains good characters for edible use. Its quality characters were compared with check variety Punjab Sarsoon. Results revealed that it contains less than 1 percent erucic acid in oil, 83.81 μ mole/g glucosinolates in oil free meal (Table 10). The quality traits recorded by N.A.R.C, Islamabad and NIFA Peshawar depicted that new variety is better than existing varieties containing sufficient amount of chemical compounds i.e., erucic acid and glucosinolates in its oil and oil free seed meal.

Table – 10: Quality characteristics of Rohi Sarsoon

Rapeseed Lines	Erucic Acid (%)	Glucosinolates (μmole/g)	Oil Content
Rohi Sarsoon	Less than 2	83.81	47.46
Punjab Sarsoon (Check)	7.62	72.16	45.48

Spot Examination and Approval

The candidate variety was evaluated by Spot Examination Committee during February 2016. The committee has recommended for submission of variety approval case to the expert sub-committee. The Expert Sub-committee approved KN-259 as new commercial variety named as Rohi Sarsoon and forwarded to Punjab Seed Council (PSC) for its final approval. The Punjab Seed Council approved variety for general cultivation in meeting held on 12 August 2016. Conclusively, Rohi Sarsoon is not only high yielding variety but also resistant to various diseases and pests with better quality traits. Due to its better adaptability, it can be substituted with already existing approved varieties.

References

Ahmad B, Mohammad S, Azam F, Ali I, Ali J and Rehman SU, 2013. Studies of genetic variability, heritability and phenotypic correlations of some qualitative traits in advance mutant lines of winter rapeseed (*Brassica napus* L.). American-Eurasian J. Agri. Environ. Sci. 13(4): 531-538.

Mustafa SZ, Yasmin S, Kisana NS and Mujahid MY, 2008. Results of the National Uniform Wheat Yield Trials (2007-08). Coordinated Wheat Barley and Triticale Programme, Pakistan Agricultural Research Council, P.O. Box 1031, Islamabad.

Pakistan Economic Survey (2015-16). Ministry of Finance Division, Economic Advisor Wing, Islamabad.

Sadat HA, Nematzadeh GA, Jelodar NB and Chapi OG, 2010. Genetic evaluation of yield and yield components at advanced generations in rapeseed (*Brassica napus* L.). African J. Agri. Res. 5(15):1958-1964.

Sarwar M, Ahmad N, Siddiqui QH, Ali A and Tofique M, 2004. Genotypic response in canola (Brassica species) against aphid (Aphidae: Homoptera) attack. The Nucleus. 41(1-4):87-92.

Singh RK and Chaudhry BD, 2004. Biometrical methods in quantitative genetic analysis. Kalyani Publishers, Ludhiana, India, p. 318.

Steel RGD, Torrie JH and Dickey DA, 1997. Principles and Procedures of Statistics. A biometrical approach. 3rd Ed., McGraw Hill Book Co., New York, USA.

22

Screening of breeding lines of *Brassica napus* L. tolerant to grain shattering

Tamoor Hussain[1]*, Muhammad Azeem Tariq[1], Ramzan Ansar[1], Muhammad Tariq[1], Ahmad Sher[2]
[1]Barani Agricultural Research Institute, Chakwal, Pakistan
[2]Department of Agronomy, BZU, Bahadar Sub Campus, Layyah, Pakistan

Abstract

Shattering causes huge losses to Brassica even if harvesting is done mechanically. To study the shattering tolerance in different rapeseeds lines, a study was conducted in split plot design at experimental area of Barani Agricultural Research Institute (BARI) Chakwal, Pakistan during two winter season 2014-15 & 2015-16 under the rainfed conditions. The experimental material comprised of 10 advanced lines including one local variety of rapeseed viz: 8CBN001, 8CBN002, 10CBN003, 10CBN005, 11CBN001, 11CBN003, 11CBN009, 11CBN011, 12CBN003 and Chakwal Sarsoon The main experimental plot was harvested at four different harvesting dates (HD) including: HD_1 = Harvesting of each advanced line at crop maturity, HD2= 10 days after the first harvesting (DAFH), HD_3= 20 DAFH, HD_4= 30 DAFH. The comparison of difference between the grain yield of each advanced line at various harvesting dates with its seed yield at first harvesting date (HD_1-HD_i) were calculated as indices of pods shattering. Combined analysis of variance depicted that all the genotypes were significantly different for HD, harvesting date and advanced lines interaction effect, year and its interaction effects to each factors. The significant difference of genotypes into harvesting dates demonstrated various level of pods shattering of all advanced lines at different harvesting dates. The amount of grain yield due to shattering losses was increased at third and fourth harvesting dates. On the basis of shattering tolerance indices, advanced line 10CBN005 was more tolerant to shattering losses and 8CBN002 was more susceptible to shattering as compared to others promising genotypes.

Keywords: Shattering tolerance, Rapeseed, Harvesting date

*Corresponding author email:
tamoorhussain@gmail.com

Introduction

Rapeseed is one of the most important oilseed crops in Pakistan and world over. For enhancement of grain yield potential of rapeseed, the important breeding strategies are the good knowledge and utilization of morphological, physiological and genetic basis of grain yield associated attributes in different climatic conditions (Bruce et al., 2002; Banga et al., 2011). Tolerance to shattering is an important attribute for rapeseed grain yield enhancement because the crop ripens and is harvested under warm environment and normally windy summer conditions (Rameeh, 2013). In rapeseed grain yield loss is usually divided into two periods, shattering before and during harvesting (chandler et al., 2005). Weather conditions prior to and during harvesting are the main factors in the field that influence the level of shattering (Tan et al., 2006). Typically grain yield losses vary from 10 to 25 percent (Price et al., 1996). Links between pods and other canopy components during windy summer conditions have also been implicit to contribute to shattering in the field. Furthermore, insect-pest and disease damage can accelerated ripening and pod shattering (Rameeh, 2013). Peng-Fei et al., (2011) in their studies reported that in brassica species marked losses of grain yield is due to shatter during maturity and harvesting. Moreover the shed seeds may remain viable during a number of years and germinate to produce plants, which represent weeds in the following crops. Shattering involves bursting of the pod valves and detachment of seed from the replum. It could take place in ripe standing crops under windy conditions once plants become in contact other plants and in windrows from the impact of harvest machinery. (Meakin and Roberts, 1990). Overseas research work on this factors explained that genetic variation for grain shattering tolerance present among rapeseed genotypes (Wen et al., 2008). Sixty eight lines of Brassica napus were studied for pods shattering resistance using a ripping method and demonstrated that ripping force ranged from 0.59 to 2.75 N existed in different Brassica napus genotypes. It was also revealed that the inheritance of shatter resistance (SR) was determined by two genes, with heritability of 50% (Peng-Fei et al., 2011). This study concurred that genetic gain can be achieved by using conventional methods of breeding in rapeseed. However, further development is required to avoid the need to windrow. Resistance in Brassica napus was recessive and mostly governed by additive genes. They observed that correlation of pods shattering tolerant with important attributes was low, signifying that it would be viable to easily introgress the shattering resistance characters into inbred lines.

Screening of rapeseed advanced lines for shattering tolerance is difficult because pods shattering attribute is also influenced by other than the genetic factors such as timing of pod senescence, maturity, method and timing of harvesting. The goal of the present study were to evaluate the presence of genetic variability for shattering tolerance in Brassica napus advanced lines and also relationship of pods shattering at different harvesting dates and the selection of lines with more tolerant to shattering characteristics which can be used by plant breeders for improvement of shattering

tolerance in their breeding program of rapeseed in future.

Material and Methods

A study was conducted to observe shattering tolerance in different rapeseed lines at the research area of BARI, Chakwal, Pakistan which is located at 32.66° latitude and 72.51° longitude with an altitude of 575ams altitude with an average rainfall less than 600 mm under rainfed conditions. The pre sowing soil analysis indicated that the soil is deficient in organic matter (0.52 %) and phosphorus (5.0 mg kg^{-1}), extractable K (110 mg kg^{-1}) and soil pH 7.8. The research site kept fallow in kharif season during both years. The field was prepared according to standard practices used in rainfed areas. The experimental material comprised of 10 advanced lines including one local variety of rapeseed viz: 8CBN001, 8CBN002, 10CBN003, 10CBN005, 11CBN001, 11CBN003, 11CBN009, 11CBN011, 12CBN003 and Chakwal Sarsoon were sown during last week of September with seed rate @ 2 kg per acre in both years in a split plot design with four replications. The main plot harvested at four different dates including HD_1 = Harvesting at crop maturity of each advance line, HD_2= 10 days after the first harvesting (DAFH), HD_3= 20 (DAFH), HD_4= 30 (DAFH). All the treatments (HD_1, HD_2, HD_3, and HD_4) were harvested according to planned different harvesting dates and threshed by hand. Each sub plot comprised of same advance lines and consisted of 5 rows of 1.8 m long with plant to plant and row to row distance was kept at 10cm and 45 cm respectively. All the plant protection measures and cultural practices are adopted same for all the plots as per requirement of the crop. The difference of seed yield (kg/ha) of each advanced lines in various dates of harvesting with compare to first harvesting date (HD_1-HD_i) were considered as indices of shattering tolerance and compute by using the formula give as Rameeh, (2013).

Shattering Resistance= (HD_1-HD_i)

Shattering Resistance (%) = [(HD_1-HD_i)/HD_1] x100

Where "HD_1" is the grain yield of each genotype in the 1st harvesting date and "HD_i" is the grain yield of each advanced line in subsequent harvesting dates. Combine analysis of variance on the basis of split plot design were calculated by using the Statistix software version 8.1.

Results and Discussion

The detail of metrological data regarding total rainfall (mm), average minimum and maximum temperature (°C), average humidity (%) and average sunshine hours is shown in figure No. 1 and 2 for cropping season 2014-15 and 2015-16 respectively. From the investigation of present study, combined analysis of variance on the basis of split plot design for grain yield represented that each advanced line was different significantly from each other's. Also different harvesting dates and harvesting dates x advanced lines interaction had significant effects on grain yield which indicated that variation in grain yield of advanced lines were varied in different harvesting dates (Table-1). The table-2 depicted that on the basis of average grain yield (kg/ha) for two year results of each advanced line in different harvesting dates, the local check variety "Chakwal Sarsoon" showed highest average value at all the harvesting dates followed by 11CBN009. However value decreased in all the advance lines at subsequent harvesting dates.

The difference of seed yield of all the advanced line in first and second harvesting date varied from 76 kg/ha to 106 kg/ha in 10CBN005 and 8CBN002 respectively (Table-3). This result indicates that the 8CBN002 is more susceptible to shattering as compared to 10CBN005. Others genotypes likes 11CBN011, 10CBN009, 8CBN001 and Chakwal Sarsoon are also susceptible to shattering. Weng et al., (2008) and Rameeh, (2013) also investigated that genetic variation are present for pods shattering resistance in *Brassica napus* lines. When compare the difference of grain yield of 1st and 3rd harvesting dates, advanced lines ranged from 122 to 186 kg/ha in 10CBN003 and 8CBN002 respectively. Highest average value was determined for the 8CBN002 followed by 11CBN009 (Table-3). This investigation revealed that 8CBN002 and 11CBN009 are more susceptible to shattering. The difference of mean value of 1st and 4th harvesting dates (HD_1-HD_4) was more varied then all the others differences, so on the basis of this consequence, the (HD_1-HD_4) index is more important source for the selection of advanced lines against the shattering tolerance. The difference of grain yield of (HD_1-HD_4) index significant varied from 154 kg/ha to 266 kg/ha in 10CBN005 & 8CBN002 respectively (Table-3). On the basis of present inference from (HD_1-HD_4) index, the 8CBN002 is more susceptible to shattering then all the other advanced lines. In addition, the percentage of grain yield shattering with comparison to 1st harvesting date revealed that grain shattering at 2nd harvesting date ranged from 8.05 to 10.04 percent in 10CBN003 & 8CBN002 advanced lines respectively. However, grain shattering in 3rd harvesting date varied from 12.23 to 17.57 percent in 10CBN003 & 8CBN002 advanced lines respectively. Furthermore, the grain shattering in 4th harvesting date ranged from 16.85 to 25.15 percent in 10CBN005 & 8CBN002 advanced lines respectively (Table-4). The present investigations are in corroborated with the observations of Price et al., (1996) who reported grain yield losses ranged from 10 to 25 percent in Brassica napus lines.

In present experiment study, the genetic variation for pods shattering tolerance was found among all advance lines of rapeseed. However different methods used to determines the inheritance of shattering resistance but delaying in harvesting date in contrast to physiological maturity is also viable technique for the assessment of shattering tolerance of rapeseed advanced lines under normal conditions in rainfed areas. As the inference from the current study, the difference of grain yield of 1st and 4th harvesting dates (HD_1-HD_4) were more varied therefore, it is useful tool for the selection of rapeseed advanced lines for the shattering tolerance. Among all the advanced lines the 10CBN005 and Chakwal Sarsoon were more tolerant to shattering then all the others advanced lines and 8CBN002 & 8CBN001 were relatively more susceptible to shattering. So the lines with more tolerant to shattering characteristics can be used by plant breeders for improvement of shattering tolerance in their breeding program of rapeseed.

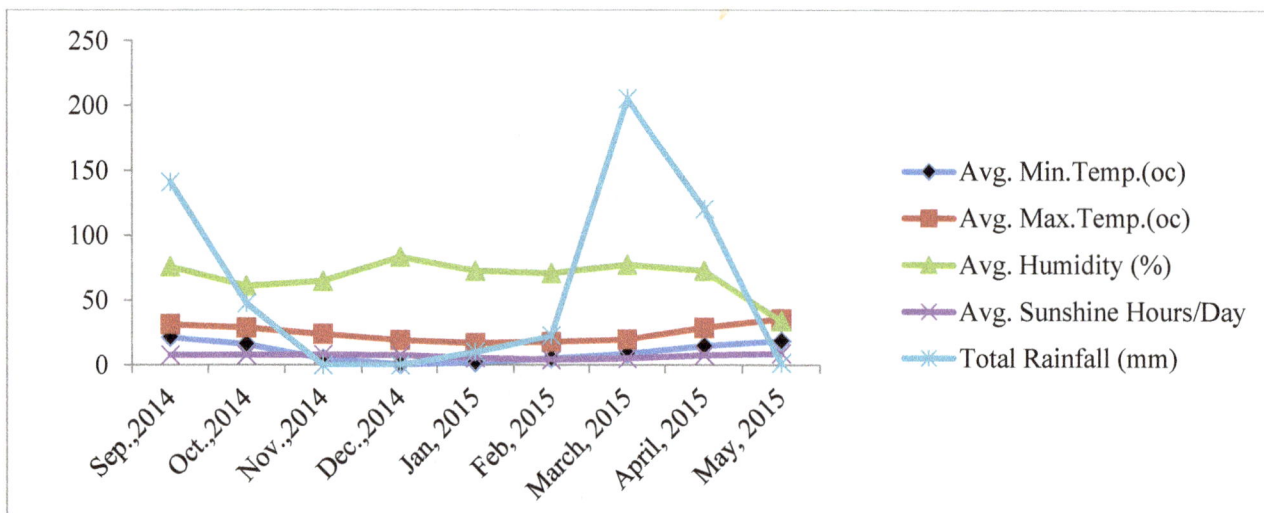

Fig 1: Metrological data during winter season 2014-15

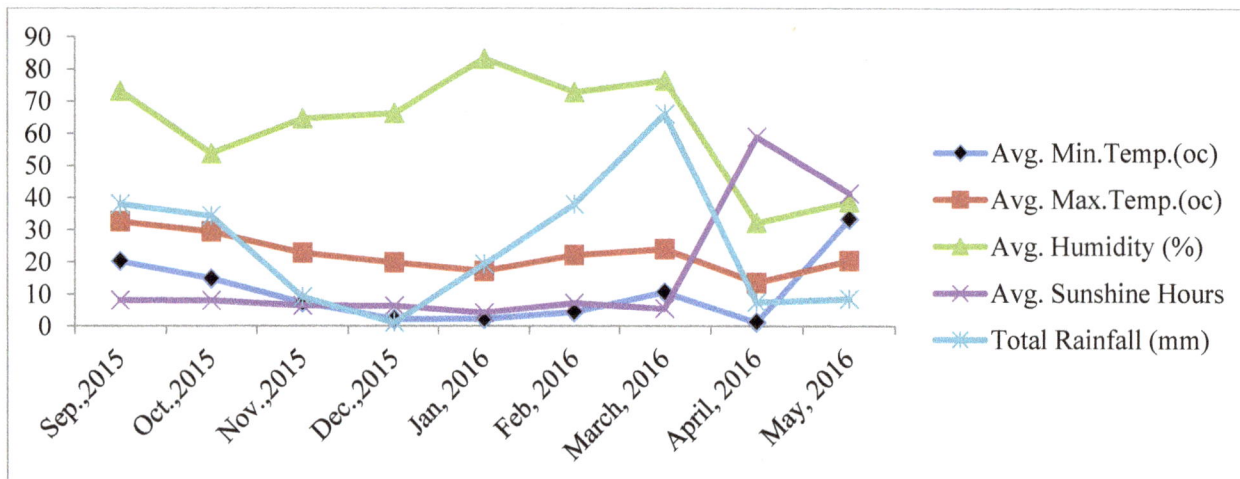

Fig 1: Metrological data during winter season 2015-16

Table-1: Results of combine analysis of variance of two years 2014-15 & 2015-16

Source of Variation	Degree of Freedom	M S	F. test	P. Value
Y	1	2936653**	1135.52	0.0000
R(Y)	6	16687**	5.67	0.0001
HD	3	647848**	250.50	0.0000
Y x HD	3	15629**	6.04	0.0006
Error 1	18	2351	-	-
PL	9	110999**	42.92	0.0000
Y x PL	9	115060**	44.49	0.0000
HD x PL	27	4881*	1.89	0.0067
Y x HD x PL	27	6633**	2.56	0.0001
Error 2	120	2771	-	-

* and ** Significant at 5% and 1% level respectively Y: (Year), R: (Replication), HD: (Harvesting date), PL: (Promising lines)

Table-2: Average Grain Yield (kg/ha) of Rapeseed advanced lines in different harvesting dates during two rabi growing seasons 2014-15 & 2015-16

Advanced Lines	1st Harvesting Date (HD$_1$)	2nd Harvesting Date (HD$_2$)	3rd Harvesting Date (HD$_3$)	4th Harvesting Date (HD$_4$)
8CBN001	966	874	825	730
8CBN002	1056	950	870	790
10CBN003	994	914	872	809
10CBN005	911	835	785	758
11CBN001	872	790	743	674
11CBN003	1025	937	885	822
11CBN009	1063	959	911	848
11CBN011	937	844	793	720
12CBN003	998	914	853	793
Chakwal Sarsoon	1097	998	956	904

HD$_1$ = Harvesting at maturity of crop of each Promising line, HD$_2$= 10 days after the first harvesting (DAFH), HD$_3$= 20 (DAFH), HD$_4$= 30 (DAFH)

Table-3: Least Significant Difference (LSD) for mean grain yield (kg/ha) shattering of rapeseed advanced lines during two rabi growing seasons 2014-15 & 2015-16

Advanced Lines	(HD$_1$-HD$_2$)	(HD$_1$-HD$_3$)	(HD$_1$-HD$_4$)
8CBN001	93**	141**	237**
8CBN002	106**	186**	266**
10CBN003	80*	122**	185**
10CBN005	76*	126**	154**
11CBN001	82**	129**	198**
11CBN003	89**	141**	204**
11CBN009	104**	152**	215**
11CBN011	94**	145**	217**
12CBN003	85**	145**	205**
Chakwal Sarsoon	99**	141**	193**

* and ** Significant at 5% and 1% level respectively HD$_1$ = Harvesting at maturity of crop of each Promising line, HD$_2$= 10 days after the first harvesting (DAFH), HD$_3$= 20 (DAFH), HD$_4$= 30 (DAFH)

Table-4: Percentage of average grain yield (kg/ha) shattering of rapeseed advanced lines during two rabi growing season 2014-15 & 2015-16

Advanced Lines	[(HD$_1$-HD$_2$)/HD$_1$]x100	[(HD$_1$-HD$_3$)/HD$_1$]x100	[(HD$_1$-HD$_4$)/HD$_1$]x100
8CBN001	9.58	14.60	24.48
8CBN002	10.04	17.57	25.08
10CBN003	8.05	12.23	18.62
10CBN005	8.34	13.83	16.85
11CBN001	9.40	14.79	22.71
11CBN003	8.63	13.71	19.85
11CBN009	9.79	14.26	20.24
11CBN011	9.98	15.42	23.16
12CBN003	8.47	14.53	20.54
Chakwal Sarsoon	8.98	12.81	17.56

HD$_1$ = Harvesting at maturity of crop of each Promising line, HD$_2$= 10 days after the first harvesting (DAFH), HD$_3$= 20 (DAFH), HD$_4$= 30 (DAFH)

Reference

Bruce DM, Farrent JW, Morgan CL and Child RD, 2002. Determining the oilseed rape pod strength needed to reduce seed loss due to pod shatter. Biosystems Eng. 81: 179-184.

Banga S, Kaur G, Grewal N, Salisbury PA and Banga SS, 2011. Transfer of resistance to seed shattering from B. carinatato B. napus. 13th International Rapeseed Congress, Prague, Zhech Republic. pp. 863-865.

Chandler JL, Corbesier P, Spielmann J, Dettendorfer D, Stahl K, Apel and Melzer S, 2005. Modulating flowering time and prevention of pod shatter in oilseed rape. Mol. Breed. 15: 87–94.

Meakin PJ and Roberts JA, 1990. Dehiscence of Fruit in Oilseed Rape (B. napus L.) I. Anatomy of pod dehiscence. J. Exp. Bot. 41 (229): 995-1002.

Peng-Fei P, Yun-chang L, De-sheng D, Ying-de L, Yu-song X and Qiong H, 2011. Evaluation and genetic analysis of pod shattering resistance in Brassica napus.13th International Rapeseed Congress, Prague, Czech Republic. pp. 617-620.

Price JS, Hobson RN, Neale MA and Bruce DM, 1996. Seed losses in commercial harvesting of oilseed rape. J. Agric. Eng. Res. 65: 183-191.

Rameeh V, 2013. Evaluation of different spring rapeseed (Brassica napus L.) genotypes for shattering tolerance. J. Oilseed Brassica. 4(1): 19-24.

Tan XL, Zhang JF and Yang L, 2006. Quantitative determination of the strength of rapeseed pod dehiscence. Trans CSAE. 22: 40-43.

Wang R, Ripley VL and Rakow G, 2007. Pod shatter resistance evaluation in cultivars and breeding lines of B. napus, B. juncea and Sinapis alba. Plant Breed. 126: 588-595.

Wen YC, Fu TD, Tu JX, Ma CZ, Shen JX and Zhang SF, 2008. Screening and analysis of resistance to silique shattering in rape (B. napus). Acta Agron Sin. 34: 163-166.

Permissions

All chapters in this book were first published in AJAB, by Life Sciences Society; hereby published with permission under the Creative Commons Attribution License or equivalent. Every chapter published in this book has been scrutinized by our experts. Their significance has been extensively debated. The topics covered herein carry significant findings which will fuel the growth of the discipline. They may even be implemented as practical applications or may be referred to as a beginning point for another development.

The contributors of this book come from diverse backgrounds, making this book a truly international effort. This book will bring forth new frontiers with its revolutionizing research information and detailed analysis of the nascent developments around the world.

We would like to thank all the contributing authors for lending their expertise to make the book truly unique. They have played a crucial role in the development of this book. Without their invaluable contributions this book wouldn't have been possible. They have made vital efforts to compile up to date information on the varied aspects of this subject to make this book a valuable addition to the collection of many professionals and students.

This book was conceptualized with the vision of imparting up-to-date information and advanced data in this field. To ensure the same, a matchless editorial board was set up. Every individual on the board went through rigorous rounds of assessment to prove their worth. After which they invested a large part of their time researching and compiling the most relevant data for our readers.

The editorial board has been involved in producing this book since its inception. They have spent rigorous hours researching and exploring the diverse topics which have resulted in the successful publishing of this book. They have passed on their knowledge of decades through this book. To expedite this challenging task, the publisher supported the team at every step. A small team of assistant editors was also appointed to further simplify the editing procedure and attain best results for the readers.

Apart from the editorial board, the designing team has also invested a significant amount of their time in understanding the subject and creating the most relevant covers. They scrutinized every image to scout for the most suitable representation of the subject and create an appropriate cover for the book.

The publishing team has been an ardent support to the editorial, designing and production team. Their endless efforts to recruit the best for this project, has resulted in the accomplishment of this book. They are a veteran in the field of academics and their pool of knowledge is as vast as their experience in printing. Their expertise and guidance has proved useful at every step. Their uncompromising quality standards have made this book an exceptional effort. Their encouragement from time to time has been an inspiration for everyone.

The publisher and the editorial board hope that this book will prove to be a valuable piece of knowledge for researchers, students, practitioners and scholars across the globe.

List of Contributors

Saima Batool
Department of Botany, Govt. Degree College for Women, Samanabad, Faisalabad, Pakistan

Shahbaz Khan and Faisal Mehmood
Department of Agronomy, University of Agriculture, Faisalabad, Pakistan

Sohail Irshad
In-Service Agricultural Training Institute, Rahim Yar Khan, Pakistan

Muhammad Nawaz
Department of Agronomy, College of Agriculture, BZU, Bahadur Campus Layyah, Pakistan

Muhammad Mazhar Iqbal
Soil and Water Testing Laboratory for Research, Chiniot, Department of Agriculture, Government of Punjab
Institute of Soil and Environmental Sciences, University of Agriculture, Faisalabad
Provincial Pesticide Reference Laboratory, Kala Shah Kaku, Sheikhupura
Institute of Soil Chemistry and Environmental Sciences, Ayub Agriculture Research Institute, Faisalabad
Department of Soil and Environmental Sciences, University of Sargodha, Sargodha

Ghulam Murtaza, Tayyaba Naz and Wasim Javed
Institute of Soil and Environmental Sciences, University of Agriculture, Faisalabad

Sabir Hussain
Department of Environmental Sciences and Engineering, Government College University Faisalabad, Faisalabad

Muhammad Ilyas and Muhammad Ashfaq Anjum
Provincial Pesticide Reference Laboratory, Kala Shah Kaku, Sheikhupura
Institute of Soil Chemistry and Environmental Sciences, Ayub Agriculture Research Institute, Faisalabad

Sher Muhammad Shahzad and Muhammad Ashraf
Department of Soil and Environmental Sciences, University of Sargodha, Sargodha

Zafar Iqbal
Department of Plant Pathology, University of Sargodha, Sargodha

Mobina Maktabdaran
MSc Student, Department of Ecology, Faculty of Agriculture, University of Birjand, Iran

Mohammad Hassan Sayyari Zohan
Associate Professor, Department of Soil Science and Engineering, Faculty of Agriculture, University of Birjand, Iran

Majid Jami Alahmadi and Golam Reza Zamani
Associate Professor, Department of Agronomy and Plant Breeding, Faculty of Agriculture, University of Birjand, Iran

Muhammad Zakirullah, Sumayya Innayat, Tariq Jan and Muhammad Ali
Agricultural Research Institute Tarnab, Peshawar, Khyber Pakhtunkhwa, Pakistan

Muhammad Arif
Directorate of Outreach, Agricultural Research, Khyber Pakhtunkhwa, Pakistan

Mehboob Alam
Department of Horticulture, The University of Agriculture, Peshawar, Khyber Pakhtunkhwa, Pakistan

Marulak Simarmata, Uswatun Nurjanah and Nanik Setyowati
Department of Agronomy, University of Bengkulu, Jalan W.R. Supratman Kandang Limun, Bengkulu 38371, Indonesia

Bindu Singh, Virendra Kumar Singh and Khalid Monowar Alam
Department of Environmental Science, Integral University, Lucknow, UP, India

Muhammad Anwar Zaka, Khalil Ahmed, Hafeezullah Rafa and Muhammad Sarfraz
Soil Salinity Research Institute, Pindi Bhattian, Punjab, Pakistan

Helge Schmeisk
Faculty of Organic Agriculture Sciences, University of Kassel, Germany

Hassan A. Hassan and Sahar S.Taha
Department of Vegetables, Faculty of Agriculture, Cairo University, Cairo, Egypt

Mohamed A. Aboelghar and Noha A. Morsy
National Authority for Remote Sensing and Space Sciences (NARSS), Cairo, Egypt

Saeed Ahmad, Muhammad Sajjad and Muhammad Arshad Hussain
Regional Agricultural Research Institute, Bahawalpur, Pakistan

Rabia Nawaz
Government Sadiq Women University, Bahawalpur, Pakistan

Muhammad Naveed Aslam
University College of Agriculture and Environmental Sciences, The Islamia University of Bahawalpur, Pakistan

Muhammad Aslam and Khola Rafique
Pest Warning and Quality Control of Pesticides, Department of Agriculture, Govt. of Punjab, Pakistan

Alfred Maroyi
Medicinal Plants and Economic Development (MPED) Research Centre, Department of Botany, University of Fort Hare, Private Bag X1314, Alice 5700, South Africa

Farheen Nazli and Muhammad Ramzan Kashif
Pesticide Quality Control Laboratory Bahawalpur-63100, Pakistan

Bushra
Institute of Soil and Environmental Sciences, University of Agriculture Faisalabad, Pakistan

Muhammad Mazhar Iqbal
Soil and Water Testing Laboratory, Chiniot, Pakistan

Fatima Bibi
Mango Research Station, Multan, Pakistan

Zafar-ul-Hye
Department of Soil Science, College of Agriculture, Bahauddin Zakariya University, Multan, Pakistan

Maqshoof Ahmad
Department of Soil and Environmental Sciences, the Islamia University of Bahawalpur, Bahawalpur-63100, Pakistan

Turhadi Turhadi, Hamim Hamim and Miftahudin Miftahudin
Department of Biology, Faculty of Mathematics and Natural Sciences, Bogor Agricultural University, Kampus IPB Darmaga, 16680 Bogor, Indonesia

Munif Ghulamahdi
Department of Agronomy and Horticulture, Faculty of Agriculture, Bogor Agricultural University, Kampus IPB Darmaga, 16680 Bogor, Indonesia

Nosheen Elahi and Abdul Majeed
Institute of Pure and Applied Biology, B. Z. University, Multan, Pakistan

Muhammad Ishaq Asif Rehmani
Department of Agronomy, Ghazi University, Dera Ghazi Khan, Pakistan
Muhammad Ahmad
Agricultural Extension Department, Govt. of Punjab, Sanghoie, Jhelum, Pakistan

Aqila Shaheen, Rabia Tariq and Abdul Khaliq
Department of Soil and Environmental Sciences, The University of Poonch, Rawalakot, Azad Jammu and Kashmir

Laila Nazirah
Doctoral Program of Agricultural Sciences, Faculty of Agriculture, Universitas Sumatera Utara, Padang Bulan, Medan 20155, Indonesia
Lecture Faculty of Agriculture, Malikussaleh University, Indonesia

Edison Purba, Chairani Hanum and Abdul Rauf
Lecture Program Study of Agriculture, Universitas Sumatera Utara, Padang Bulan, Medan 20155 Indonesia

Ahmed M. Abbas and Sabah Hammad
Department of Botany, Faculty of Science, South Valley University, Qena 83523, Egypt

Wagdi Saber Soliman
Horticulture Department, Faculty of Agriculture and Natural Resources, Aswan University, Aswan 81528, Egypt

Abdul Rehman, Aurangzeb, Rafi Qamar, Jamshaid Qamar and Farwa Hassan
Department of Agronomy, University College of Agriculture, University of Sargodha, Pakistan

Atique-ur-Rehman
Department of Agronomy, Bahauddin Zakariya University, Multan, Pakistan

Muhammad Shoaib
Maize and Millets Research Institute, Yousufwala, Sahiwal, Pakistan

Nwokwu Gilbert Nwogboduhu
Department of Crop Production and Landscape Management, Faculty of Agriculture and Natural Resources Management, Ebonyi State University, Abakaliki, Nigeria

Muhammad Aslam Nadeem, Hafiz Muhammad Zia Ullah, Abdul Majid Khan, Fida Hussain, Muhammad Jamil and Idrees Ahmad
Oilseeds Research Station, Khanpur, District Rahim Yar Khan, Pakistan

Tamoor Hussain, Muhammad Azeem Tariq, Ramzan Ansar and Muhammad Tariq
Barani Agricultural Research Institute, Chakwal, Pakistan

Ahmad Sher
Department of Agronomy, BZU, Bahadar Sub Campus, Layyah, Pakistan

Index